# THE IMMUNOLOGY OF HOST—ECTOPARASITIC ARTHROPOD RELATIONSHIPS

# The Immunology of Host–Ectoparasitic Arthropod Relationships

Edited by

Stephen K. Wikel
*Department of Entomology*
*Oklahoma State University*
*USA*

CAB INTERNATIONAL

CAB INTERNATIONAL
Wallingford
Oxon OX10 8DE
UK

Tel: +44 (0)1491 832111
Fax: +44 (0)1491 833508
E-mail: cabi@cabi.org
Telex: 847964 (COMAGG G)

© CAB INTERNATIONAL 1996. All rights reserved. No part of this publication may be reproduced in any form or by any means, electronically, mechanically, by photocopying, recording or otherwise, without the prior permission of the copyright owners.

A catalogue record for this book is available from the British Library.

ISBN 0 85199 125 4

Typeset by Solidus (Bristol) Limited
Printed and bound in the UK by Biddles Ltd, Guildford

# Contents

| | | |
|---|---|---|
| **Contributors** | | vii |
| **Preface** | | ix |
| **Acknowledgements** | | xi |
| 1 | **Immunology of the Skin**<br>*Stephen K. Wikel* | 1 |
| 2 | **Mouthparts and Feeding Mechanisms of Haematophagous Arthropods**<br>*Douglas K. Bergman* | 30 |
| 3 | **Salivary Gland Physiology of Blood-feeding Arthropods**<br>*John R. Sauer, Alan S. Bowman, Janis L. McSwain and Richard C. Essenberg* | 62 |
| 4 | **Pharmacology of Haematophagous Arthropod Saliva**<br>*Donald E. Champagne and Jesus G. Valenzuela* | 85 |
| 5 | **Arthropod Modulation of Host Immune Responses**<br>*Stephen K. Wikel, Rangappa N. Ramachandra and Douglas K. Bergman* | 107 |
| 6 | **Digestion and Fate of the Vertebrate Bloodmeal in Insects**<br>*Michael J. Lehane* | 131 |

| 7 | **Immune Responses to Fleas, Bugs and Sucking Lice**<br>*Carl J. Jones* | 150 |
|---|---|---|
| 8 | **Immune Responses to Mosquitoes and Flies**<br>*R. Mark Sandeman* | 175 |
| 9 | **Immunology of the Tick–Host Interface**<br>*Stephen K. Wikel* | 204 |
| 10 | **Immunology of Scabies**<br>*Larry G. Arlian* | 232 |
| 11 | **Immune Responses to Mange Mites and Chiggers**<br>*William J. Wrenn* | 259 |
| 12 | **Immunological-based Control of Blood-feeding Arthropods**<br>*Stephen K. Wikel, Douglas K. Bergman and Rangappa N. Ramachandra* | 290 |
| 13 | **A Synthesis of Current Concepts Regarding the Immunology of the Host–Arthropod Interface**<br>*Stephen K. Wikel* | 316 |
| **Index** | | 319 |

# Contributors

**Larry G. Arlian**
Department of Biological Sciences and Department of Microbiology and Immunology
Wright State University
Dayton, Ohio 45435
USA

**Douglas K. Bergman**
Department of Entomology
Oklahoma State University
Stillwater, Oklahoma 74078
USA

**Alan S. Bowman**
Department of Entomology
Oklahoma State University
Stillwater, Oklahoma 74078
USA

**Donald E. Champagne**
Department of Veterinary Science and Center for Insect Science
University of Arizona
Tucson, Arizona 85721
USA

**Richard C. Essenberg**
Department of Biochemistry and Molecular Biology
Oklahoma State University
Stillwater, Oklahoma 74078
USA

**Carl J. Jones**
Department of Veterinary Pathobiology
College of Veterinary Medicine
University of Illinois
Urbana, Illinois 61801
USA

**Michael J. Lehane**
School of Biological Sciences
University of Wales
Bangor LL57 2UW
UK

**Janis L. McSwain**
Department of Entomology
Oklahoma State University
Stillwater, Oklahoma 74078
USA

**Rangappa N. Ramachandra**
Department of Entomology
Oklahoma State University
Stillwater, Oklahoma 74078
USA

**R. Mark Sandeman**
School of Agriculture

Faculty of Science and
Technology
La Trobe University
Bundoora, Victoria
Australia 3083
**John R. Sauer**
Department of Entomology
Oklahoma State University
Stillwater, Oklahoma 74078
USA
**Jesus G. Valenzuela**
Department of Biochemistry and
Center for Insect Science
University of Arizona

Tucson, Arizona 85721
USA
**Stephen K. Wikel**
Department of Entomology
Oklahoma State University
Stillwater, Oklahoma 74078
USA
**William J. Wrenn**
Department of Biology
University of North Dakota
Grand Forks, North Dakota
58202
USA

# Preface

There has been an increasing interest in recent years in the immunology of the host–ectoparasitic arthropod interface. The number of different arthropod–host associations being examined is expanding. The complex interactions of host and arthropod are being more extensively defined and new avenues of investigation explored. Tremendous advances in cellular and molecular immunology coupled with enhanced capabilities in immunogen purification and biotechnology provide an increasing array of tools for characterization of these associations. Immunological control of blood-feeding arthropods has progressed from a possibility to a reality, and further advances in this exciting approach to control will certainly occur. Major advances in characterization of the interactions of cells, cytokines and immunogens in the skin provide invaluable information for investigators studying interactions of arthropods at the host cutaneous interface. Fundamental knowledge of the immunology of host–ectoparasitic arthropod interactions will enhance understanding of how arthropods successfully infest a host and act as vectors of disease-causing agents.

This is the first book to draw together in one source an extensive treatment of the emerging field that I refer to as 'immunoentomology'. The first six chapters serve as 'foundation information' for examining the immunology of the host–arthropod interface. Chapter 1, Immunology of the Skin, is intended to impart a fundamental core of information regarding the rapid advances in cutaneous immunobiology. Chapters focusing on salivary gland physiology, pharmacology of haematophagous arthropod saliva and arthropod modulation of host immune responses provide a fascinating insight into the diverse array of pharmacological and biochemical activities associated with arthropod saliva. Ectoparasitic arthropods have developed complex countermeasures to host haemostatic, inflammatory and immune defences. Fate of the bloodmeal is discussed in Chapter 6. The immunology of specific ectoparasitic arthropod–host relationships is examined in Chapters 7 to 11. An applied extension of basic studies is the development of immunological-based control of arthropod species of human and veterinary public health importance. This topic is discussed in Chapter 12. The final chapter provides an overview of selected findings and proposed avenues for future investigation.

Hopefully, information presented in this book will serve as a catalyst for new avenues of investigation and encourage the entrance of new researchers into this field of study. A wealth of exciting research opportunities awaits us!

Stephen K. Wikel

# Acknowledgements

I express my sincere appreciation to my colleagues whose discussions and valuable comments have been of great help in the preparation of this book: Douglas Bergman, Alan Bowman, Kim Burnham, Jack Dillwith, Melanie Palmer, Rangappa Ramachandra, John Sauer and William Wrenn. The excellent guidance of Tim Hardwick and the staff of CAB INTERNATIONAL was invaluable and made the preparation of this book an enjoyable experience. The most special thank you is extended to my wife Nancy for her suggestions, encouragement and support. Her love is a blessing which makes all things possible.

# 1

# Immunology of the Skin

## Stephen K. Wikel

*Department of Entomology, Oklahoma State University, Stillwater, Oklahoma 74078, USA*

## INTRODUCTION

Arthropod–host immune interactions which occur at the cutaneous interface are complex and dynamic. An understanding of the immunology of these associations can only be achieved by careful analysis of the interplay between arthropod and host factors. Arthropod-associated elements that affect the host include: immunogenic molecules and pharmacologically active components of saliva; duration of exposure to arthropod immunogens; and number of infesting arthropods. Host factors influencing immunocompetence include: genetic background; age; underlying diseases; diet; infections; and drug therapies. Interactions of ectoparasitic arthropods and the animals they parasitize through time have resulted in stimulation of host humoral and cellular immune responses to infestation and development of arthropod countermeasures to those defences.

Ectoparasitic arthropod interactions with the host occur at the cutaneous interface (Fig. 1.1). A detailed description of the skin is beyond the scope of this chapter. Readers are referred to any histology text for a description of the microscopic anatomy of the skin. Skin physiology, biochemistry and molecular biology have been thoroughly addressed in the excellent text edited by Goldsmith (1991). The skin is a truly wondrous organ that adapts to a potentially continuously changing array of stresses, including desiccation, physical insults, chemical irritants and microorganisms. The epidermis is a keratinizing epithelium that covers the body with openings for glands and

© CAB INTERNATIONAL 1996. From Wikel, S.K. (ed.) *The Immunology of Host–Ectoparasitic Arthropod Relationships.* CAB INTERNATIONAL, Wallingford.

hair follicles. Epidermis terminates at orifices in mucocutaneous junctions (Odland, 1991). The major cell of the epidermis is the keratinocyte, which in recent years has been recognized as a source of biological response modifiers. The epidermis is characterized by the absence of vasculature and continuous replacement of cells. Dermis is described by Odland (1991) as a moderately dense connective tissue containing collagen, elastic fibres, nerves and microcirculatory elements of the blood vasculature and lymphatics.

This book is structured so that the reader can acquire an understanding of the complex immunological interactions of ectoparasitic arthropods with their hosts. Significant advances have been made in this field of investigation

**Fig. 1.1.** Scanning electron micrograph of two female *Amblyomma americanum* attached to skin. Mouthparts are inserted into the skin to approximately the level of the dermal–epidermal junction where a 'pool' is created in the tissues of the host upon which the tick feeds to obtain a bloodmeal.

during recent years. Contributors to this book have provided an up-to-date analysis of the immunological interactions of hosts and arthropods. Chapters addressing arthropod factors influencing these interactions include: mouthpart structure and feeding mechanisms; salivary gland physiology; pharmacology of haematophagous arthropod saliva; arthropod modulation of host immune responses; and digestion and fate of the vertebrate bloodmeal. Chapters focusing on specific arthropods provide in-depth information concerning these and other aspects of particular arthropod–host relationships.

## OVERVIEW OF THE IMMUNE RESPONSE

The immune system is an integrated series of interactions among a complex array of cells and soluble mediators, which have reached their greatest complexity among higher vertebrates. The immune system is composed of intricately orchestrated pathways of innate and acquired (adaptive) immunity, which provide protection against disease-causing organisms and environmental insults. Desired outcome of the immune defences is elimination or neutralization of the threat to homeostasis. The immune system does not function as an isolated entity. Complex links exist with the clotting, fibrinolytic, nervous and endocrine systems.

The objective of this section is to provide an overview of the general aspects of the immune response prior to focusing on a more detailed analysis of the immunology of the skin. A comprehensive description of immune regulatory and effector pathways is beyond the scope of this book. Readers requiring more general information about the fundamentals of contemporary immunology should consult the superb textbook edited by Paul (1993).

### Innate Immunity

Innate immune defences are considered to be non-specific factors that represent a primary line of defence. The intact skin provides a barrier against disease-causing organisms and environmental insults. Cutaneous fatty acids, lactic acid and sebaceous secretions are additional elements inhibitory to microorganisms. The normal bacterial flora present on skin also contributes to the control of potential disease-causing microbes. Additional non-specific defences include mucus, tears, saliva and urine. Lysozyme present in many secretions is a non-specific defence element.

Once the skin is disrupted an array of cells and soluble mediators contribute to innate immunity. Acute-phase proteins and complement, particularly the alternative pathway, are first lines of defence. Activated complement components can destroy microorganisms and abnormal cells; attract cells to a site of injury or focus of microorganisms; and, facilitate the uptake of foreign material by macrophages and neutrophils. Interferons are cytokines which also contribute to innate defence.

In addition to complement component C3b, phagocytosis can also be initiated in the non-immune host by interaction of the 'foreign' material with carbohydrate moieties on the phagocyte surface. Upon phagocytosis the engulfed material is enclosed in a vacuole, which fuses with a lysosome to form a phagolysosome. Activated phagocytes produce the oxygen compounds superoxide anion, hydrogen peroxide, singlet oxygen and hydroxyl radicals, which are highly toxic to microorganisms. Nitric oxide plays an important role in intracellular and extracellular defences, as well as a variety of other physiological and pathological processes (Lyons, 1995).

Eosinophils, basophils and mast cells are white blood cells which contribute to innate defences. Eosinophils are phagocytic and possess surface receptors for complement component C3b, which can facilitate uptake of injurious and/or infectious material by phagocytosis. Mast cells and basophils can be degranulated by the complement-derived anaphylatoxins C3a and C5a, resulting in the release of a plethora of biologically active mediators.

The natural killer (NK) cell is another important component of innate immunity. NK cells are often referred to as large granular lymphocytes, which are neither B nor T lymphocytes. The surface receptors of NK cells predominantly have specificity for virus and tumour antigens. An important aspect of this component of cellular defence is that target antigens are recognized by the NK cell surface receptor independent of the major histocompatibility antigens of the target. NK cells release perforin, which acts to form transmembrane channels in target cells. Host cytokines play an important role in the activation of these cells.

These elements and others comprise the first line of defence, innate immunity. The differences between innate and acquired (adaptive) defences are not always distinct. Many of the same cellular elements are involved in both aspects of immune defences.

## Acquired (Adaptive) Immunity

Environmental substances and microorganisms that escape the innate defences are the target of the acquired (adaptive) immune response. The goal of the acquired immune response is the elimination of the 'foreign' material (immunogen) that poses a threat to the host. Acquired immunity is a learned response characterized by 'specificity' for the sensitizing immunogen and memory, the ability to rapidly develop a heightened response on subsequent exposure to the immunogen. An important aspect of the immune response (simplified) is the ability to distinguish self from non-self. The principal cellular elements of the acquired immune response are antigen-presenting cells (macrophages, B lymphocytes, Langerhans cells, dendritic cells) and antigen-specific B and T lymphocytes. Each B and T lymphocyte possesses a specific surface receptor that recognizes a distinct chemical configuration. The term immunogen refers to a molecule that is capable of inducing a host immune response. A microorganism might be composed of hundreds to thousands of immunogens. The term antigen is often assumed to mean the same as immunogen; however, it refers to the ability of a molecule to interact with the products of an immune response. An immunogen actually determines which clones of B and/or T lymphocytes will be activated due to their ability to recognize unique structural regions of the immunogen.

An immunogen is processed and presented by an antigen-presenting cell, which presents the immunogen to a helper T lymphocyte ($T_H$) with a specific surface receptor for a region of that immunogen, which is called an epitope or antigenic determinant. The immunogen is presented to helper T lymphocytes ($CD4^+$) in association with a major histocompatibility molecule on the surface of the antigen-presenting cell. Helper T lymphocytes receive secondary activation signals through receptor–ligand interactions between themselves and the antigen-presenting cell, as well as immunoregulatory signals in the form of cytokines. Cytokines are soluble molecules that regulate the functions of many cell types through an intricate web of interactions. Helper T lymphocytes act just as their name indicates. They provide signals essential for the development of $CD8^+$ lymphocytes into cytotoxic effector lymphocytes and for B lymphocytes to differentiate into end-stage plasma cells that elaborate antibodies. Helper T lymphocytes are also the effectors of the delayed type hypersensitivity response that results in the accumulation of macrophages that isolate and hopefully destroy an immunogen. A unique form of delayed type hypersensitivity elicited by arthropods is cutaneous basophil hypersensitivity.

Activated B lymphocytes ultimately become plasma cells producing antibodies. Each immunoglobulin isotype (IgM, IgG, IgD, IgA, IgE) has unique biological properties. IgM and IgG activate the classical pathway of complement, which generates a vast array of biological activities. Coating of microorganisms or other immunogens with IgG facilitated phagocytosis by macrophages and granulocytes through a process referred to as opsonization. Macrophages, neutrophils and eosinophils possess receptors for the Fc region of antibodies that are important in the host defence activities of these cells. Mast cells and basophils possess Fc receptors for IgE, which is the basis for the importance of this isotype and these cells in immediate (allergic) hypersensitivity.

A primary antibody response might require days to weeks to develop. A primary cell-mediated response will develop within days. The duration post-sensitization before reactivity is noted depends upon the sensitivity of the detection assay. A secondary, or anamnestic, antibody response will occur more rapidly than the primary response resulting in a higher titre of antibody, higher affinity antibody and be predominantly of the IgG isotype. The initial stage of a primary response consists mainly of IgM. Memory B and T lymphocytes are responsible for the ability of an animal to develop a heightened response upon re-exposure to an immunogen. This heightened and rapid reactivity upon secondary exposure to immunogens is the fundamental basis for the concept of immunization.

A key element of all immune responses is the exquisite degree of regulation. Many microorganisms, multicellular parasites and tumours have developed countermeasures to host immune defences. Ectoparasitic arthropods are no exception. We can more fully understand the biology of ectoparasitic arthropods by examining their relationship with the host immune defences, particularly those at the cutaneous interface with the arthropod.

## THE SKIN IMMUNE SYSTEM

The purpose of this chapter is to provide fundamental information essential to establishing a basic understanding of immune regulatory and effector functions of the skin. Investigators interested in the immunology of arthropod–host interactions need to be aware of the rapidly expanding base of knowledge about the immunology of the skin. Further general information can be obtained by consulting reviews of cutaneous immunology (Streilein, 1983; Bos and Kapsenberg, 1993; Salmon et al., 1994), particularly the comprehensive text on the subject edited by Bos (1990).

New findings about the immunology of the skin have been accumulating at an explosive rate during the past few years (Bos and Kapsenberg 1993; Salmon et al., 1994). The unique features of cutaneous immune function and their interactions with other elements of the immune system were highlighted by Streilein (1978), when he developed the concept of skin-associated lymphoid tissues (SALT). Results of continuing investigations led to further refinement of the cellular basis of SALT (Streilein, 1990). Cells comprising SALT interact in the development and expression of cutaneous immunity. Lymph nodes draining the skin are integral components of SALT, due to their role in providing a site for interaction of immunocompetent lymphocytes and antigen-presenting cells, particularly Langerhans cells that have migrated from the epidermis. Functions attributed to SALT are the development of primary immune responses, expression of secondary immune responses to previously encountered immunogens and selective avoidance of immune reactivity to normal skin components (Streilein, 1990). The physiological and immunological roles of the skin occur in a balanced manner.

The term skin immune system (SIS) was proposed by Bos and Kapsenberg (1986) to describe the complex immunological interactions of cells and humoral factors in normal skin. Cutaneous immune responses involve both resident and infiltrating cells including: Langerhans cells, keratinocytes, dendritic epidermal T lymphocytes, melanocytes, mast cells, tissue macrophages, monocytes, granulocytes and endothelial cells. The complex interactions of these cells involve adhesion molecules and biological response modifiers: cytokines, colony stimulating factors, growth factors, prostaglandins, leukotrienes and complement components (Bos and Kapsenberg, 1993). Research defining the roles of soluble molecules as regulators and effectors of cutaneous immune function is providing extremely valuable insights into normal immune function of the skin and responses to insults.

A unique aspect of the skin immune system is that the epidermis has no direct access to the blood or lymphatics. However, an intricate web of cytokines coordinates immunologic interactions among keratinocytes, Langerhans cells, epidermotrophic T lymphocytes, cells of the dermis and regional lymph nodes (Bos and Kapsenberg, 1993; Salmon et al., 1994). These biological response modifiers (cytokines) are taken up by the superficial vasculature of the upper dermis and distributed systemically (Salmon et al., 1994). These interactions result in direction of cells of the inflammatory/immune responses to the skin.

The intricate interactions of cytokines, resident cells and cells migrating into the skin are beginning to be elucidated. The highest accumulation in normal skin of cells related to the immune response are found in the perivascular region of post-capillary venules in the papillary and deep dermis (Bos and Kapsenberg, 1993). These regions become more densely populated with cells during cutaneous inflammation. Endothelial cells of the dermal microvasculature are key participants in cutaneous immune responses through their elaboration of cytokines, growth factors and chemokines, which are involved in trafficking of leukocytes into the dermis and epidermis. Adhesion molecules expressed on the endothelial cell surface include: intercellular adhesion molecule-1 (ICAM-1); vascular cell adhesion molecule-1 (VCAM-1); E-selectin; and, P-selectin (Bauer and Caughman, 1994). The expression of adhesion molecules is modulated by the actions of interleukin-1 (IL-1), tumour necrosis factor alpha (TNF-$\alpha$) and interferon-gamma (IFN-$\gamma$) levels (Ruszczak et al., 1990; Detmar et al., 1992). Leukocytes' integrins adhere to endothelial cell adhesion molecules to form stable complexes on the luminal surface of the microvasculature (Butcher, 1991). Chemotactic gradients and endothelial cell retraction contribute to leukocyte migration into the cutaneous interstitium.

## Langerhans Cells

Langerhans cells are key elements that direct immune responses in the skin through their ability to process and present antigen, as well as interact with their cellular neighbours through an array of soluble mediators. Langerhans cells occur in the epidermis as a dendritic network of cells whose processes extend between keratinocytes as an efficient 'web' for trapping foreign material introduced into or on to the skin (Wolff and Stingl, 1983). Langerhans cells, comprising 2–4% of epidermal cells, are bone-marrow derived and they have modest phagocytic capability. In addition to the epidermis, they occur in the dermis, lymph nodes and thymus (Wolff and Stingl, 1983). Distinguishing features of Langerhans cells are cytoplasmic organelles referred to as Birbeck granules (Kapsenberg et al., 1990) and positive histochemical staining for adenosine triphosphatase (Juhlin and Shelly, 1977).

Langerhans cells are important antigen-presenting cells that can migrate from their suprabasal position in the epidermis, where they interact with immunogens, to the draining lymph node for presentation of immunogen to T lymphocytes (Kapsenberg et al., 1990; Larsen et al., 1990). Langerhans cells in the afferent lymph are referred to as veiled cells (Bos and Kapsenberg, 1986). Upon entering the draining lymph node, Langerhans cells function as dendritic antigen-presenting cells, stimulating naive or memory T lymphocytes (Salmon et al., 1994). Langerhans cells express surface class II major histocompatibility antigens important in the presentation of antigen to specifically reactive T lymphocytes (Shimada et al., 1987). They are the only cells in normal epidermis that express these markers (Salmon et al., 1994). Langerhans cells bear the B7-1 surface molecule, which complexes with the CD28 and CTLA-4 molecules on the surface of $T_H1$ lymphocytes (Larsen et al., 1992). Expression of B7-1 correlates with the *in vitro* transition of Langerhans cells into highly active antigen-presenting cells. Langerhans cells function in antigen presentation in a manner similar to macrophages with 'processed' antigen presented in association with class II major histocompatibility antigen to T lymphocytes. These specific interactions result in generation of effector and memory T lymphocytes (Streilein et al., 1990). Langerhans cells are essential for the development of cutaneous delayed type hypersensitivity responses (Toews et al., 1980), which are mediated by $T_H1$ lymphocytes (Mossman and Coffman, 1989).

Additional surface receptors of Langerhans cells include those for the Fc piece of IgG and IgE, as well as complement component C3 (Wolff and Stingl, 1983; Kapsenberg et al., 1990). Langerhans cells express intercellular adhesion molecule-1 (ICAM-1) and lymphocyte function associated antigen-3 (LFA-3), which facilitate interactions with T lymphocytes (Teunissen, 1992). The presence of E-cadherin on the Langerhans cell surface is involved in adherence *in vitro* to keratinocytes (Tang et al., 1993). Loss of E-cadherin might facilitate the movement of Langerhans cells from the epidermis (Bauer and Caughman, 1994). Langerhans cells express heat-stable antigen, which acts as a co-stimulatory molecule in $T_H1$-lymphocyte-mediated immune reactions in the skin (Enk and Katz, 1994). This co-stimulatory molecule for $CD4^+$ T-lymphocyte proliferation is also expressed by B lymphocytes (Bruce et al., 1981).

Cytokines play a critical role in orchestration of immune responses (Kroemer et al., 1993). Langerhans cells provide important cytokines for regulation of cutaneous immune responses. Interleukin-1 (IL-1) enhances viability of Langerhans cells in culture and increases their ability, when applied with granulocyte/monocyte-colony-stimulating factor, to activate allogeneic T lymphocytes (Heufler et al., 1988). An early indicator of Langerhans cell activation is an increase in the amount of interleukin-1-β (IL-1-β), which occurs within 15 minutes of *in vivo* exposure to contact allergens (Enk et al., 1993a).

Langerhans cells do not function in isolation from dendritic epidermal T lymphocytes and keratinocytes. Activated keratinocytes produce interleukin-10 (IL-10), which inhibits the ability of Langerhans cells to support antigen-induced $T_H1$-lymphocyte proliferation (Enk et al., 1993b). Interleukin-10 does not inhibit the ability of Langerhans cells to provide co-stimulatory proliferation signals to $T_H2$ lymphocytes. $T_H1$ lymphocytes are the effector cells of delayed type hypersensitivity reactions, while $T_H2$ lymphocytes provide helper signals for B lymphocytes (Mossman and Coffman, 1989). $T_H1$ lymphocytes can act as helper cells for differentiation of B lymphocytes, and both helper T-lymphocyte populations enhance production of cytotoxic T lymphocytes (Coffman et al., 1988). Interleukin-1-α inhibits the ability of Langerhans cells to present antigen, in part through tumour necrosis factor alpha (TNF-α) (Grabbe et al., 1994). Keratinocytes produce both IL-1-α and TNF-α (Bos and Kapsenberg, 1993). Tumour necrosis factor alpha maintains the *in vitro* viability of murine Langerhans cells, but does not induce their functional maturation (Koch et al., 1990). The same cytokine of dermal origin may provide a signal for Langerhans cell migration (Cumberbatch and Kimber, 1992).

An intriguing association between Langerhans cell antigen-presenting capabilities and the nervous system was described (Hosoi et al., 1993). Calcitonin gene-related peptide-positive nerve axons contact Langerhans cells within the epidermis. Approximately 70–80% of Langerhans cells were in contact with calcitonin gene-related peptide-positive epidermal nerves. In addition, calcitonin gene-related peptide was found in association with the surface of Langerhans cells. Calcitonin gene-related peptide inhibited the ability of Langerhans cells to present the antigen chicken ovalbumin. Langerhans cells may be subject to neurological modulation. Can stress, including ectoparasite infestation, impact cutaneous immune competence?

## Keratinocytes

Keratinocytes, which comprise 90% of the epidermis, are no longer simply considered just an inert barrier essential for maintaining homeostasis (Chu et al., 1990; Salmon et al., 1994). A complex network of interactions involving cells and soluble mediators occurs within the skin. The keratinocyte has emerged as an immunologically important cell involved in expression of class II major histocompatibility complex antigens and elaboration of cytokines (Bos and Kapsenberg, 1993). These properties of keratinocytes correlate with a role in antigen presentation and regulation of T-lymphocyte responses.

Class II major histocompatibility antigens were first described for human keratinocytes involved in a graft versus host reaction (Lampert *et al.*, 1981). T-lymphocyte-derived interferon-γ (IFN-γ) activates human keratinocytes to transiently express class II major histocompatibility antigens (Volc-Platzer *et al.*, 1985). The role of major histocompatibility class-II-positive keratinocytes as antigen-presenting cells is becoming more clearly defined. Keratinocytes expressing class II major histocompatibility antigens were reported to be unable to present antigens to naive T lymphocytes; however, they were capable of activating memory and effector T cells (Salmon *et al.*, 1994). In addition, major histocompatibility complex class-II-positive keratinocytes were capable of specifically inhibiting cutaneous cell-mediated immune reactions (Krueger and Stingl, 1989).

Recent studies provide new insights into the role of keratinocytes as antigen-presenting cells. T-lymphocyte activation, proliferation and cytokine elaboration require occupancy of the antigen-specific receptor and co-stimulatory signals (Mueller *et al.*, 1989; Jenkins, 1992). Absence of the co-stimulatory signal after antigen receptor occupancy results in development of anergy, specific unresponsiveness, rather than activation (Jenkins, 1992). Co-stimulatory signals are provided by surface molecules of the antigen-presenting cells (Mueller *et al.*, 1989; Jenkins, 1992; Nickoloff and Turka, 1994). Keratinocytes expressing major histocompatibility complex class II markers are capable of providing co-stimulatory signals to purified T lymphocytes (Nickoloff *et al.*, 1993). Keratinocyte co-stimulatory signals are not provided through B7-3 acting as a ligand for T-lymphocyte CD28 (Nickoloff and Turka, 1994). Co-stimulatory surface molecular interactions remain to be identified for keratinocytes and T lymphocytes.

T-lymphocyte cytokine responses that are induced by keratinocytes differ from those elicited by 'professional' antigen-presenting cells, such as macrophages, dendritic cells and activated B lymphocytes (Nickoloff and Turka, 1994). Monocytes acting as antigen-presenting cells induced the elaboration of $T_0$ cytokine profile of IL-2, IL-4 and IFN-γ. T-lymphocyte cytokines induced by keratinocytes consisted of normal levels of IL-2 and IL-4; however, IFN-γ production was deficient (Nickoloff and Turka, 1994). Differences in IFN-γ production were attributed to levels of interleukin-12 available to drive the transition from a $T_0$ to a $T_H1$ cell. Keratinocytes lack the ability to provide an adequate IL-12 signal for this transition. In addition, keratinocytes can elaborate IL-10 (Enk and Katz, 1992). Interleukin-12 contributes to the generation of $T_H1$ lymphocytes rather than $T_H2$ cells (Trinchieri, 1995). Interleukin-10 inhibits $T_H1$-lymphocyte immune functions (Moore *et al.*, 1993). Antigen presentation by keratinocytes would appear to deviate immune responses towards the $T_H2$-lymphocyte pathway with apparent anergy of $T_H1$-lymphocyte function (Nickoloff and Turka, 1994). The repertoire of immune responses arising from antigen presentation by Langerhans cells or keratinocytes might be different. $T_H1$ lymphocytes are the mediators of delayed type hypersensitivity reaction, provide help to cytotoxic T cells and elaborate cytokines such as IL-2 and IFN-γ (Mossman and Coffman, 1989).

Cytokines provide a link between keratinocytes and dendritic epidermal T lymphocytes. Keratinocyte-derived interleukin-7 (IL-7) is a growth factor for dendritic epidermal T lymphocytes (Matsue *et al.*, 1993). Expression of IL-7 by keratinocytes is regulated by IFN-γ, which is derived from T lymphocytes (Ariizumi *et al.*, 1995). A positive feedback loop exists consisting of dendritic epidermal T-cell secretion of IFN-γ inducing production of IL-7 by keratinocytes, which in turn acts as a growth factor for epidermal T cells. Activated gamma/delta T lymphocytes derived from the skin elaborate keratinocyte growth factor (Boismenu and Havran, 1994). Cutaneous T lymphocytes bearing the alpha/beta antigen receptor, as well as lymphoid T cells with either antigen receptor structure, did not express the mitogen for keratinocytes.

Keratinocytes produce a complex array of cytokines with diverse biological activities. Given that 90% of the epidermis is composed of keratinocytes, these cells represent a significant source of biological response modifiers. Keratinocytes do not constitutively produce most of these cytokines *in vivo*, but their expression is induced by factors such as irritants, ultraviolet radiation and injury (Bos and Kapsenberg, 1993). Keratinocyte-derived cytokines contribute significantly to cutaneous immune reactions through autocrine/paracrine roles, induction of adhesion molecules and generation of chemotactic gradients. Cytokines elaborated by keratinocytes are described in Table 1.1 (Luger and Schwarz, 1990; Enk and Katz, 1992; Bos and Kapsenberg, 1993; Ariizumi *et al.*, 1995).

Activated keratinocytes can be induced by IFN-γ and TNF to express

**Table 1.1.** Cytokines elaborated by keratinocytes.

| |
|---|
| Interleukins: IL-1-α, IL-1-β, IL-6, IL-7, IL-8, IL-10, IL-12 (low levels) |
| Interferons: IFN-α, IFN-β |
| Tumour necrosis factor: TNF-α |
| Growth factors: Platelet derived and fibroblast |
| Transforming growth factors: TGF alpha and beta |

(*Source*: Adapted from Luger and Schwarz, 1990; Enk and Katz, 1992; Bos and Kapsenberg, 1993; and Ariizumi *et al.*, 1995.)

ICAM-1, which is involved in the movement of cells participating in immune and inflammatory responses into the epidermis (Griffiths et al., 1989; Singer et al., 1990). T-lymphocyte ligand for keratinocyte ICAM-1 is LFA-1, since antibodies to either of these molecules inhibit cell migration (Singer et al., 1990). Cellular trafficking into the dermis is dependent in part upon endothelial cell adhesion molecule interaction with appropriate receptors on migrating leukocytes. Endothelial cells express low levels of ICAM-1; however, levels are increased following induction by IL-1, TNF or IFN-γ (Ruszczak et al., 1990; Detmar et al., 1992). Cytokines provide a possible signal pathway between endothelial cells and keratinocytes. Additional receptor–ligand interactions occur between endothelial cell leukocyte adhesion molecule and cutaneous lymphocyte-associated antigen, which facilitate T-cell movement into the skin (Picker et al. 1991).

## T Lymphocytes

The roles of T lymphocytes in the skin have been more clearly defined in recent years (Sim, 1995). Epidermotropic T lymphocytes and murine dendritic epidermal T cells are important cutaneous immune regulatory and effector elements. Results of investigations performed on mouse skin are not always directly applicable to human cutaneous immune function. The presence of dendritic epidermal T cells has not been conclusively established for humans (Salmon et al., 1994).

Bos et al. (1987) reported that the vast majority of T lymphocytes are located in the dermal perivascular units of normal human skin. Approximately 90% of T lymphocytes are localized around the post-capillary venules of the dermis and 5% of T cells occur in the epidermis of normal human skin (Bos et al., 1990). Intraepithelial T lymphocytes are predominantly of the CD8 lineage (Bos and Kapsenberg, 1990), while $CD4^+$ T lymphocytes are mainly of the memory, $CDw29^+$, immunophenotype (Bos et al., 1989). Presence of memory cells assures a rapid secondary immune response upon exposure to the sensitizing immunogen. An additional marker associated with T lymphocytes in human skin was an epitope of a 200 kilodalton (kDa) cell surface glycoprotein binding with monoclonal antibody HECA-452 (Picker et al., 1990). This marker occurred on 16% of peripheral blood T lymphocytes of both CD4 and CD8 subpopulations. Examination of inflammatory skin lesions revealed that 85% of T lymphocytes were HECA-452 positive. The HECA-452 reactive epitope was associated with memory T cells. Human dermis contains 7–9% gamma/delta and 80% alpha/beta receptor-bearing T lymphocytes, as determined by use of monoclonal antibody probes (Bos et al., 1990). Approximately 18–29% of epidermal T cells are gamma/delta receptor positive, while 60% possess alpha/beta receptors (Bos et al., 1990).

A population of dendritic cells located in the basal layer of the murine epidermis consists of Thy-1$^+$ T lymphocytes (Bergstresser et al., 1983; Tschachler et al., 1983). Dendritic epidermal T cells are bone-marrow derived and epidermotropic (Tigelaar et al., 1990). Dendritic epidermal T cells isolated from skin selectively migrate to skin and thymus upon intravenous infusion (Cruz et al., 1990). Antibodies reactive with lymphocyte-function-associated antigen-1 (LFA-1) block T-lymphocyte migration to the epidermis (Shiohara et al., 1991).

Dendritic epidermal T cells can be distinguished by the presence of CD3, CD45, the lack of CD4 and CD8, and the unique expression of the gamma/delta receptor for antigen (Bergstresser et al., 1993). Immunophenotyping of dendritic epidermal T lymphocytes reveals an absence of CD4 and CD5 markers with a limited presence of CD8 (Tschachler et al., 1983, 1989). Murine dendritic epidermal T lymphocytes can be distinguished from Langerhans cells by their lack of expression of class II (Ia) major histocompatibility antigens (Bergstresser et al., 1993). A unique feature of dendritic epidermal T cells is expression of a non-polymorphic, gamma/delta T-cell receptor (Koning et al., 1987; Havran et al., 1989a). Gamma/delta receptor T lymphocytes in the skin possess a very limited repertoire of antigen recognition specificities (Allison and Havran, 1991; Bergstresser et al., 1993). Dendritic epidermal T lymphocytes recognize self-antigens presented by cultured keratinocytes in a manner not restricted by major histocompatibility antigens (Havran et al., 1991). This homogeneity of antigen receptors is thought to reflect a population of T lymphocytes reactive with common antigens, stress (heat shock) proteins, induced in the skin by a variety of insults (Asarnow et al., 1988). Dendritic epidermal T cells might function as a primitive defence system within the epidermis (Tigelaar et al., 1990).

Upon stimulation, dendritic epidermal T cells produce mRNAs for the following cytokines: IL-1-$\alpha$, IL-2, IL-3, IL-6, IL-7, IFN-$\gamma$, TNF-$\alpha$, TNF-$\beta$ and granulocyte/monocyte colony stimulating factor (Bergstresser et al., 1993). They do not produce IL-4 (Havran et al., 1989b). Dendritic epidermal T cells proliferate in vitro in the presence of concanavalin A and IL-2 (Nixon-Fulton et al., 1986). Keratinocyte-derived IL-7 supports the growth of dendritic epidermal T cells (Matsue et al., 1993). Gamma/delta dendritic epidermal T lymphocytes express keratinocyte growth factor, indicating that these cells might recognize and nurse injured epithelial cells (Boismenu and Havran, 1994). These cells do not live in isolation from their neighbours in the epidermis.

Epidermal T lymphocytes of humans do not possess a dendritic morphology. A population of human T cells equivalent to the murine dendritic epidermal T cells has not been described (Salmon et al., 1994). However, skin-associated gamma/delta T cells of humans may be important in maintaining skin immune function (Bergstresser et al., 1993). The proportion of human cutaneous T lymphocytes with these two receptor populations does not differ significantly from the proportion of cells with similar receptors in the lymphoid tissues (Bos *et al.*, 1990).

Possibly, gamma/delta T lymphocytes resident in epithelia have functions distinct from T cells in the circulation and lymphoid tissues possessing gamma/delta or alpha/beta receptors for antigen. Dendritic epidermal T lymphocytes are associated with resistance to vaccinia virus infection (Ikeda *et al.*, 1991) and tumour cells *in vitro* (Nixon-Fulton *et al.*, 1988), indicating a possible role as cytotoxic effector cells. Dendritic epidermal T cells stimulated with IL-2 acquire properties of natural killer (NK) cells and can mediate cytotoxicity in a non-major histocompatibility complex dependent manner (Nixon-Fulton *et al.*, 1988). Dendritic epidermal T cells in normal skin express mRNA for perforin, the pore-forming protein produced by cytotoxic lymphocytes that forms transmembrane channels in target cells (Kobata *et al.*, 1990).

## Mast Cells and Basophils

The importance of mast cells and basophils in immediate hypersensitivity (type I) reactions and inflammation is well established and the topic of excellent reviews (Siraganian, 1988; Kitamura, 1989; Van Loveren *et al.*, 1990; Schwartz, 1993; Bochner, 1995; McNeil and Austen, 1995). Mast cells, basophils and IgE are the hallmark components of allergic reactions. Although mast cells and basophils have distinct differences, a number of similarities exist between these cells. Mast cells and basophils are two distinct cell lineages (Valent and Bettelheim, 1992). Immunophenotyping with monoclonal antibodies revealed unique and shared CD antigens between basophils and mast cells (Valent and Bettelheim, 1992). Mast cells and basophils both express surface receptors for IgE. Murine and rat mast cells occur as two morphologically and functionally distinct subsets in specific tissues (McNeil and Austen, 1995). The first subset is referred to as connective tissue mast cells occurring in the skin, the submucosa of the small intestine and serosal surfaces. The mucosal mast cell subset occurs in the mucosa and lamina propria of the small intestine. Evidence exists to support the occurrence of human mast cell subpopulations (McNeil and Austen, 1995). Mast cells are resident in tissues, while basophils are circulating granulocytes. However, basophils do accumulate in tissues as components of cell-mediated reactions known as cutaneous basophil hypersensitivity (CBH) responses, which are a form of delayed type hypersensitivity (Askenase *et al.*, 1979). The CBH response is frequently induced by arthropod feeding (Wikel, 1982, 1996).

Mast cells and basophils contain a variety of mediators that have important roles in inflammatory and immune responses. Granules of human basophils, mucosal mast cells and connective tissue mast cells contain the biogenic amine, histamine (Schwartz, 1993). Mouse and rat mucosal mast cells contain histamine, while connective tissue mast cell granules of both species contain histamine and serotonin (Schwartz, 1993). Human mucosal and connective tissue mast cells both contain the proteoglycans heparin and chondroitin sulphate E, while basophils contain chondroitin sulphate A (Schwartz, 1993). Mouse and rat mucosal mast cells contain chondroitin sulphate, and connective tissue mast cells contain heparin.

Mast-cell-derived bioactive molecules are classified as preformed secretory granule mediators, newly synthesized membrane-lipid-derived mediators and cytokines (Schwartz, 1993). Preformed mediators of human mast cells and basophils include biogenic amines, proteoglycans, neutral proteases and acid hydrolases, while mediators synthesized *de novo* upon activation include prostaglandin, leukotriene, platelet activating factor and adenosine (Schwartz, 1993). Mast-cell-derived cytokines (Table 1.2) possess pro-inflammatory, mitogenic/growth factor and immunomodulatory activities (Schwartz, 1993). These mediators possess a diverse array of biological activities, including chemotactic properties. Leukotriene $B_4$ is chemotactic for neutrophils (Schwartz, 1993). Histamine is chemotactic for eosinophils (Clark *et al.*, 1975).

Release of mast cell and basophil mediators occurs through the complexing of the antigen-reactive site of Fc-receptor-bound IgE molecules with bivalent or multivalent antigens (Ishizaka and Ishizaka, 1978). The number of basophil and mast cell Fc receptors for IgE is estimated to be $10^4$ to $10^6$ per cell (Conrad *et al.*, 1975; Malveaux *et al.*, 1978). Also, complement C3a and C5a are anaphylatoxins, which cause the degranulation of basophils and mast cells (Dvorak *et al.*, 1981). The biochemical bases of basophil and mast cell degranulation have been reviewed by Schwartz (1993).

Histamine acts through cell-specific receptors designated $H_1$, $H_2$ and $H_3$ (Schwartz, 1993; NcNeil and Austen, 1995). The effects of histamine depend upon the predominant receptor type on the target cell population (McNeil and Austen, 1995). Histamine binding to $H_1$ or $H_2$ receptors results in separate pathways of signal transduction (Falus and Meretey, 1992). Histamine interaction with $H_1$ receptors induces positive effects, while stimulation of the

**Table 1.2.** Mast-cell-derived cytokines.

| | |
|---|---|
| Pro-inflammatory cytokines: | TNF-α, IL-1-α, IL-1-β, IL-6 and macrophage inflammatory protein family |
| Mitogenic cytokines/growth factors: | IL-3, IL-4, IL-5, Il-10 and GM-CSF |
| Immunomodulatory cytokines: | IL-1-α, Il-1-β, IL-4, IL-10 and IFN-γ |

(*Source*: Adapted from Schwartz, 1993.)

$H_2$ receptor has a negative effect. Stimulation of the $H_1$ receptor results in increased synthesis of complement component C3, while $H_2$ receptor stimulation suppresses complement and acute-phase protein production by monocytes, macrophages and hepatocytes (McNeil and Austen, 1995). Histamine stimulates IL-6 synthesis by B lymphocytes, but inhibits production of TNF-α and IL-1 by macrophages, and reduces T-lymphocyte production of IL-2 and IFN-γ (McNeil and Austen, 1995). Interaction of histamine with $H_2$ receptors is immunomodulatory by inhibiting secretion by cytotoxic lymphocytes, neutrophils and basophils, as well as enhancing the suppressor activity of T lymphocytes (Melmon et al., 1981). Histamine stimulation of $H_1$ receptors results in the classic signs associated with an allergic reaction: vasodilation, increased vascular permeability, increased mucus secretion, and contraction of bronchial and gastrointestinal smooth muscles. Cytolytic lymphocytes are induced by histamine to release a chemoattractant for monocytes, T lymphocytes and eosinophils (McNeil and Austen, 1995).

Basophils and mast cells produce a variety of lipid-derived molecules with biological activities. Arachidonic acid is metabolized along the cyclo-oxygenase pathway to produce prostaglandins and thromboxanes or by the lipoxygenase pathway to yield leukotrienes (Siraganian, 1988; Schwartz, 1993; McNeil and Austen, 1995). Prostaglandin $D_2$ is the predominant prostaglandin produced by human and rodent mast cells (Urade et al., 1987). Presence of prostaglandin $D_2$ is used to distinguish mast cell from basophil-mediated responses (MacGlashan and Lichtenstein, 1983). Prostaglandin $D_2$ is chemokinetic for human neutrophils; enhances leukotriene $B_4$ accumulation of the same cell type; and inhibits platelet aggregation (Schwartz, 1993).

Leukotrienes $C_4$, $D_4$ and $E_4$ are the biological response modifiers previously identified as the slow reacting substance of anaphylaxis (SRS-A) (McNeil and Austen, 1995). These molecules possess biological activities very similar to prostaglandin $D_2$, providing significant pro-inflammatory stimuli. Leukotriene $B_4$ causes leukocyte adherence to endothelial cells and transmigration to the extravascular compartment.

Mast cells produce cytokines with pro-inflammatory, mitogenic and immunomodulatory actions (Gordon et al., 1990; Romagnani, 1992; McNeil and Austen, 1995). Mast-cell-derived pro-inflammatory cytokines are TNF-α, IL-1-α, IL-1-β and IL-6. Pro-inflammatory properties of these molecules include: chemotactic activity for leukocytes, induction of expression of endothelial cell adhesion molecules and stimulation of hepatic cell synthesis of acute-phase proteins. Levels of pro-inflammatory cytokine mRNAs peak approximately 30 to 60 minutes after activation of the Fc receptor for IgE (Gordon et al., 1990). Mitogenic cytokines/haematopoietic growth factors produced by mast cells are IL-3, IL-4, IL-5 and GM-CSF. The immunomodulatory cytokines of mast cell origin are IL-1-α, IL-1-β, IL-4, IL-10 and IFN-γ. Mast cell production of IL-4 is thought to be associated with the induction of T-lymphocyte helper responses of the $T_H2$ subset (Romagnani, 1992).

Mast cells and basophils play important roles in host immune responses to ectoparasitic arthropods. An understanding of their role in this web of immune interactions provides new insights into host–ectoparasitic arthropod relationships.

## Eosinophils

Eosinophils are blood cells slightly larger than neutrophils, which are characterized by a bilobed nucleus and granules rich in cationic proteins that are stained by eosin (Spry, 1993; Gleich et al., 1995). Eosinophils can accumulate in tissues during inflammatory and/or allergic reactions, and their numbers in the circulation can increase greatly. Eosinophils secrete cytokines, lipids and peptides in response to immunological stimuli (Spry, 1993; Weller, 1993; Gleich et al., 1995; Wardlaw et al., 1995). Eosinophils appear to cooperate with other elements of the immune system in protection against potentially harmful insults (Spry, 1993). Eosinophils are bone-marrow-derived cells that circulate in the blood and migrate to tissue in response to chemoattractants, adhesion molecules and cytokines (Walsh et al., 1993). Tissue damage arising from eosinophils occurs in allergic and granulomatous hypersensitivity states (Spry, 1993). Prior to the discovery of the pro-inflammatory role of eosinophils, their function was thought primarily to be counteracting the effects of mast-cell-derived mediators released during anaphylaxis (Gleich et al., 1995).

Cytokines are important in eosinopoiesis. The primary eosinopoietic factors are IL-3, IL-5 and GM-CSF (Sanderson et al., 1988). Eosinophils can be activated by IL-2 (Silberstein et al., 1989) and IL-5 (Sanderson et al., 1988), while IFN-γ activates in a delayed manner (Valerius et al., 1990). Synergistic action of IL-3 and IFN-γ resulted in expression of class II major histocompatibility antigen by eosinophils, while TNF synergized equally with IL-3, IL-5 and GM-CSF for maximal induction of eosinophil intercellular adhesion molecule-1 (Hansel et al., 1992). Eosinophils are capable of phagocytosis, processing/presenting antigen and providing accessory functions for interacting with T lymphocytes (Hansel et al., 1992). The significance of eosinophils in immunoregulation remains to be determined.

Eosinophils express surface receptors for IgG (Henson, 1969; Tai and Spry, 1976), IgE (Capron and Capron, 1987), IgA (Kulczycki, 1984), complement component C3b (Fischer et al., 1986) and cytokines (Wardlaw et al., 1995). The number of CR1 receptors for C3b is increased by histamine, eosinophil chemotactic factor of anaphylaxis, leukotriene $B_4$ and other biological response modifiers (Fearon, 1985). Surface binding to C3b receptors causes eosinophil degranulation. Eosinophils possess receptors for the anaphylatoxins C3a and C5a (Glovsky et al., 1979); however, the physiological significance of these receptors has not been determined.

The movement of eosinophils into tissues involves expression of adhesion molecules by activated vascular endothelial cells and their interaction with receptors on the eosinophil surface (Walsh et al., 1993). Both E-selectin (also known as ELAM-1) and ICAM-1 are expressed by endothelial cells and are involved in binding eosinophils and their subsequent migration into the extravascular compartment (Kyan-Aung et al., 1991). In addition, endothelium expression of vascular cell adhesion molecule-1 (VCAM-1) contributed to the adherence and migration of eosinophils (Walsh et al., 1993). Eosinophil surface receptors for specific adhesion molecules are LFA-1 and CR3 (Mac-1) binding ICAM-1; VLA-4 interacting with VCAM-1; and CD-18 with E-selectin (Kyan-Aung et al., 1991; Walsh et al., 1993).

Eosinophil transmigration of the vascular endothelium and accumulation in tissues at foci of inflammatory/immune reactions requires chemotactic signals. Platelet activating factor (PAF) and C5a are highly chemotactic for eosinophils (Wardlaw et al., 1986; Walsh et al., 1993). Leukotriene $B_4$ ($LTB_4$) is slightly chemoattractive for human eosinophils (Wardlaw et al., 1986) and highly active in attracting guinea pig eosinophils (Sehmi et al., 1991). Chemoattractive activity of eosinophil chemotactic factor of anaphylaxis (ECF-A) is attributable to $LTB_4$ and 8(S), 15(S)-diHETE (Sehmi et al., 1991). Histamine is chemotactic for eosinophils by acting upon the $H_2$ receptor (Clark et al., 1975).

Constituents of eosinophil granules are responsible for the cytotoxic potential, pro-inflammatory properties and the role of these cells in immune responses (Gleich and Adolphson, 1986; Venge, 1993; Weller, 1993). Eosinophils possess two types of granules (Venge, 1993). The first granule type contains peroxidase and is characterized by a crystalloid bar of major basic protein. The second granule population is peroxidase negative. Both types of granules contain eosinophil cationic protein and eosinophil protein X, also known as eosinophil-derived neurotoxin (Venge, 1993). Lipid mediators derived from eosinophils contribute to the array of inflammatory and immune response roles of these cells (Weller, 1993).

Major basic protein is an important constituent of eosinophil granules, comprising over 50% of the granular protein of guinea pig eosinophils (Gleich et al., 1973). In addition, major basic protein is also found in basophils (Ackerman et al., 1983). Although lacking enzymatic activity, the primary role of major basic protein is to damage or kill parasites and normal human cells (Spry, 1993). The following activities are attributed to eosinophil major basic protein: inhibition of C3b; release of histamine from rat mast cells; killing of bacteria; activation of neutrophils and alveolar macrophages; and induction of neutrophil superoxide production (Gleich et al., 1992, 1995; Venge, 1993).

Composed primarily of basic amino acids, eosinophil cationic protein is a major constituent of the granule core (Venge, 1993). This molecule is a potent cytotoxin recognized for its ability to kill mammalian and non-mammalian cells, particularly helminth parasites (Venge, 1993). Additional properties attributed to eosinophil cationic protein include: killing of bacteria; neurotoxicity; causing release of basophil histamine; release of histamine from rat mast cells; blocking T-lymphocyte proliferation; and inhibition of delayed type hypersensitivity (Gleich et al., 1992, 1995; Venge, 1993).

The third key protein to consider is eosinophil peroxidase, which in the presence of a halide and hydrogen peroxide acts as a cytotoxic effector (Spry, 1993). Activities linked to eosinophil peroxidase include: killing of microorganisms and helminths; release of histamine from rat mast cells; inhibition of neutrophil phagocytosis mediated through IgG and C3b; inhibition of neutrophil chemotaxis; increased adhesiveness of neutrophils; and inhibition of lipid mediators such as leukotrienes (Gleich et al., 1992, 1995; Venge, 1993). Many properties of eosinophil peroxidase are considered anti-inflammatory (Venge, 1993).

Eosinophil protein X and eosinophil-derived neurotoxin were purified by separate groups and subsequently found to be identical molecules (Slifman et al., 1989; Venge, 1993). Biological activities attributed to this molecule are potent neurotoxicity; inhibition of T-lymphocyte proliferative response to mitogen; and cytotoxicity considerably less than that of eosinophil cationic protein (Gleich and Adolphson, 1986; Venge, 1993).

Eosinophils are capable of producing lipid mediators with a diverse array of biological activities (Weller, 1993; Gleich et al., 1995). Eosinophils can synthesize platelet-activating factor by acetylating lysophospholipid (Lee et al., 1984). Other cells capable of producing platelet-activating factor are basophils, neutrophils, monocytes, macrophages, endothelial cells and platelets (Gleich et al., 1995). Biological activities of platelet-activating factor include: induction of aggregation and secretion by platelets; aggregation and stimulation of neutrophils; heightened metabolic activity of macrophages; and increased cutaneous vascular permeability (Gleich et al., 1995). Eosinophils predominantly metabolize arachidonic acid by the lipoxygenase pathway with the preferential production of leukotrienes $C_4$ and $D_4$, which have strong pro-inflammatory properties (Gleich et al., 1995). Stimulated guinea pig eosinophils release thromboxane $B_2$, 5-hydroxyeicosatetraenoic acid (5-HETE) and leukotriene $B_4$ (Gleich et al., 1995). Eosinophil arachidonic acid metabolism through the cyclo-oxygenase pathway results in the production of prostaglandin $D_2$, prostaglandin $E_2$ and thromboxane $B_2$ (Weller, 1993).

In addition to being influenced by the actions of cytokines, eosinophils elaborate a number of these important biological response modifiers, which can affect a variety of cell types (Wardlaw et al., 1995). The following cytokines produced by eosinophils were detected at either the mRNA and/or protein levels: GM-CSF, TGF-α, TGF-β, IL-1, IL-3, IL-5, IL-6 and IL-8 (Gleich et al., 1995). Interleukin-5 might act in an autocrine manner to stimulate inflammation associated with tissue eosinophilia.

Eosinophils are frequently encountered at attachment sites of ectoparasitic arthropods on a variety of host species (Wikel, 1982, 1996). The importance of these cells in immunity to arthropods will be more fully understood as the basic knowledge of the function and interactions with other cells becomes more clearly defined.

## Neutrophils

The biology of neutrophils has been the topic of recent reviews (Hellewell and Henson, 1993; Rosenberg and Gallin, 1995). Neutrophils represent 55–60% of haematopoietic cells in the bone marrow (Bainton, 1992). Maturation of neutrophils occurs through a series of stages: the pluripotent stem cell, myeloid stem cell, myeloblast, promyelocyte, myelocyte, metamyelocyte, band and mature neutrophil. Neutrophils have a very short half-life and comprise approximately two-thirds of circulating leukocytes (Rosenberg and Gallin, 1995). A variety of signal events control motility, adherence to endothelium, transmigration from the vasculature, phagocytosis and degranulation. Neutrophils secrete a number of molecules that contribute to the inflammatory response. These cells are highly phagocytic and contain two types of granules characterized by a variety of enzymes that degrade engulfed foreign or altered 'self' material. The cytokines G-CSF, GM-CSF and IL-3 affect the maturation of neutrophils (Rosenberg and Gallin, 1995). In fact, the neutrophil once considered to be only an effector of host defences contributes to the regulation of immune responses (Lloyd and Oppenheim, 1992).

Plasma membrane molecules include receptors for the Fc region of IgG, complement component C3b, adhesion molecules and chemoattractants (Hellewell and Henson, 1993; Rosenberg and Gallin, 1995). Chemoattractants for neutrophils are C5a, the formylated peptide fMLP, $LTB_4$, IL-8, macrophage inflammatory peptides and platelet-activating factor. Selectins are proteins thought to contribute to the slowing and rolling of neutrophils prior to endothelial cell adherence and transmigration (Rosenberg and Gallin, 1995). L-selectin occurs on all leukocytes. E-selectin (ELAM-1) mediates neutrophil adhesion to activated endothelial cells. P-selectin is involved in neutrophil interactions with platelets and endothelial cells. Adhesion molecule interactions involve neutrophil surface Mac-1 with its ligand on the endothelium known as ICAM-1 (Rosenberg and Gallin, 1995). Levels of expression of ICAM-1 are up regulated by IL-1, TNF-$\alpha$ and IFN-$\gamma$, which occur during inflammatory and immune responses.

Phagocytosis is facilitated by coating material to be engulfed with C3b and/or IgG, a process known as opsonization. Intracellular killing of microorganisms involves two pathways: (i) the respiratory burst yielding reactive oxygen-dependent factors; and (ii) the oxygen-independent system consisting of antimicrobial proteins (Rosenberg and Gallin, 1995). Phagocytosed material is enclosed in a phagosome that fuses with granules to expose the engulfed material to enzymes and products of the respiratory burst, which include superoxide anion, hydrogen peroxide, hypochlorite and hydroxyl radical (Hellewell and Henson, 1993; Rosenberg and Gallin, 1995). Nitric oxide may be an important factor in intracellular killing. Oxygen-independent microbicidal elements are lysozyme, lactoferrin, bacterial permeability increasing protein, defensins, cathepsin G, azurocidin/CAP37 and cationic proteins.

The concept that neutrophils were end stage cells incapable of protein synthesis has been discarded (Lloyd and Oppenheim, 1992). Neutrophils can synthesize proteins relevant to effector function as well as molecules with immunoregulatory properties. Neutrophils are capable of producing the following cytokines: IL-1-$\beta$, TNF-$\alpha$, GM-CSF, M-CSF, IL-6, IL-8 and IL-1 receptor antagonist (Lloyd and Oppenheim, 1992). How can neutrophils affect the afferent limb of the immune response? Interleukin-1-$\beta$ mRNA is transcribed within 2 to 6 hours of stimulation (Marucha *et al.*, 1990). Possible functions attributed to neutrophil-derived IL-1 during the inflammatory response include: stimulation of synthesis and release of acute-phase proteins; augmentation of B- and T-lymphocyte activation; induction of IL-6, IL-8 and GM-CSF; stimulation of further IL-1 elaboration; and induction of endothelial cell adhesion molecules (Lloyd and Oppenheim, 1992). These cytokine activities may augment neutrophil production and accumulation. Production of TNF-$\alpha$ could augment the activities of IL-1 (Lloyd and Oppenheim, 1992). Action of IL-6 at the inflammatory focus might be the induction of B-lymphocyte maturation. Neutrophil elaboration of IL-8 might result in the further accumulation of these cells at the inflammatory focus due to the chemoattractant properties of this cytokine. In addition to these pro-inflammatory cytokines, production of the IL-1 receptor antagonist by neutrophils acts as a negative regulator of the inflammatory response (Lloyd and Oppenheim, 1992)

The contribution of granulocytes to inflammatory and immune responses is indeed complex. The importance of these pathways to the arthropod–host interface is just beginning to be understood.

## ACKNOWLEDGEMENTS

Research of the author is supported by US Department of Agriculture, Oklahoma Center for Advancement of Science and Technology, Pfizer Animal Health and Oklahoma Agricultural Experiment Station Project OKLO2174. Appreciation is expressed to Dr Kim Burnham, Dr Melanie Palmer, Dr Rangappa Ramachandra and Nancy Wikel for critical review of the manuscript.

## REFERENCES

Ackerman, S.J., Kephart, G.M., Habermann, T.M., Greipp, P.R. and Gleich, G.J. (1983) Localization of eosinophil granule major basic protein in human basophils. *Journal of Experimental Medicine* 158, 946–961.

Allison, J.P. and Havran, W.L. (1991) The immunobiology of T cells with invariant gamma/delta antigen receptors. *Annual Review of Immunology* 9, 679–705.

Ariizumi, K., Meng, Y., Bergstresser, P.R. and Takashima, A. (1995) IFN-gamma-dependent IL-7 gene regulation in keratinocytes. *Journal of Immunology* 154, 6031–6039.

Asarnow, D.M., Kuziel, W.A., Bonyhadi, M., Tigelaar, R.E., Tucker, P.W. and Allison, J.P. (1988) Limited diversity of gamma/delta antigen receptor genes of Thy-1$^+$ dendritic epidermal cells. *Cell* 55, 837–847.

Askenase, P.W., Graziano, F. and Worms, M. (1979) Immunobiology of cutaneous basophil reactions. *Monographs in Allergy* 14, 222–235.

Bainton, D.F. (1992) Developmental biology of neutrophils and eosinophils. In: Gallin, J.I., Goldstein, I.M. and Snyderman, R. (eds), *Inflammation: Basic Principles and Clinical Correlates*, 2nd edn. Raven Press, New York, pp. 303–324.

Bauer, J.W. and Caughman, S.W. (1994) Cytokines, neuropeptides, and other factors in cutaneous immune responses. *Western Journal of Medicine* 160, 181–183.

Bergstresser, P.R., Tigelaar, R.E., Dees, J.H. and Streilein, J.W. (1983) Thy-1 antigen-bearing dendritic cells populate murine epidermis. *Journal of Investigative Dermatology* 81, 286–288.

Bergstresser, P.R., Cruz, P.D., Jr and Takashima, A. (1993) Dendritic epidermal T cells: lessons from mice for humans. *Journal of Investigative Dermatology* 100, 80s–83s.

Bochner, B.S. (1995) Basophils. In: Frank, M.M., Austen, K.F., Claman, H.N. and Unanue, E.R. (eds), *Samter's Immunological Diseases*, 5th edn. Little, Brown and Co., Boston, pp. 185–204.

Boismenu, R. and Havran, W.L. (1994) Modulation of epithelial cell growth by intraepithelial gamma/delta T cells. *Science* 266, 1253–1255.

Bos, J.D. (ed.) (1990) *Skin Immune System (SIS)*. CRC Press, Boca Raton, Florida, 483pp.

Bos, J.D. and Kapsenberg, M.L. (1986) The skin immune system. Its cellular constituents and their interactions. *Immunology Today* 7, 235–239.

Bos, J.D. and Kapsenberg, M.L. (1990) Lymphocyte subpopulations of the skin immune system. In: Bos, J.D. (ed.), *Skin Immune System (SIS)*. CRC Press, Boca Raton, Florida, pp. 89–108.

Bos, J.D. and Kapsenberg, M.L. (1993) The skin immune system: progress in cutaneous biology. *Immunology Today* 14, 75–78.

Bos, J.D., Zonnevold, I., Das, P.K., Krieg, S.R., van der Loos, Ch. M. and Kapsenberg, M.L. (1987) The skin immune system (SIS): distribution and immunophenotype of lymphocyte subpopulations in normal human skin. *Journal of Investigative Dermatology* 88, 569–573.

Bos, J.D., Hagenaars, C., Das, P.K., Krieg, S.R., Voorn, W.J. and Kapsenberg, M.L. (1989) Predominance of memory cells (CD4$^+$, CDw29$^+$) over naive T cells (CD4$^+$, CD45R$^+$) in both normal and diseased skin. *Archives of Dermatology Research* 281, 24–30.

Bos, J.D., Teurissen, M.B.M., Cairo, I., Krieg, S.R., Kapsenberg, M.L., Das, P.K. and Borst, J. (1990) T-cell receptor gamma/delta bearing cells in normal human skin. *Journal of Investigative Dermatology* 94, 37–42.

Bruce, J., Symington, F.W., McKean, T.J. and Sprent, J. (1981) A monoclonal antibody discriminating between subsets of T and B cells. *Journal of Immunology* 127, 2446–2501.

Butcher, E.C. (1991) Leukocyte-endothelial cell recognition: three (or more) steps to specificity and diversity. *Cell* 67, 1033–1036.

Capron, M. and Capron, A. (1987) The IgE receptor of human eosinophils. In: Kay, A.B. (ed.), *Allergy and Inflammation*. Academic Press, London, pp. 151–159.

Chu, T.C., Morris, J.F. and McLelland, J. (1990) Keratinocyte. In: Bos, J.D. (ed.), *Skin Immune System (SIS)*. CRC Press, Boca Raton, Florida, pp. 75–87.

Clark, R.A.F., Gallin, J.I. and Kaplan, A.P. (1975) Selective chemotactic activity of histamine. *Journal of Experimental Medicine* 142, 1462–1476.

Coffman, R.L., Seymour, B.W., Lebman, D.A., Hiraki, D.D., Christiansen, J.A., Shrader, B., Cherwinski, H.M., Savelkoul, H.F.J., Finkelman, F.D., Bond, M.W. and Mossmann, T.R. (1988) The role of helper T cell products in mouse B cell differentiation and isotype regulation. *Immunological Reviews* 102, 5–28.

Conrad, D.H., Bazin, H., Sehon, A.H. and Froese, A. (1975) Binding parameters of the interaction between rat IgE and rat mast cell receptors. *Journal of Immunology* 114, 1688–1691.

Cruz, P.D., Jr, Tigelaar, R.E. and Bergstresser, P.R. (1990) Langerhans cells that migrate to skin after intravenous infusion regulate the induction of contact hypersensitivity. *Journal of Immunology* 144, 2486–2492.

Cumberbatch, M. and Kimber, I. (1992) Dermal tumor necrosis factor-alpha induces dendritic cell migration to draining lymph nodes, and possibly provides one stimulus for Langerhans' cell migration. *Immunology* 75, 257–263.

Detmar, M., Tenorio, S., Hettmannsperger, U., Ruszczak, Z. and Orfanos, C.E. (1992) Cytokine regulation of proliferation and ICAM-1 expression of human dermal microvascular endothelial cells in vitro. *Journal of Investigative Dermatology* 98, 147–153.

Dvorak, A.M., Lett-Brown, M.A., Theuson, D.O. and Grant, J.A. (1981) Complement-induced degranulation of human basophils. *Journal of Immunology* 126, 523–528.

Enk, A.H. and Katz, S.I. (1992) Identification and induction of keratinocyte-derived IL-10. *Journal of Immunology* 149, 92–95.

Enk, A.H. and Katz, S.I. (1994) Heat-stable antigen is an important costimulatory molecule on epidermal Langerhans cells. *Journal of Immunology* 152, 3264–3270.

Enk, A.H., Angeloni, V.L., Udey, M.C. and Katz, S.I. (1993a) An essential role for Langerhans cell-derived IL-1 beta in the initiation of primary immune responses in skin. *Journal of Immunology* 150, 3698–3704.

Enk, A.H., Angeloni, V.L., Udey, M.C. and Katz, S.I. (1993b) Inhibition of Langerhans cell antigen-presenting function by IL-10. A role for IL-10 in induction of tolerance. *Journal of Immunology* 151, 2390–2398.

Falus, A. and Meretey, K. (1992) Histamine: an early messenger in inflammatory and immune reactions. *Immunology Today* 13, 154–156.

Fearon, D.T. (1985) Human complement receptors for C3b (CR1) and C3d (CR2).

*Journal of Investigative Dermatology* 85, 53s–57s.

Fischer, E., Capron, M., Prin, L., Kusnierz, J.P. and Kazatchkine, M.D. (1986) Human eosinophils express CR1 and CR3 complement receptors for cleavage fragments of C3. *Cellular Immunology* 97, 297–306.

Gleich, G.J. and Adolphson, C.R. (1986) The eosinophilic leukocyte: structure and function. *Advances in Immunology* 39, 177–253.

Gleich, G.J., Loegering, D.A. and Maldonado, J.E. (1973) Identification of a major basic protein in guinea pig eosinophil granules. *Journal of Experimental Medicine* 137, 1459–1471.

Gleich, G.J., Adolphson, C.R. and Leiferman, K.M. (1992) The eosinophil. In: Gallin, J.J., Goldstein, I.M. and Snyderman, R. (eds), *Inflammation: Basic Principles and Clinical Correlates*, 2nd edn. Raven Press, New York, pp. 663–700.

Gleich, G.J., Kita, H. and Adolphson, C.R. (1995) Eosinophils. In: Frank, M.M., Austen, K.F., Claman, H.N. and Unanue, E.R. (eds), *Samter's Immunologic Diseases*, 5th edn. Little, Brown and Co., Boston, pp. 205–245.

Glovsky, M.M., Hugli, T.E., Ishizaka, T., Lichtenstein, L.M. and Erickson, B.W. (1979) Anaphylatoxin-induced histamine release from human leukocytes: studies of C3a leukocyte binding and histamine release. *Journal of Clinical Investigation* 64, 804–811.

Goldsmith, L.A. (ed.) (1991) *Physiology, Biochemistry and Molecular Biology of the Skin*, 2nd edn. Oxford University Press, Oxford, 1529pp.

Gordon, J.R., Burd, P.R. and Galli, S.J. (1990) Mast cells as a source of multifactorial cytokines. *Immunology Today* 11, 458–464.

Grabbe, S., Bruvers, S. and Granstein, R.D. (1994) Interleukin $1\alpha$ but not transforming growth factor beta inhibits tumor antigen presentation by epidermal antigen-presenting cells. *Journal of Investigative Dermatology* 102, 67–73.

Griffiths, C.E.M., Voorhees, J.J. and Nickoloff, B.J. (1989) Characterization of intercellular adhesion molecule-1 and HLA-DR expression in normal and inflamed skin: modulation by recombinant gamma-interferon and tumor necrosis factor. *Journal of the American Academy of Dermatology* 20, 617–629.

Hansel, T.T., DeVries, I.J.M., Carballido, J.M., Braun, R.K., Carballido-Perrig, N., Rihs, S., Blasen, K. and Walker, C. (1992) Induction and function of eosinophil intercellular adhesion molecule-1 and HLA-DR. *Journal of Immunology* 149, 2130–2136.

Havran, W.L., Grell, S., Duwe, G., Kimura, J., Wilson, A., Kruisbeek, A.M., O'Brien, R.L., Born, W., Tigelaar, R.E. and Allison, J.P. (1989a) Limited diversity of T-cell receptor gamma-chain expression of murine thy-$1^+$ dendritic epidermal cells revealed by $V_v3$-specific monoclonal antibody. *Proceedings of the National Academy of Sciences (USA)* 86, 4185–4189.

Havran, W.L., Poenie, M., Tigelaar, R.E., Tsien, R.Y. and Allison, J.P. (1989b) Phenotypic and functional analysis of gamma/delta T cell receptor-positive murine dendritic epidermal clones. *Journal of Immunology* 142, 1422–1428.

Havran, W.L., Chien, Y.-H. and Allison, J.P. (1991) Recognition of self antigens by skin-derived T cells with invariant gamma/delta antigen receptors. *Science* 252, 1430–1432.

Hellewell, P.G. and Henson, P.M. (1993) Neutrophils and their mediators. In: Lachmann, P.J., Peters, K., Rosen, F.S. and Walport, M.J. (eds), *Clinical Aspects of Immunology*, 5th edn. Blackwell Scientific Publications, Oxford, pp. 505–522.

Henson, P.M. (1969) The adherence of leukocytes and platelets induced by fixed IgG or complement. *Immunology* 16, 107–121.

Heufler, C., Koch, F. and Schuler, G. (1988) Granulocyte/macrophage colony-stimulating factor and interleukin-1 mediate the maturation of murine epidermal Langerhans cells into potent immunostimulatory cells. *Journal of Experimental Medicine* 167, 700–705.

Hosoi, J., Murphy, G.F., Egan, C.L., Lerner, E.A., Grabbe, S., Asahina, A. and Granstein, R.D. (1993) Regulation of Langerhans cell function by nerves containing calcitonin gene-related peptide. *Nature* 363, 159–162.

Ikeda, S., Tominaga, T. and Nishimura, C. (1991) Thy-$1^+$ asialo $GM_1^+$ dendritic epidermal cells in skin defense mechanisms of vaccina virus-infected mice. *Archives of Virology* 117, 207–218.

Ishizaka, T. and Ishizaka, K. (1978) Triggering of histamine release from mast cells by divalent antibodies against IgE receptors. *Journal of Immunology* 120, 800–805.

Jenkins, M.K. (1992) The role of cell division in the induction of clonal anergy. *Immunology Today* 13, 69–73.

Juhlin, L. and Shelly, W.B. (1977) New staining techniques for the Langerhans cells. *Acta Dermatologia Venerologia* 57, 289–296.

Kapsenberg, M.L., Teunissen, M.B.M. and Bos, J.D. (1990) Langerhans cells: a unique subpopulation of antigen presenting dendritic cells. In: Bos, J.D. (ed.), *Skin Immune System (SIS)*. CRC Press, Boca Raton, Florida, pp. 109–124.

Kitamura, Y. (1989) Heterogeneity of mast cells and phenotypic change between subpopulations. *Annual Review of Immunology* 7, 59–76.

Kobata, T., Shinkai, Y., Ligo, Y., Kawasaki, A., Yagita, H., Ito, S., Shimada, S., Katz, S.I. and Okumura, K. (1990) Thy-1 positive dendritic epidermal cells contain a killer protein perforin. *International Immunology* 2, 1113–1116.

Koch, F., Heufler, C., Kampgen, E., Schneeweiss, D., Brock, G and Schuler, G. (1990) Tumor necrosis factor alpha maintains the viability of murine epidermal Langerhans' cells in culture, but in contrast to granulocyte/macrophage colony-stimulating factor, without inducing their functional maturation. *Journal of Experimental Medicine* 171, 159–171.

Koning, F., Stingl, G., Yokoyama, W.M., Yamada, H., Maloy, W.L., Tschachler, E., Shevach, E.M. and Coligan, J.E. (1987) Identification of a T3-associated gamma/delta T cell receptor on thy-$1^+$ dendritic epidermal cell lines. *Science* 236, 834–837.

Kroemer, G., deAlboran, I.M., Gonzalo, J.A. and Martinez-A.C. (1993) Immunoregulation by cytokines. *Critical Reviews in Immunology* 13, 163–191.

Krueger, G.G. and Stingl, G. (1989) Immunology/inflammation of the skin – A 50 year perspective. *Journal of Investigative Dermatology* 92, 32s–51s.

Kulczycki, A. (1984) Human neutrophils and eosinophils have structurally distinct Fc-alpha receptors. *Journal of Immunology* 133, 849–855.

Kyan-Aung, U., Haskard, D.O., Poston, R.N., Thornhill, M.H. and Lee, T.H. (1991) Endothelial leukocyte adhesion molecule-1 and intercellular adhesion molecule-1 mediate the adhesion of eosinophils to endothelial cells *in vitro* and are expressed by endothelium in allergic cutaneous inflammation *in vivo*. *Journal of Immunology* 146, 521–528.

Lampert, I.A., Smitters, A.J. and Chisholm, P.M. (1981) Expression of Ia antigen on epidermal keratinocytes in graft-versus-host disease. *Nature* 293, 149–150.

Larsen, C., Steinman, R.M., Witmer-Pack, M., Hankins, D.F., Morris, P.J. and Austyn, J.M. (1990) Migration and maturation of Langerhans cells in skin transplants and explants. *Journal of Experimental Medicine* 172, 1483–1493.

Larsen, C., Ritchie, S., Pearson, T., Linsley, P. and Lowry, R. (1992) Functional expression of the costimulatory molecule, B7/BB1, on murine dendritic cell populations. *Journal of Experimental Medicine* 176, 1215–1220.

Lee, T.-C., Lenihan, D.J., Malone, B., Roddy, L.L. and Wasserman, S.I. (1984) Increased biosynthesis of platelet-activating factor in activated human eosinophils. *Journal of Biological Chemistry* 259, 5526–5530.

Lloyd, A.R. and Oppenheim, J.J. (1992) Poly's lament: the neglected role of the polymorphonuclear neutrophil in the afferent limb of the immune response. *Immunology Today* 13, 169–172.

Luger, T.A. and Schwarz, T. (1990) Epidermal cell-derived cytokines. In: Bos, J.D. (ed.), *Skin Immune System (SIS)*. CRC Press, Boca Raton, Florida, pp. 257–291.

Lyons, C.R. (1995) The role of nitric oxide in inflammation. *Advances in Immunology* 60, 323–371.

MacGlashan, D.W., Jr and Lichtenstein, L.M. (1983) Studies of antigen binding on human basophils. I. Antigen binding and functional consequences. *Journal of Immunology* 130, 2330–2336.

Malveaux, F.J., Conroy, M.C., Adkinson, N.F. and Lichtenstein, L.M. (1978) IgE receptors on human basophils: relationship to serum IgE concentration. *Journal of Clinical Investigation* 62, 176–181.

Marucha, P.T., Zeff, R.A. and Kreutzer, D.L. (1990) Cytokine regulation of IL-1 beta gene expression in the human polymorphonuclear leukocyte. *Journal of Immunology* 145, 2932–2937.

Matsue, H., Bergstresser, P.R. and Takashima, A. (1993) Keratinocyte-derived IL-7 serves as a growth factor for dendritic epidermal T cells in mouse skin. *Journal of Immunology* 151, 6012–6019.

McNeil, H.P. and Austen, K.F. (1995) Biology of the mast cell. In: Frank, M.M., Austen, K.F., Claman, H.N. and Unanue, E.R. (eds), *Samter's Immunologic Diseases*, 5th edn. Little, Brown and Co., Boston, pp. 185–204.

Melmon, K.L., Rocklin, R.E. and Rosenkranz, R.P. (1981) Autacoids as modulators of the inflammatory and immune response. *American Journal of Medicine* 71, 100–106.

Moore, K.W., O'Garra, A., de Waal Malefyt, R., Vieira, P. and Mossmann, T.R. (1993) Interleukin-10. *Annual Review of Immunology* 11, 165–190.

Mossmann, T.R. and Coffman, R.L. (1989) TH1 and TH2 cells: differential patterns of lymphokine secretion lead to different functional properties. *Annual Review of Immunology* 7, 145–173.

Mueller, D.L., Jenkins, M.K. and Schwartz, R.H. (1989) Clonal expansion versus functional clonal inactivation: a costimulatory signalling pathway determines the outcome of T cell antigen receptor occupancy. *Annual Review of Immunology* 7, 445–480.

Nickoloff, B.J. and Turka, L.A. (1994) Immunological functions of non-professional antigen-presenting cells: new insights from studies of T-cell interactions with keratinocytes. *Immunology Today* 15, 464–469.

Nickoloff, B.J., Mitra, R.S., Green, J., Zheng, X.-G., Shimizu, Y., Thompson, C. and Turka, L.A. (1993) Accessory cell function of keratinocytes for superantigens.

Dependence on lymphocyte function-associated antigen-1/intercellular adhesion molecule-1 interaction. *Journal of Immunology* 150, 2148–2159.

Nixon-Fulton, J.L., Bergstresser, P.R. and Tigelaar, R.E. (1986) Thy-1$^+$ epidermal cells proliferate in response to concanavalin A and interleukin-2. *Journal of Immunology* 136, 2776–2786.

Nixon-Fulton, J.L., Hackett, J., Jr, Bergstresser, P.R., Kumar, V. and Tigelaar, R.E. (1988) Phenotypic heterogeneity and cytotoxic activity of Con A and IL-2 stimulated cultures of mouse Thy-1$^+$ epidermal cells. *Journal of Investigative Dermatology* 91, 62–68.

Odland, G.F. (1991) Structure of the skin. In: Goldsmith, L.A. (ed.), *Structure of the Skin*, 2nd edn. Oxford University Press, Oxford, pp. 3–62.

Paul, W.E. (ed.) (1993) *Fundamental Immunology*, 3rd edn. Raven Press, New York, 1490pp.

Picker, L.J., Michie, S.A., Rott, L.S. and Butcher, E.C. (1990) A unique phenotype of skin-associated lymphocytes in humans. Preferential expression of the HECA-452 epitope by benign and malignant T cells at cutaneous sites. *American Journal of Pathology* 136, 1053–1068.

Picker, L.J., Kishimoto, T.K., Smith, C.W., Warnock, R.A. and Butcher, E.C. (1991) ELAM-1 is an adhesion molecule for skin-homing T cells. *Nature* 349, 796–799.

Romagnani, S. (1992) Induction of $T_H1$ and $T_H2$ responses: a key role for the 'natural' immune response? *Immunology Today* 13, 379–381.

Rosenberg, H.F. and Gallin, J.I. (1995) Neutrophils. In: Lachmann, P.J., Peters, K., Rosen, F.S. and Walport, M.J. (eds), *Clinical Aspects of Immunology*, 5th edn. Blackwell Scientific Publications, Oxford, pp. 505–522.

Ruszczak, A.B., Detmar, M., Imcke, E. and Orfanos, C.E. (1990) Effects of rIFN-alpha, rIFN-beta, and rIFN-gamma on the morphology, proliferation and cell surface antigen expression of human dermal microvascular endothelial cells in vitro. *Journal of Investigative Dermatology* 95, 693–699.

Salmon, J.K., Armstrong, C.A. and Ansel, J.C. (1994) The skin as an immune organ. *Western Journal of Medicine* 160, 146–152.

Sanderson, C.J., Campbell, H.D. and Young, I.G. (1988) Molecular and cellular biology of eosinophil differentiation factor (interleukin-5) and its effects on human and mouse B cells. *Immunological Reviews* 102, 29–50.

Schwartz, L.B. (1993) Mast cells and basophils and their mediators. In: Lachmann, P.J., Peters, K., Rosen, F.S. and Walport, M.J. (eds), *Clinical Aspects of Immunology*, 5th edn. Blackwell Scientific Publications, Oxford, pp. 549–593.

Sehmi, R., Cromwell, O., Taylor, G.W. and Kay, A.B. (1991) Identification of guinea pig eosinophil chemotactic factor of anaphylaxis as leukotriene $B_4$ and 8(S), 15 (S)-dihydroxy-5,9,11,13 (Z,E,Z,E)-eicosatetraenoic acid. *Journal of Immunology* 147, 2276–2283.

Shimada, S., Caughman, S.W., Sharrow, S.O., Stephany, D. and Katz, S.I. (1987) Enhanced antigen presenting capacity of cultured Langerhans cells is associated with markedly increased expression of Ia antigen. *Journal of Immunology* 139, 2551–2555.

Shiohara, T., Moriya, N., Saizawa, K., Gotoh, C., Yagita, H. and Nagashima, M. (1991) Evidence for involvement of lymphocyte function-associated antigen 1 in T cell migration to epidermis. *Journal of Immunology* 146, 840–845.

Silberstein, D.S., Schoof, D.D., Rodrick, M.L., Tai, P.-C., Spry, C.J.F., David, J.R. and

Eberlein, T.J. (1989) Activation of eosinophils in cancer patients treated with IL-2 and IL-2 generated lymphokine-activated killer cells. *Journal of Immunology* 142, 2162–2167.

Sim, G.-K. (1995) Intraepithelial lymphocytes and the immune system. *Advances in Immunology* 58, 297–343.

Singer, K.H., Le, P.T., Denning, S.M., Whichard, L.P. and Haynes, B.F. (1990) The role of adhesion molecules in epithelial-T-cell interactions in thymus and skin. *Journal of Investigative Dermatology* 94, 85s–90s.

Siraganian, R.P. (1988) Mast cells and basophils. In: Gallin, J.I., Goldstein, I.M. and Snyderman, R. (eds), *Inflammation: Basic Principles and Clinical Correlates*. Raven Press, New York, pp. 513–542.

Slifman, N.R., Venge, P., Peterson, C.G.B., McKean, D.J. and Gleich, G.J. (1989) Human eosinophil-derived neurotoxin and eosinophil protein X are likely the same protein. *Journal of Immunology* 143, 2317–2322.

Spry, C.J.F. (1993) Eosinophils and their mediators. In: Lachmann, P.J., Peters, K., Rosen, F.S. and Walport, M.J. (eds), *Clinical Aspects of Immunology*, 5th edn. Blackwell Scientific Publications, Oxford, pp. 523–548.

Streilein, J.W. (1978) Lymphocyte traffic, T cell malignancies and the skin. *Journal of Investigative Dermatology* 71, 167–171.

Streilein, J.W. (1983) Skin-associated lymphoid tissues (SALT): origins and functions. *Journal of Investigative Dermatology* 80, 12s–16s.

Streilein, J.W. (1990) Skin associated lymphoid tissues (SALT): the next generation. In: Bos, J.D. (ed.), *Skin Immune System (SIS)*. CRC Press, Boca Raton, Florida, pp. 25–48.

Streilein, J.W., Grammer, S.F., Yoshikawa, T., Demidem, A. and Vermer, M. (1990) Functional dicotomy between Langerhans cells that present antigen to naive and to memory/effector T-lymphocytes. *Immunological Reviews* 117, 159–182.

Tai, P.C. and Spry, C.J.F. (1976) Studies on blood eosinophils. I. Patients with transient eosinophilia. *Clinical and Experimental Immunology* 24, 415–422.

Tang, A., Amagai, M., Granger, L., Stanley, J. and Udley, M. (1993) Adhesion of epidermal Langerhans cells to keratinocytes mediated by E-cadherin. *Nature* 361, 82–85.

Teunissen, M.C.M. (1992) Functional role of adhesion molecules LFA-3 and ICAM-1 on cultured human epidermal Langerhans cells in antigen specific T-cell activation. *Journal of Investigative Dermatology* 99, 77s–79s.

Tigelaar, R.E., Lewis, J.M. and Bergstresser, P.R. (1990) TCR gamma/delta$^+$ dendritic epidermal T cells as constituents of skin-associated lymphoid tissue. *Journal of Investigative Dermatology* 94, 58s–63s.

Toews, G.B., Bergstresser, P.R. and Streilein, J.W.(1980) Epidermal Langerhans cell density determines whether contact hypersensitivity or unresponsiveness follows skin painting with DNFB. *Journal of Immunology* 124, 445–453.

Trinchieri, G. (1995) Interleukin-12: A proinflammatory cytokine with immunoregulatory functions that bridge innate resistance and antigen-specific adaptive immunity. *Annual Review of Immunology* 13, 251–276.

Tschachler, E., Schuler, G., Hutterer, J., Leibl, H., Wolff, K. and Stingl, G. (1983) Expression of Thy-1 antigen by murine epidermal cells. *Journal of Investigative Dermatology* 81, 282–285.

Tschachler, E., Steiner, G., Yamada, H., Elbe, A., Wolff, K. and Stingl, G. (1989)

Dendritic epidermal T cells: activation requirements and phenotypic characterization of proliferating cells. *Journal of Investigative Dermatology* 92, 763–768.

Urade, Y., Fujimoto, N., Ujihara, M. and Hayaishi, O. (1987) Biochemical and immunological characterization of rat spleen prostaglandin D2 synthetase. *Journal of Biological Chemistry* 262, 3820–3825.

Valent, P. and Bettelheim, P. (1992) Cell surface structures on human basophils and mast cells: biochemical and functional characterization. *Advances in Immunology* 52, 333–423.

Valerius, T., Repp, R., Kalden, J.R. and Platzer, E. (1990) Effects of IFN on human eosinophils in comparison with other cytokines. A novel class of eosinophil activators with delayed onset of action. *Journal of Immunology* 145, 2950–2958.

Van Loveren, H., Teppema, J.S. and Askenase, P.W. (1990) Skin mast cells. In: Bos, J.D. (ed.), *Skin Immune System (SIS)*. CRC Press, Boca Raton, Florida, pp. 171–193.

Venge, P. (1993) Human eosinophil granule proteins: structure, function and release. In: Smith, H. and Cook, R.M. (eds), *Immunopharmacology of Eosinophils*. Academic Press, London, pp. 43–55.

Volc-Platzer, B., Leibl, H., Luger, T., Zahn, G. and Stingl, G. (1985) Human epidermal cells synthesize HLA-DR alloantigens in vitro upon stimulation with gamma-interferon. *Journal of Investigative Dermatology* 85, 16–19.

Walsh, G.M., Wardlaw, A.J. and Kay, A.B. (1993) Eosinophil accumulation, secretion and activation. In: Smith, H. and Cook, R.M. (eds), *Immunopharmacology of Eosinophils*. Academic Press, London, pp. 73–89.

Wardlaw, A.J., Moqbel, R., Cromwell, O. and Kay, A.B. (1986) Platelet activating factor. A potent chemotactic and chemokinetic factor for human eosinophils. *Journal of Clinical Investigation* 78, 1701–1706.

Wardlaw, A.J., Moqbel, R. and Kay, A.B. (1995) Eosinophils: biology and role in disease. *Advances in Immunology* 60, 151–266.

Weller, P. (1993) Human eosinophil granule proteins: structure, function and release. In: Smith, H. and Cook, R.M. (eds), *Immunopharmacology of Eosinophils*. Academic Press, London, pp. 43–55.

Wikel, S.K. (1982) Immune responses to arthropods and their products. *Annual Review of Entomology* 27, 21–48.

Wikel, S.K. (1996) Host immunity to ticks. *Annual Review of Entomology* 41, 1–22.

Wolff, K. and Stingl, G. (1983) The Langerhans cell. *Journal of Investigative Dermatology* 80, 17s–21s.

# 2

# Mouthparts and Feeding Mechanisms of Haematophagous Arthropods

## Douglas K. Bergman

*Department of Entomology, Oklahoma State University, Stillwater, Oklahoma 74078, USA*

### INTRODUCTION

In order to understand better the host's immune responses to haematophagous arthropods, a thorough understanding of the processes of blood-feeding must first be attained. The following chapters deal with immune responses to lice, bugs, fleas, mosquitoes, flies, mites and ticks. In this chapter, aspects of the structure of feeding apparatuses and feeding behaviour of haematophagous arthropods are described to provide the reader with a background with which to understand better the immune responses by the host. Descriptions of mouthparts, probing and feeding behaviour, duration of feeding, feeding repetition and saliva deposition are included. Most published descriptions of arthropod mouthparts describe their orientation relative to their positions in the resting arthropod. This author, as did Sutcliffe and Deepan (1988), takes the liberty to describe them as it works best for himself, describing the position and orientation of mouthparts relative to their positions in the feeding arthropod. Saliva components are not treated here, but are detailed in subsequent chapters.

Taxonomic positions of haematophagous arthropods and notes concerning blood-feeding in each taxon are given in Table 2.1. Hematophagy occurs in four orders of insects, one of which (Siphonaptera) is composed only of obligate ectoparasites. The arachnid order Parasitformes contains obligate blood-feeding mites and ticks. Although mites in the arachnid order Acariformes are not truly haematophagous, they are included in the table

© CAB INTERNATIONAL 1996. From Wikel, S.K. (ed.) *The Immunology of Host–Ectoparasitic Arthropod Relationships*. CAB INTERNATIONAL, Wallingford.

because immune responses to these mites are addressed in subsequent chapters. These mites, including chiggers (Trombiculidae) and those causing scabies (Sarcoptidae), consume lymph, lysed tissue and occasionally blood or its components.

## DERIVATION OF ARTHROPOD MOUTHPARTS

Insect mouthparts are derived from paired appendages associated with three gnathal segments of ancestral arthropods (Snodgrass, 1935). The *mandibles* and *maxillae* are paired appendages arising from the first and second gnathal segments, respectively. In insects with mandibulate, or basic chewing-type, mouthparts, the mandibles are rigid cutting and grinding jaws, while maxillae retain segmentation characteristic of limbs. The *cardo* of a maxilla articulates with the head and bears the *stipes*. Attached to the stipes are the inner *lacinia*, middle *galea*, and outer maxillary *palp*. The third gnathal segment bears paired appendages that are fused to form the *labium*. In mandibulate insects, the labium has two sets of lobes, *glossae* and *paraglossae*, corresponding with maxillary laciniae and galeae, respectively, and a pair of palps. The preoral cavity is closed in the back by the labium and in front by the *labrum*. The labrum has a lobe, called the epipharynx, on its inside surface that forms the roof of the mouth (opening of the *pharynx*). Mandibulate insects also have a fleshy, median *hypopharynx* situated behind the mouth and between the mandibles and maxillae. In these insects, saliva empties into the preoral cavity from the common salivary duct opening behind the hypopharynx and in front of the labium. In haematophagous insects, these structures are highly modified, adapted to facilitate this form of feeding.

Arachnid *chelicerae* are derived from a segment (postantennal) that has no appendages in insects. Arachnid *pedipalps* arise from the segment that bears mandibles in insects. Insect maxillae and labium arise from segments bearing the first two pairs of arachnid legs.

## HAEMATOPHAGOUS ARTHROPODS

### Phthiraptera

Lice are excellent examples of the relationship between the development of specialized mouthparts and blood-feeding. Chewing lice (suborder Mallophaga) have modified mandibulate mouthparts that are used to scrape skin or feather surfaces. The lice then feed on debris generated by these activities. This may include dried blood from scarified skin resulting from the host scratching. Typical mallophagan mouthparts include heavily sclerotized mandibles bounded in front by the labrum–epipharynx and in the back with a

Table 2.1. Haematophagous arthropods (Phylum Arthropoda).

| | Class Insecta (Hexapoda) |
|---|---|
| Order Phthiraptera | |
|     Suborder Mallophaga | Chewing lice; feed on skin or feather scrapings, few species feed on blood |
|     Suborder Rhynchophthirina | Ectoparasites of elephants; 2 species |
|     Suborder Anoplura | Sucking lice; blood-feeding parasites of eutherian mammals; $c.$ 450 species (67% on rodents) |
| Order Hemiptera | |
|     Family Cimicidae | Bedbugs |
|     Family Reduviidae | Subfamily Triatominae, kissing bugs; $c.$ 155 species |
|     Family Polyctenidae | Permanent blood-feeding ectoparasites of bats |
| Order Siphonaptera | Fleas; adults feed on blood of mammals (74% on rodents) and birds |
| Order Diptera | |
|     Suborder Nematocera | |
|         Family Psychodidae | Subfamily Phlebotominae, sand flies; both sexes feed on nectar, females feed on blood |
|         Family Culicidae | Subfamilies Anophelinae and Culicinae, mosquitoes; both sexes feed on nectar, females feed on blood |
|         Family Ceratopogonidae | Blood-feeding midges in four genera; both sexes feed on nectar, females feed on blood |
|         Family Simuliidae | Blackflies; females feed on blood |
|     Suborder Brachycera | |
| | Infraorder Tabanomorpha |
|         Family Rhagionidae | Snipe flies; most species insect predators, few blood-feeding species; adults feed on nectar |
|         Family Tabanidae | Important genera include *Chrysops* (deer flies), *Tabanus* (horse flies) and *Haematopota* (clegs); both sexes feed on nectar, females feed on blood |
| | Infraorder Muscamorpha |
|         Family Muscidae | Few haematophagous species include *Stomoxys* spp. and *Haematobia* spp., both sexes feed on blood |
|         Family Glossinidae | *Glossina* spp., tsetse flies; both sexes feed on blood only |
|         Family Hippoboscidae | Ectoparasitic as adults; includes keds |
|         Family Streblidae | Ectoparasites of bats, bat flies |
|         Family Nycteribiidae | Ectoparasites of bats (also called bat flies) |

**Class Arachnida, Subclass Acari**

| | | |
|---|---|---|
| Order Acariformes | | |
| Suborder Acaridida (=Astigmata) | | |
| | Family Sarcoptidae | Includes mites *Sarcoptes scabiei* and *Notoedres cati*; feed predominantly on lymph and lysed tissue, occasionally on blood |
| Suborder Actinedida (=Prostigmata) | | |
| | Family Trombiculidae | Larvae (chiggers) feed on liquefied host tissues; few blood cells may be found in gut |
| Order Parasitiformes | | |
| Suborder Mesostigmata | | |
| | Family Dermanyssidae | Includes haematophagous mites *Dermanyssus gallinae* and *Liponyssoides sanguineus* |
| | Family Macronyssidae | Includes *Ornithonyssus* spp. and *Chryptonyssus robustipes* |
| | Family Laelapidae | Mite ectoparasites of rodents |
| Suborder Ixodida | | |
| | Family Ixodidae | Hard ticks (*c.* 650 species) |
| | Family Argasidae | Soft ticks (*c.* 145 species) |

maxillary–labial plate where the maxillae are fused to the lateral edges of the labium (Buxton, 1940).

Blood-feeding species of chewing lice have developed specialized mouthparts. *Menacanthus stramineus* (Nitzsch) (family Menoponidae) pierces young developing feathers of chickens with pointed mandibles and feeds on blood resulting from these wounds (Wilson, 1933). Another species, *Trinoton anserinum* (F.), in the same family (Menoponidae), feeds on the blood of swans using enlarged mandibles and maxillae with a serrated cutting surface (Cohen et al., 1991). *Ricinus* spp. (family Ricinidae) feed on blood of hummingbirds and passerine birds using mandibles that are elongated and pointed (Clay, 1949). A related genus in the same family, *Trochiloecetes* spp., also feeds on blood of hummingbirds with highly modified mouthparts. These lice have a stylet-like structure of hypopharyngeal origin and mandibles with piercing blades that lie alongside the median stylet-like structure.

The two species in the single genus, *Haematomyzus* (family Haematomyzidae), in the suborder Rhyncophthirina feed on the blood of thick-skinned elephants and wart hogs. They have cutting mandibles on the end of a lengthened oral tube that lacerate tissues within skin folds (Ferris, 1931). Blood resulting from this action is then sucked up.

Blood-feeding is accomplished by sucking lice (suborder Anoplura) with highly modified mouthparts. When the haustellum (an extrudable membranous portion of the preoral cavity) is everted, teeth around its apex cut into the skin and anchor the louse in the epidermis of the host during feeding (Lavoipierre, 1967). The stylet bundle or fascicle, that rests in a sac below the alimentary canal when the louse is not feeding, is then extended through the opening in the haustellum into the dermis. The stylets probe 50–100 μm into tissues in different directions until a suitable blood vessel is found. The vessel is then penetrated by the tip of the stylet bundle and feeding begins. Both nymphs and adults feed from the lumen of venules. Lavoipierre (1967) noted that lice attached to the host in a few minutes and immediately began to suck up blood. Feeding was completed in 10 to 15 minutes.

The mouthparts of sucking lice are so highly modified and reduced that homologies with structures in other insects are ill-defined (Buxton, 1940; Stojanovich, 1945; Ferris, 1951). The haustellum is a membranous portion of the labrum. The stylet bundle is derived from postmandibular segments that have been retracted into a sac in the head. The stylet bundle is composed of ventral, medial and dorsal stylets (Fig. 2.1a). Embryonic and morphological evidence indicates that the ventral stylets are derived from the labial segment. Lavoipierre (1967) states that the ventral stylet assists and guides penetration of the dermis. The medial stylet is an extension of the salivary duct with its opening at the apex of the stylet bundle. This salivary stylet lies in a groove in the ventral stylet. Above the salivary stylet and ventral stylet is the dorsal stylet. The dorsal stylet is composed of two closely associated appendages of the maxillary segment. The close aposition of these appendages forms a hollow tube that serves as the food canal (Lavoipierre, 1967).

Both nymphal and adult sucking lice feed exclusively on blood. Human body and head lice (*Peduculus humanus*) have life spans of *c*. 30 days (Buxton, 1940; Busvine, 1948; Culpepper, 1948; Flemings and Ludwig, 1964). Presenting blood two times each day was sufficient to maintain colonies. Similar percentages of nymphs matured to adults whether they were fed one to three times each day (Gooding, 1963).

## Hemiptera (Heteroptera)

Three groups of true bugs feed exclusively on the blood of vertebrate hosts (Schuh and Slater, 1995). These are the families Cimicidae and Polyctenidae and the tribe Triatominae in the family Reduviidae. Although hematophagy evolved independently in each of these groups, the modification of mouthparts remains the same, as it does for predacious and plant-feeding relatives. The maxillae are modified in all hemipteran insects to oppose each other closely and form separate salivary and food canals (Fig. 2.1b). The mandibles form a ring that partially encircles the maxillae. The tips of the mandibles are usually serrated and are used for cutting into plant or animal tissues. These

Mouthparts and Feeding Mechanisms of Haematophagous Arthropods 35

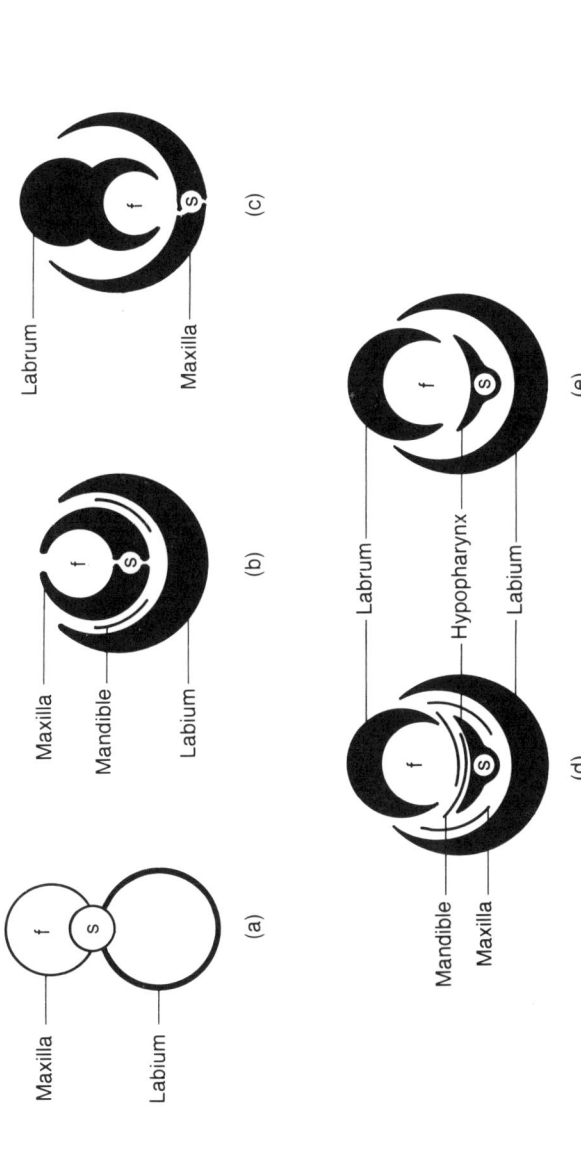

**Fig. 2.1.** Schematic representations of anopluran (a), hemipteran (b), siphonapteran (c), nematoceran (c) and muscamorphan (e) mouthparts with reference to the derivation of mouthparts and the positions of food (f) and salivary (s) canals.

stylets form the feeding fascicle and lie in the trough-like three- or four-segmented labium.

Cimicid nymphs and adults are temporary ectoparasites primarily of birds and bats (Usinger, 1966). However, two species, the common temperate bedbug, *Cimex lectularius* L., and the common tropical bedbug, *C. hemipterus* (F.), are often associated with humans. Bedbugs feed by first probing tissues with the fascicle (Dickerson and Lavoipierre, 1959a). The labium does not puncture the skin, but lip-like structures on its end grasp the fascicle and support it as it is thrust into host tissues. In the act of piercing the skin, the mandibles are alternately thrust forward and retracted, cutting into the skin slightly in front of the maxillae. Probing continues in different directions until a suitable blood vessel is located. The vessel is then penetrated by the maxillae alone. Probing may be take some time, but engorgement is rapid.

Mated *C. lectularius* females ingest *c.* 7.8 mg of blood during the bloodmeal and feed to repletion in 5 to 10 minutes (Usinger, 1966). They then remain quiescent off the host for 5 to 6 days, digesting the blood and developing their eggs. In this manner, young adult females feed approximately once per week and oviposit continually under temperate conditions. Nymphs feed within 24 hours of moulting and at shorter intervals than adults. Last (fifth) instars ingest *c.* 7.1 mg of blood during the bloodmeal. Dolling (1991) stated that in temperate London, *C. lectularius* develops from egg to adult in about 10 weeks and adults may live several years under normal indoor conditions.

The Polyctenidae are permanent ectoparasites of bats (Schuh and Slater, 1995). Phylogenetically, they are closely related to Cimicidae (Schuh and Stys, 1991). Little is known of the biology of these obscure insects. Each species has an extremely narrow host range, limited to a particular host genus and often to a single host species (Marshall, 1981). No single bat species is parasitized by more than one species of Polyctenidae. Polyctenids are born alive and the three postnatal nymphal instars and adults never leave the host. Adults require frequent bloodmeals and do not live more than *c.* 24 hours off the host under conditions of temperature and humidity simulating host microenvironments.

Although all reduviids are predators, only the Triatominae are haematophagous (Schuh and Slater, 1995). These important vectors of human disease evolved from nest parasites of birds and both nymphs and adults utilize a wide array of hosts. Lavoipierre *et al.* (1959) described the mouthparts and feeding behaviour of *Rhodnius prolixus* Stal, *Triatoma infestans* (Klug) and *T. protracta* (Uhler). The fascicle, containing the maxillary and mandibular stylets, quickly penetrated the outer layer of skin. The mandibles then ceased their saw-like penetrating activity and anchored the fascicle into the upper skin layer with their recurved apical teeth. The maxillae then continued to penetrate the dermis and probe in all directions until a suitable vessel was penetrated. Once a suitable vessel was penetrated, the left maxilla withdrew relative to the right maxilla, the spine-like tip on the left maxilla bent outwards at a 90 degree angle and ingestion commenced. Friend and Smith (1971) observed *R. prolixus* feeding on artificial diet through a membrane and found that the withdrawal of the left maxilla relative to the right exposed or opened the food canal and that the outwards-bent tip of the left maxilla served to anchor the maxillae during feeding. In artificial diets, salivation was continuous during probing, but was not observed at all during ingestion. The authors postulated that saliva was continuously sucked back up with diet during ingestion on the basis that regurgitated diet 'looked different' from regular diet.

Triatominae are long-lived, requiring large and frequent bloodmeals. *Triatoma barberi* Usinger completed feeding in 10 to 24 minutes, on average, depending upon size and developmental stage (Zarate *et al.*, 1984). Fifth instars and female adults fed an average of 24 and 15 minutes, respectively. In general, feeding began soon after exposure to the host. In the laboratory, *T. barberi* nymphs develop to adults in 250 days and feed 266 times, on average (Zarate, 1983). Male and female adults lived an average of 178 and 222 days, respectively. Adult triatominae reared outdoors in Argentina lived for up to 6 months feeding every 12 to 22 days (Carcavallo and Martinez, 1972). First and fifth instar *T. dimidiata* (Latreille) ingested an average of 5 and 175 mg of blood per meal, respectively (Zeledon *et al.*, 1970). In this study, adult female *T. dimidiata* lived an average of 561 days and ingested an average of 6745 mg of blood in 32 feedings (approximately 211 mg of blood per feeding).

## Siphonaptera

Fleas are a monophyletic group of wingless ectoparasitic insects evolved from a winged ancestor that also gave rise to the order Diptera (Marshall, 1981). Adult fleas feed exclusively on the blood of mammals and birds. Most (74%) flea species are associated with rodents. Fleas (*Pulex irritans* L., *Xenopsylla cheopis* (Rothschild), *Ctenocephalides felis* (Bouche), and *Atyphloceras multidentatus* (Fox)) fed by puncturing small blood vessels (capillaries) near the surface of the skin (Lavoipierre and Hamachi, 1961). The host's skin was penetrated by the fascicle composed of three stylets. As reviewed by Askew (1971), paired blade-like maxillary laciniae form a trough that holds and is enclosed by a single, deeply-grooved stylet that is derived from the epipharynx. The maxillary laciniae cut into host tissues and anchor the fascicle during feeding with recurved teeth located at their apex (Deoras and Prasad, 1967). Only the tip of the epipharynx enters the lumen of capillaries and blood is sucked up through the groove in its surface into the food canal formed by this groove and the interior surfaces of the maxillary laciniae (Fig. 2.1c). The maxillary laciniae remain outside the vessel, periodically emitting saliva through a small salivary canal formed by their close apposition.

Adult fleas complete feeding in 2 to 10 minutes (Rothschild, 1975). Females imbibe more blood (0.18 mg) per feeding than males (0.10 mg). Adult fleas are long-lived, with or without the availability of a bloodmeal. Rothschild and Clay (1957) asserted that the northern rat flea, *Nosopsyllus fasciatus* (Bosc), may survive 17 months without feeding under experimental conditions. The human flea, *Pulex irritans*, survived approximately the same amount of time on a diet of human blood. Marshall (1981) reviewed the longevity of ten flea species and found that the maximum longevity for adults of seven of these species was 1 year or more.

An exception to the above scenario is found in fleas in the family Tungidae. These mammalian entoparasites, including *Tunga penetrans* L. (human chigoe flea), burrow into the dermis of their hosts. In an experimental infestation of mice with *Tunga monositus* Barnes and Rodovsky, fleas became deeply embedded in ear tissues within 24 hours and fed initially on tissue exudates as well as connective tissue (Lavoipierre *et al.*, 1979). Fleas began to feed on blood from vessels 14 days after the initial infestation. Fleas then began to reproduce and lived 2 to 3 months.

## Diptera

Flies are a highly diverse group of insects whose larvae develop in many different moist, semi-aquatic and aquatic environments. However, all adult flies feed on liquids (McAlpine, 1981; Smith, 1985). Some feed directly on fluids such as blood or nectar, while others first liquify dry nutrients with saliva. In general, the liquid diet is imbibed through a food canal closed anteriorly (or dorsally) by the labrum (labrum–epipharynx) and posteriorly (or ventrally) by the hypopharynx and labium. The hypopharyngeal stylet usually contains the salivary canal. Mandibles, usually flat, blade-like structures, are found in females of some species. Mandibles in males of these species are vestigial. Maxillary palpi are present in all flies. Maxillary stylets, derived from maxillary laciniae, are present in some species. The labium is highly developed in flies. It most cases, it forms the trough or sheath in which the other mouthparts reside. On its distal end, it bears labella (derived from labial palpi). These structures are very diverse, ranging from the grasping, clamp-like structures that direct the piercing activity of mosquito stylets to large, sponge-like lobes bearing pseudotracheae that direct fluids into the food canal of the house fly, *Musca domestica* L.

Crosskey (1993a) provides an excellent overview of current dipteran systematics. Species of haematophagous flies occur in both suborders of Diptera, Nematocera and Brachycera. Blood-feeding flies in the suborder Nematocera include the Psychodidae, Ceratopogonidae, Simuliidae and Culicidae and have similar mouthpart arrangements. The Nematocera are closely related phylogenetically and represent the oldest form of blood-feeding found in modern dipterans (Downes, 1971). The feeding fascicle or syntrophium (structure that pierces the skin of the host) in females is composed of well-developed labral, mandibular, maxillary and hypopharyngeal stylets (Jobling, 1976). The fascicle or syntrophium lies in a trough in the labium. The latter mouthpart does not penetrate host tissues.

Haematophagous species in the suborder Brachycera are found in two infraorders, Tabanomorpha and Muscamorpha. Blood-feeding flies in the infraorder Tabanomorpha are found in two families, Tabanidae and Rhagionidae. These families retain the original mandibular blood-feeding mechanism, sexual dimorphism of mouthparts and feeding behaviour of nematocerans (Downes, 1971). The fascicle or syntrophium of female Tabanomorpha contains a well-developed labrum, blade-like mandibular and maxillary stylets and a stylet-like hypopharynx. However, the labium of Tabanomorpha bears large, sponge-like labellar lobes that do not enter the wound created by the fascicle.

Blood-feeding flies in the infraorder Muscamorpha developed a new method of acquiring blood, primarily from mammals, following the loss of mandibles, maxillary laciniae and the original mode of biting and cutting (Downes, 1971). Acquisition of the bloodmeal is facilitated in these flies by prestomal teeth. The prestomal cavity, or prestomen, lies between the labellar lobes in front of the opening of the food canal (Snodgrass, 1935; West, 1951). In the Muscidae, it is bordered on either side by five prestomal teeth attached to a U- or V-shaped sclerite that opens anteriorly.

Flies with non-piercing mouthparts have prestomal teeth that are delicate and blade-like. They are used to scratch or scrape nutrients off a surface that can then be sucked up in saliva or regurgitated contents of the crop. Facultative haematophagic muscids, such as *Musca autumnalis* De Geer, *Hydrotaea irritans* (Fallen) and *Morellia hortorum* (Fallen), may feed on blood oozing from wounds, some of which may have been caused by blood-feeding stomoxyins or tabanids (Greenberg and Povolny, 1971). These species suck blood off skin surfaces with their sponging/lapping muscamorphan mouthparts exemplified by those found in the house fly.

Some facultative haematophagous muscids, such as *Musca sorbens* Weidemann or *Musca tempestiva* Fallen, use their prestomal teeth to scratch or scrape wounded tissues, feeding on any exudates that form. In the case of *Musca crassirostris* Stein, enlarged prestomal teeth cause a bleeding wound on which the fly feeds (Crosskey, 1993b). Finally, the prestomal teeth associated with the piercing mouthparts of the muscid subfamily Stomoxyinae are well developed and born on the end of a rigid, strongly sclerotized labium (Greenberg, 1973; Zumpt, 1973). When the greatly reduced labella are everted or opened, the prestomal teeth are exposed and are used to cut into host tissues. Similar mouthparts and feeding habits are found in tsetse flies (family Glossinidae), louse flies (family Hippoboscidae) and bat flies (families Streblidae and Nycteribiidae). Like all flies, these haematophagous species imbibe fluids (blood) through a labral food canal that is closed posteriorly/ dorsally by the hypopharynx. Although bat flies in the families Streblidae and Nycteribiidae have labial cutting teeth similar to the other haematophagous flies in the infraorder Muscamorpha, there is some question as to whether their mode of feeding has developed from a common origin or independently (Downes, 1971).

## Nematocera

The mouthparts of female hematophagous sandflies (family Psychodidae, subfamily Phlebotominae), biting midges (family Ceratopogonidae) and blackflies (family Simuliidae) are similar in arrangement, size, shape and function. The mouthparts of sandflies and their function were described by Lewis (1975), Jobling (1976) and Brinson et al. (1993). The mouthparts of ceratopogonids, primarily *Culicoides* spp., and their function were described by Jobling (1928a), McKeever et al. (1988) and Sutcliffe and Deepan (1988). The mouthparts of blackflies and their function were described by Nicholson (1945) and Sutcliffe and McIver (1984), and reviewed by Wenk (1981) and Crosskey (1990).

The labrum is the first mouthpart in the piercing fascicle to contact host tissues. In all three groups, the labrum is pointed at its apex, more broadly in Ceratopogonidae and Simuliidae than in Psychodidae. Sandfly labra terminate in large, blunt peg-like sensilla, while those of both biting midges and blackflies terminate with two upward-directed tricuspid teeth. The labra in these groups are deeply guttered and range in length from *c.* 0.170 mm in *Culicoides sanguisuga* (Coquillet) (Sutcliffe and Deepan, 1988) to *c.* 0.350 mm in sandflies in the genera *Lutzomyia* and *Phlebotomus* (Brinson et al., 1993).

The mandibles of these flies are broad, flat and heavily sclerotized blades. They are approximately the same length as the labrum. The distal margins are serrated with numerous small teeth. In *C. sanguisuga* and sandflies, only the outside margin is serrated. In blackflies, both distal margins are serrated. The mandibles overlap each other and lie directly behind the labrum, closing the food canal posteriorly (Fig. 2.1d). The mandibles of Ceratopogonidae and Simuliidae are held in this position by interlocking indentations and protuberances occurring midlength on each mandible. In blackflies, the middle of each mandible has a forward-projecting raised portion that articulates with a corresponding indentation in the overlying mandible or with the labral gutter. In *C. sanguisuga*, the mandibles are held in place by medial backward-projecting knobs articulating with either the corresponding indentation in the mandible behind it or with the salivary groove in the hypopharynx. There are no such interlocking mechanisms in sandflies.

Maxillary laciniae in sandflies, biting midges and blackflies are long, slender structures with upward-pointing (retrorse or recurved) teeth on their apical spear-shaped edges. In sandflies and *Culicoides* spp., the curved laciniae close the food canal laterally and clip the feeding fascicle together by their close apposition with labrum and hypopharynx. In blackflies, the distal portion of the laciniae are more broadly triangular with retrorse teeth on both medial and lateral edges.

The hypopharynx in sandflies, biting midges and blackflies is a blade-like structure, bearing teeth on its distal edges and the salivary canal in a keel along the middle of the posterior margin. The salivary canal is completely enclosed proximally and opens into a groove distally.

The mouthparts of the biting fascicle are sheathed in the well-developed and deeply-grooved labium. The labium terminates in two relatively large, two-segmented labellar lobes. These modified labial palps bear numerous spines, setae and other sensory organs.

The following sandfly feeding scenario was proposed by Lewis (1975). First, labral sensory structures locate a crevice in the host's skin. The laciniae are then thrust into the outer skin layer, anchoring the sandfly to its host. The mandibles then cut into the host's skin with primarily repeated protraction/retraction (saw-like) motions. After the fascicle has penetrated the dermis to a depth of approximately one-half its length (c. 0.170 mm), blood from lacerated capillaries and other small vessels is sucked up through the labral food canal. Saliva is injected into the wound through the hypopharynx. The labium serves to support and guide the fascicle during feeding, but does not participate in creating the wound nor does it penetrate it. Johnson and Hertig (1961) observed female *Phlebotomus* spp. feeding in the laboratory and found them to be restless feeders, landing on the host, probing with the proboscis and then leaving the host three or more times before feeding. Once begun, feeding was completed in 5 minutes or less.

Ceratopogonid and simuliid feeding is similar to feeding by psychodids, but incorporates some specialized adaptations. After *Culicoides sanguisuga* selects a suitable biting site, the labella retract and the distal tip of the syntrophium is pushed down against the skin (Sutcliffe and Deepan, 1988). Hypopharyngeal teeth engage the skin and are deflected backward while the tricuspid teeth on the labrum engage the skin as it rolls forward. This action stretches the skin between the tips of the labrum and hypopharynx. The mandibles then cut and saw into the taut skin. The main action is protraction/retraction of the mandibles, into and out of tissues. Protraction (insertion) is associated with the tips of the mandibles closing and retraction (withdrawal) is associated with the tips of the mandibles opening slightly laterally. Cutting occurs principally during this latter motion.

The laciniae function to draw the syntrophium into host tissues. After the wound is opened by the mandibles, the laciniae are inserted into host tissues where the recurved or retrorse teeth on their distal tips become embedded. The laciniae are then retracted, drawing the syntrophium deeper into the wound. The mandibles then lacerate newly accessible tissue and the process is repeated until the syntrophium is fully drawn-down and feeding on haemorrhaging blood begins. The fully inserted syntrophium is surrounded by the spread labellar lobes, directing escaping blood towards the syntrophium.

Blackfly mouthparts function to obtain the bloodmeal in a similar manner (Sutcliffe and McIver, 1984). However, the motion of blackfly mandibles may be more scissor-like during the protraction/retraction cycle, with cutting occurring primarily on retraction/adduction (closing). In this case, the teeth on the inside edge would be used primarily for scissor-like cutting. The teeth on the outside edge of the mandible may be used to slice into tissues during the protraction/abduction (opening) part of the cycle. Blackfly mouthparts may penetrate host tissues up to $c$. 0.4 mm and feeding is completed in 4 to 5 minutes (Wenk, 1981).

Mosquitoes have evolved specialized nematoceran mouthparts, adapted for piercing rather than cutting (Downes, 1971). Although the mouthparts of female mosquitoes are long, very thin and closely adjoined, they are similar in arrangement and function to those of females in the families Psychodidae, Ceratopogonidae and Simuliidae. Robinson (1939) and Lee (1974) described the mouthparts of female *Anopheles maculipennis* (Meigen) and *Aedes aegypti* (L.). The food canal is formed by the close apposition anteriorly of circular labral tissues. In *Ae. aegypti*, these tissues are fused apically to form a complete tube. The apex of the labrum is bevelled in both species, resembling the tip of a hypodermic needle. Very thin, paired mandibles lie behind the labrum and in front of the hypopharynx. As in other nematocerans, the mandibles serve to close both the food canal in the labrum and the salivary canal in the hypopharynx, thus separating blood intake from salivary discharge. The salivary canal runs along the midline of the hypopharynx, forming a midrib along its posterior surface. In anopheline mosquitos, blade-like tips of mandibular stylets bear fine teeth on the apical one-half of their outer edges. These stylets in *Ae. aegypti* have broadened tips that bear no teeth. Behind and on either side of the hypopharynx lie the paired maxillary stylets (laciniae). The distal tips of the laciniae are spear-shaped with recurved teeth on their extreme end. During penetration of the host's skin, protractor and retractor muscles drive each maxillary stylet alternatively up and down, first cutting into and then gripping host tissues. As in other nematocerans, this action pulls the remainder of the fascicle into and through the host's dermis. The maxillary palpi raise and lower relative to the movement of individual maxillary lacinia (Jones and Pilitt, 1973). The maxillary palpi are at rest during feeding.

The behaviour of female *Ae. aegypti* feeding on mouse ears (Griffiths and Gordon, 1952) was similar to those feeding on the foot webbing of a frog (Gordon and Lumsden, 1939). The skin surface was first probed by the labella. When a feeding site was selected, the labellum guided and supported the fascicle as it entered the skin. The fascicle was found to be very flexible, sometimes bending 90 degrees to penetrate tissues parallel to the skin surface. Salivation was observed for mosquitoes feeding in the mouse ear as 'puffs' of clear liquid discharged primarily during probing.

Gordon and Lumsden (1939) observed mosquitoes feeding both directly from capillaries and from haemorrhages formed by the laceration of capillary beds by the penetrating actions of the fascicle. Clements (1992) pointed out that the mosquito fascicle is too large to enter capillaries, so mosquitoes must feed from larger arterioles or venules. Mosquitoes feeding from pooled blood required as much as 10 minutes to completely engorge while mosquitoes feeding directly from vessels engorged in only 3 minutes. When feeding occurred from a vessel, the fascicle penetrated the lumen of the vessel and no haemorrhaging occurred in surrounding tissues. On ten occasions where the entire feeding process was well documented, pool feeding was observed on three occasions and vessel feeding in the remaining seven occasions. Feeding from both vessels and hemorrhages during the same feeding episode was observed on occasion. O'Rourke (1956) found that the time required for female *Ae. aegypti* to complete feeding was distributed bimodally, with peaks at *c.* 2.2 and 4.5 minutes. The author postulated that the faster and slower feeding times represented vessel and pool feeding, respectively. On this basis, *Ae. aegypti* females fed from a vessel *c.* 60% of the time and from pooled blood *c.* 40% of the time.

Haematophagous nematocerans feed on both warm- and cold-blooded vertebrates (Downes, 1971). Female sandflies feed on a wide range of vertebrates, including many species of mammals, birds, reptiles and even Amphibia. Even though sandflies may have broad host ranges, they are opportunistic feeders, feeding primarily on the most readily available host (De Colmenares *et al.*, 1995; Morrison *et al.*, 1993b). Hosts of female biting midges and blackflies appear to be more restricted, feeding primarily on warm-blooded hosts (Downes, 1971). Some species of *Culicoides* feed either on mammals or birds, while other species feed on a wide range of both. In Newfoundland, *Simulium* spp. were attracted to a variety of mammals, including humans (McCreadie *et al.*, 1994). In general, female mosquitoes feed on blood from a wide array of mammals and birds, with a few species feeding on reptiles, amphibia and fish (Downes, 1971). This is true for both the family as a whole and individual genera. In Missouri, 8.2% of the female *Aedes albopictus* (Skuse) collected had fed on human blood, 55.8% had fed on a variety of other mammals and 16.9% had fed on birds (Savage *et al.*, 1993). Host utilization is more limited in other species. Most (99%) *Anopheles quadrimaculatus* Say females collected in North Carolina had fed on white-tailed deer or horses (Robertson *et al.*, 1993). Scott *et al.* (1993a) found that 88% of the *Ae. aegypti* collected around a rural village in Thailand had fed on human blood alone. Although some species prefer certain hosts, they also may display 'opportunistic' feeding. In Egypt, 79% of anthropophilic female *Culex pipiens* L. collected from inside houses had fed on human blood (Gad *et al.*, 1995). However, 35% of those collected from animal sheds had fed on sheep or goats.

Sandflies and biting midges imbibe less than 0.5 µl of blood during each blood meal (Braverman, 1994). Blackflies may ingest up to 3.3 µl of blood per meal. Mean amounts of blood imbibed by four species of *Aedes* feeding on restrained rabbits ranged from 5.0 to 11.2 µl per feeding episode (Klowden and Lea, 1979). Both feeding success and amounts of blood ingested were significantly reduced in unrestrained rabbits, demonstrating the importance of host defensive responses, such as grooming and shaking, in regard to the success of haematophagous arthropods in obtaining a complete bloodmeal. Amounts of blood ingested by other mosquito species, per feeding episode, were similar (Clements, 1992).

Some species of sandflies are autogenous (Braverman, 1994), however, continued egg production in these species and all egg production in anautogenous species require blood. It has been commonly reported that most sandfly species are gonotrophically concordant, requiring and taking only one bloodmeal per batch of eggs matured (Lane, 1993). However, it is becoming increasingly evident that many sandflies may commonly feed more than once per gonotrophic cycle. Guzman *et al.* (1994) found that female *Phlebotomus duboscqi* Neveu-Lemaire fed twice during the first gonotrophic cycle. In Columbia, *c.* 2 to 13% of female *Lutzomyia longipalpus* (Lutz and Neiva) captured near animal pens and in the latter stages of egg development contained blood in their guts (Ferro *et al.*, 1995). In the laboratory, Elnaiem *et al.* (1992) observed *L. longipalpus* re-engorging on hamsters one, two and four days after initially feeding and before ovipositing. On day four, almost half of the females re-fed. Both female biting midges and black flies are generally gonotrophically concordant (Braverman, 1994). Many species of ceratopogonids are autogenous (Linley, 1983), while autogeny is rare in the family Simuliidae (Anderson, 1987; Crosskey, 1990).

It was also generally believed that female mosquitoes took only one bloodmeal per gonotrophic cycle (host location through egg deposition). However, it is also becoming evident that many species feed to repletion more than once while maturing a single clutch of eggs. Female *Ae. aegypti* fed *c.* 24 hours after emerging from pupae as adults and again 2 to 5 hours later (Jones and Pilitt, 1973). Xue *et al.* (1995) found that older *Ae. aegypti* females fed more times and required more blood to initiate egg development. Half of the female *Ae. aegypti* collected within houses in Puerto Rico had ingested multiple bloodmeals and most had fed twice on consecutive days (Scott *et al.*, 1993b). The authors hypothesized that most of the wild females studied had fed twice during each gonotrophic cycle.

After feeding to repletion, three species of *Anopheles* showed host-seeking behaviour during the same gonotrophic cycle (Klowden and Briegel, 1994). Wekesa *et al.* (1995) examined female *Anopheles freeborni* Aitken collected in California and found that almost 10% had ingested two or three bloodmeals during a single gonotrophic cycle.

The longevity of female blood-feeding nematocerans depends on environmental conditions. Most laboratory-reared *Phlebotomus* spp. females died after producing the first batch of eggs (Johnson and Hertig, 1961). However, those that survived lived around 18 more days. El Sawaf *et al.* (1994) reported that female *P. papatasi* (Scopoli) and *P. langeroni* Nitzulescu survived, on average, 10 days at 25°C. Ferro *et al.* (1995) reported that half of laboratory-reared female *L. longipalpus* survived at least one gonotrophic cycle with some surviving four gonotrophic cycles. Average female longevity at 25°C was 25 days with a maximum of 38 days. Data on female longevity in the field are limited. Morrison *et al.* (1993a) recaptured marked wild and laboratory-reared *L. longipalpus* of different ages up to 8 days after their release in Columbia.

Female ceratopogonids may survive three to four gonotrophic cycles (Mullens and Schmidtmann, 1982). In Israel, *Culicoides* spp. fed every 3 to 5 days, depending on time of year (Braverman *et al.*, 1985). Assuming that they fed once per gonotrophic cycle and survived three gonotrophic cycles, female biting midges in this area may live up to 15 days.

Female blackflies may survive two to three gonotrophic cycles in northern Europe (*c.* 18 days) and up to five gonotrophic cycles in equatorial Africa (*c.* 23 days) (Crosskey, 1990). Marked blackflies have been recaptured 62 to 85 days after their release and maintained in the laboratory up to 63 days (Wenk, 1981).

Female *Ae. aegypti* maintained on a diet of sugar and blood lived up to 35 days ($LT_{50}$ = 29 days) (Day *et al.*, 1994). Marked female *Ae. aegypti* released in Thailand were recaptured for 12 days. The longevity of marked female *Ae. aegypti* in tropical Kenya was at least 9 days (Trpis *et al.*, 1995). In a temperate region (northern California), female *Aedes dorsalis* (Meigen) survived up to 15 days, or *c.* two gonotrophic cycles (Kramer *et al.*, 1995).

**Brachycera**

*Tabanomorpha*

The mouthparts of female horse and deer flies (family Tabanidae) are similar in form and arrangement to those of nematocerans (Lavoipierre, 1965). The mouthparts and blood-feeding mechanisms of snipe flies (family Rhagionidae) are the same as those found in tabanids (Downes, 1958). However, compared to nematocerans, tabanomorphan mouthparts are much larger and stouter, as are the flies themselves. The labrum of *Tabanus nigrovittatus* Macquart is over 2 mm long (Stoffolano and Yin, 1983). In addition, the tip of the labrum is blunt and does not bear apical teeth. Near its base, the sides of the labrum are almost fused into a complete tube. At the apex, the sides of the labrum open to form a wide channel at the opening of the food canal.

As in nematocerans, the mandibles of horse flies are blade- or scimitar-like, with fine teeth serrating one or both edges, depending on the genera to which the fly belongs. Unlike nematocerans, the salivary canal extends all the way to the tip of the tabanid hypopharynx as a closed tube. The labellar lobes of both sexes of tabanids are large and well developed. They bear small grooves or channels (pseudotracheae) on their ventral surfaces. These small channels collect into a main canal that opens near the tip of the sheathed labrum.

Female *Haematopota pluvialis* (L.) plunge the fascicle into host tissues with a thrusting action of the head and thorax (Dickerson and Lavoipierre, 1959b). The mandibles and maxillary laciniae extensively lacerate host dermis as the fly twists its head from side to side. Mandibles cut into host tissues with scissor-like actions while the laciniae saw or rasp tissues as they are rapidly protracted and retracted. The labrum and hypopharynx do not appear to participate in cutting host tissues. When a large pool of haemorrhaging blood forms, the mandibles and maxillae cease their movement and the fly feeds, sucking blood up through the labrum, engorging in 3 to 10 minutes.

The role of labellar pseudotracheae in tabanid feeding is not well understood. The pseudotracheae do not collect blood or convey it to the food canal during feeding. Neither Dickerson and Lavoipierre (1959b) nor Stoffolano and Yin (1983) observed labella resting on skin surfaces while flies fed on blood. Rather, they were retracted as is common in some nematocerans. Tabanidae also lack prestomal teeth, precluding the use of pseudotracheae to collect skin exudates resulting from scarification of the skin. It may be that these specialized labellar lobes have developed to facilitate the collection of dried sugar substrates, such as dried honeydew or plant exudates, that have been rehydrated by saliva. Increasing evidence of the importance of aphid honeydew in fly nutrition supports this conclusion (Kniepert, 1980).

Tabanids are large flies and require large amounts of blood for egg production. The deer fly *Chrysops callidus* Osten Sacken required 22 µl of blood to feed to repletion (Hollander and Wright, 1980). Larger horse flies, such as *Tabanus atratus* L., require more than 0.5 ml of blood. Due to the large amounts of blood required and the invasiveness of the bite stimulating host defences, more than one landing or feeding episode is needed for most tabanids to become fully engorged. Hollander and Wright (1980) estimated that the eight species of tabanids they sampled needed approximately ten landings on a host to complete one bloodmeal. It is interesting to note that from 6 to 13 nl of host blood may remain on tabanid mouthparts following each interrupted feeding episode (Knaus *et al.*, 1993).

Tabanids and rhagionids feed predominantly on larger mammals, with occasional records of individual species feeding on turtles, frogs or birds (Downes, 1971; Chainey, 1993). Many species are autogenous. In these species, follicular development during the first gonotrophic cycle occurs without females feeding on either blood or sugar, using nutrients stored from larval feeding (Bosler and Hansens, 1974). Subsequent gonotrophic cycles require flies to feed on blood. Female flies must feed on sugar as well as blood to survive. Flies fed on blood alone lived only 5 days in the laboratory (Wilson, 1967). Female *Tabanus lineola* F., offered multiple opportunities to feed on blood and a sugar solution, lived up to 31 days, took up to four bloodmeals and produced up to four egg masses. Similarly, *T. nigrovittatus* offered blood and a sugar solution lived up to 28 days under laboratory conditions, with an average survival of approximately 10 days (Magnarelli and Stoffolano, 1980). Most flies completed two gonotrophic cycles. In the field, most *T. abactor* Philip females, marked while feeding on a restrained cow, were recaptured 3 to 4 days later (Cooksey and Wright, 1987). Therefore, these flies matured and deposited a batch of eggs, and sought a new bloodmeal, during this time interval. A few marked flies in this study in Oklahoma and a similar one with *Tabanus* spp. in Alabama (Thornhill and Hays, 1972) were recaptured 23 to 24 days after their release.

## Muscamorpha

Haematophagous flies in the infraorder Muscamorpha with piercing mouthparts belong to the families Muscidae (subfamily Stomoxyinae), Glossinidae, Hippoboscidae, Streblidae and Nycteribiidae. Both males and females in these families feed using a piercing organ, alternatively termed the fascicle, syntrophium or proboscis, composed of the labrum, hypopharynx and labium. As in all dipterans, the food canal is formed by the labrum. In this case, it is closed by the hypopharynx (Fig. 2.1e). The salivary canal extends through the hypopharynx to its tip near the opening of the food canal. Labrum and hypopharynx are sheathed in the labium, which bears reduced labellar lobes and the rigid prestomal teeth.

Blood-feeding muscids in subfamily Stomoxyinae include, most notably, flies in the genera *Stomoxys*, *Haematobia* and *Haematobosca* (Zumpt, 1973). The labial mentum, or haustellum, of these flies is elongated, approximately 2 mm in length for stable flies, *Stomoxys calcitrans* (L.) and proportionately shorter in smaller *Haematobia* and *Haematobosca* spp. The labium is strongly sclerotized and tapers to a point (Greenberg, 1973). At the apex of the proboscis, the stable fly has five pairs of large prestomal teeth on a V-shaped sclerite. At rest, the proboscis is held parallel with the fly's long axis, its tip extending forward of the head. When feeding, it is directed down and slightly forward of perpendicular. The labella are retracted or opened and the prestomal teeth are everted and exposed. The proboscis is driven into the host tissues by repeated thrusts of the fly's head. Dermal tissues are lacerated by the prestomal teeth and the fly imbibes blood from the resulting haematoma (Lavoipierre, 1965).

The overall length of proboscis of tsetse flies (*Glossina* spp.) is approximately twice that of stable flies, ranging from 3.2 mm in *G. palpalis* (Robineau-Desvoidy) to 4.2 mm in *G. morsitans* Westwood (Jobling, 1933). The distal portion of the proboscis is very thin, with a diameter of about 0.060 mm. The labrum and hypopharynx both terminate in sharp, bevelled points. The labella are prominent in tsetse flies, with rows of fine denticles, like those on a rasp, associated with the prestomal teeth. With the labellum at rest, the prestomal teeth lie on the inside edges of the labellum, near its apex. The maxillary palpi are as long as the proboscis. At rest, their concave inner surfaces are applied to either side of the proboscis, sheathing the narrow part of the labium and the labellum.

Gordon et al. (1956) observed microscopically the behaviour of *G. morsitans* feeding on ears of guinea pigs. Flies alighted on the host with the labium enclosed in the maxillary palpi and held parallel to the skin surface. A suitable feeding site was selected where the tarsal claws could grip the host's skin. The labium was then freed from the palpi and directed downwards to lie perpendicular to the skin surface, with the labella contacting the skin surface. Penetration was accomplished by repeated and rapid eversion/inversion of the prestomal teeth and rasps on the inner surfaces of the labella. Each eversion of the labella first cut into and anchored the labium in host tissues and then pulled the labium forward, deeper into the dermis. Once the skin surface was penetrated, few forcible thrusts of the fly's head were required for further penetration. Feeding commenced when the laceration of capillaries or small venules resulted in the formation of a small haemorrhage. Blood from the haemorrhage was immediately and rapidly sucked into the labrum while the labella remained everted. Penetration and feeding was associated with intermittent ejections of saliva from the hypopharynx. When the blood flow from the initial haemorrhage became reduced, tsetse flies partially withdrew their proboces and penetrated a new area, creating another haemorrhage on which to feed. This process may be repeated a number of times until flies become fully engorged.

Jobling (1933) repeatedly notes the similarity between the mouthparts and feeding mechanism of tsetse flies and flies in the family Hippoboscidae. The proboscis of the sheep ked, *Melophagus ovinus* (L.) is long, thin and strongly sclerotized (Jobling, 1926). It is sheathed between maxillary palpi that extend past its tip. The labrum and hypopharynx are very stylet-like, terminating in sharp points. Two serrated ridges of small labellar denticles are associated with the prestomal teeth. During feeding, penetration of host tissues is rapid, accomplished by the repeated eversion and inversion of the prestomal teeth (Nelson and Petrunia, 1969). Occasionally, the proboscis is rotated back and forth, as the insect turns its head from one side to the other. The proboscis probes tissues until a suitable venule is penetrated. Nelson and Petrunia (1969) observed that the everted prestomal teeth were used to anchor the tips of the proboscis within the venule, from which the ked fed. These authors observed no attempts of flies to feed on small haemorrhages. Periodic and intermittent salivation was observed during feeding.

Jobling (1928b, 1929) also described the mouthparts of both families of bat flies (Streblidae and Nycteribiidae) and compared them with those of the Hippoboscidae. He found that the mouthparts of the two families of bat flies resembled each other more closely than they did those of hippoboscids. The probosces of bat flies, especially those of the Streblidae, are shorter and stouter than those of hippoboscids or tsetse flies. The maxillary palpi of bat flies do not sheath the proboscis.

The labrum of the Nycteribiidae is very short, extending less than one-quarter the length of the proboscis. In the distal three-quarters of the proboscis, the food canal is formed by the labral gutter alone. The food canal is completely enclosed by interlocking folds of lateral labral membranes. The labella are membranous with their inner apical surfaces covered with rows of small teeth. At rest, these small teeth are directed forwards or sideways. When the labella are everted, the points of these small teeth are directed backwards. Eversion of the labella also exposes six flat, triangular prestomal teeth and two longer, claw-like inner teeth. The numerous small teeth may function to pull the proboscis into host tissues, that are lacerated by the prestomal and inner teeth.

The labrum of the Streblidae is also very short, with the labial gutter again forming most of the food canal. The hypopharynx in these flies is also shortened, extending only a little beyond the distal tip of the labrum. Forward of the ends of the labrum and hypopharynx, the dorsal salivary canal is separated from the ventral food canal by interlocking inward projections of the labial gutter. Compared to the total length of the proboscis, the labella are relatively long. Their sclerotized sides taper inwards to a serrated point. Jobling (1929) could not identify any prestomal teeth, but the labella were armed with ridges of denticles and two inner teeth.

Stable flies, both males and females, must feed on blood to reproduce (Meola *et al.*, 1977). In nature, both sexes may also feed on nectar (Jones *et al.*, 1985). Doing so may allow them to survive, but it does not increase their reproductive capacity or longevity (Jones *et al.*, 1992). Stable flies feed on most mammals and some large reptiles, but prefer larger mammals such as cattle and horses (Greenberg, 1973). Stable flies imbibe approximately 26 mg of blood per meal. Female stable flies, feeding in cages on a restrained steer, fed approximately two times a day for around 4 minutes each time (Harris *et al.*, 1974). In the field, stable flies preferred feeding on the forelegs of cattle and required more than 30 minutes to complete feeding on unrestrained cattle (Lysyk, 1995). The author attributed the difference between feeding times on restrained and unrestrained cattle to the effects of host defensive behaviour. Stable flies may live up to 29 days in the field (Greenberg, 1973).

Adult horn flies, *Haematobia irritans* (L.), are obligate haematophagous ectoparasites of cattle, remaining on their host at all times, except when disturbed or ovipositing (Zumpt, 1973). Female horn flies caged on a restrained steer spent, on average, 163 minutes feeding per day (Harris *et al.*, 1974). Female and male flies fed approximately 38 and 24 times per day, respectively, for around 4 minutes each time. Based on analyses of gut contents of flies feeding on unrestrained cattle in the field, the authors estimated that horn flies fed at least once each hour (24 times per day). Burg *et al.* (1994) reported that 50% of female *H. irritans* fed citrated bovine blood in the laboratory lived up to 14 days. Males had shorter life spans.

Tsetse flies, both male and female, also feed exclusively on blood (Buxton, 1955). As a group, tsetse flies feed primarily on large mammals, especially wild ungulates. However, they may also feed on birds, including ostriches, and reptiles. Jordan (1993) reviewed tsetse fly host utilization extensively. Some species have defined preferences for certain hosts, but the feeding habits of all species are partially determined by what is available and the nutritional status of the fly. Frequency of feeding and amount of blood ingested per meal are also determined by the nutritional status of tsetse flies, as well as other factors including reproductive status and age (Buxton, 1955). Female tsetse flies are viviparous, each nurturing a single larva with secretions from a specialized accessory gland until it is deposited in a fully developed prepupal state. This interlarval period or pregnancy cycle takes 9 days. Female *Glossina morsitans morsitans* Westwood and *G. pallidipes* Austen reproduced best when fed four and five times per interlarval period, respectively (Langley and Stafford, 1990). Buxton (1955) reviewed reports of tsetse flies imbibing anywhere from 5 to 80 mg of blood per meal. Tobe and Davey (1972) found that female *G. austeni* Newstead of the same physiological age, fed until they reached a constant weight, irrespective of reproductive status. Prior to the first ovulation, female flies fed to a weight of approximately 50 mg by consuming about 30 mg of blood. After the first ovulation, female flies fed to a constant weight of approximately 80 mg, consuming less than 20 mg of blood during the period of maximal larval growth (days 6 to 9 of the pregnancy cycle) and up to 60 mg of blood following larviposition.

Female *G. palpalis* and *G. morsitans submorsitans* Newstead females lived an average of 84 and 72 days, respectively, with some individuals living up to 154 days (Buxton, 1955). Female *G. pallidipes* lived an average of 105 days (maximum 163 days). Based on reports of numerous mark, release and recapture studies, Buxton (1955) concluded that *G. morsitans* males may live a month in the field and females twice that long.

The Hippoboscidae are obligate haematophagous ectoparasites of birds and mammals (Maa, 1969; Marshall, 1981). Of 197 species, 153 infest birds and 44 infest mammals. Mammalian ectoparasites include 30 species of *Lipoptene*, infesting pigs, deer, cattle and their relatives, and *M. ovinus*, the sheep ked. Most hippoboscids display some degree of host specificity. Hippoboscids are viviparous, depositing mature larvae on or around the host just before they pupate. Most species of the Hippoboscidae engorge in 5 to 30 minutes and *M. ovinus* consumes 3 to 15 mg of blood per meal (Bequaert, 1952). Intervals between bloodmeals vary from 36 hours for *M. ovinus* to 6 days for *Stilbometopa impressa* (Bigot) (Marshall, 1981). Female hippoboscids are long-lived. Female *M. ovinus* may live up to 165 days, but few live more than 100 days. Male longevity is generally substantially less than that of females.

Flies in the families Streblidae and Nycteribiidae are haematophagous ectoparasites of bats (Marshall, 1981). Streblids are found mainly in the tropics and subtropics of the Old World and nycteribiids occur primarily in the tropics and subtropics of the New World. In Australia, most species of bat flies are found on gregarious or cave-dwelling bats and are host specific, infesting only one genus of bats (Maa, 1971). Conversely, one species of bat may harbour numerous species of bat flies.

Bat flies are viviparous, depositing larvae at regular intervals in the host's roost (Marshall, 1981). After emerging from pupae, adults feed as soon as they reach a host and at hourly intervals, thereafter, throughout their lives. Female nycteribiids may live up to 180 days.

## Acari

Only those acarines that feed regularly on vertebrate blood and require that blood for normal growth and development will be covered in this section. Haematophagous acarines fitting this description include ectoparasitic mites in the families Macronyssidae and Dermanyssidae (order Parasitiformes, suborder Mesostigmata) and the tick families Argasidae and Ixodidae (order Parasitiformes, suborder Ixodida). Acarines in these families have mouthparts capable of penetrating unbroken skin. Other ectoparasitic mites, such as *Psoroptes* spp. (suborder Astigmata) and mesostigmatid species in the family Laelapidae, have mouthparts that abrade, scrape or rasp host tissues (DeLoach and Wright, 1981; Radovsky, 1994). Tissues affected may include wounds from other sources. These mites feed on the skin secretions caused by these actions. Such skin secretions may contain blood or its components (Rafferty and Gray, 1987).

Tick feeding apparatuses and mechanisms have been extensively described. The following summary was prepared from the reviews of Kemp *et al.* (1982) and Sonenshine (1991). The mouthparts and feeding of mesostigmatid mites have received less attention, and the following is derived from observations of the mouthparts of the adult female tropical rat mite, *Ornithonyssus bacoti* (Hirst) (Gorirossi, 1950), feeding mechanism of *Chiroptonyssus robustipes* (Ewing) (Lavoipierre and Beck, 1967) and the reviews of Radovsky (1985, 1994).

The principal mouthparts of all acarines, relative to feeding, are the chelicerae and hypostome. The chelicerae are paired protractible/retractible cylindrical tubes with apical cheliceral digits. In the Ixodida, the cheliceral digits have laterally-directed cutting edges. With outward, opposable movements of the chelicerae, the cheliceral digits cut or tear into host tissues. In contrast, the cheliceral digits of mesostigmatid mites have one movable and one fixed digit and are capable of a scissor-like motion with which they can grasp, tear or pierce tissues. Mites in the family Dermanyssidae have long, slender, stylet-like chelicerae that fit together to form a tube. After the chelicerae pierce the skin and possibly a venule, blood is directed through this tube to the oral cavity.

The hypostome is an unpaired structure beneath the chelicerae that directs blood into the oral cavity and may be used by the acarine to anchor to its host. The inner or dorsal surface of the hypostome is concave and has a V- or U-shaped gutter extending posteriorly/anteriorly along its midline. Along with the ventral sides of the chelicerae, this channel forms a canal that is used for both the intake of blood and the output of saliva. In the Ixodidae, blood intake and salivary output are separated by a pharyngeal valve that closes off the alimentary canal during salivation and a long period of time between each of these events. In rapid feeding argasids, an elongated, blade-like labrum lies along the entire length of the hypostomal channel. The posterior end of the labrum is attached to the dorsal surface of the hypostome. When elevated, the labrum closes off the salivarium, opens the hypostomal channel and diverts blood into the pharynx. When depressed, it completely blocks the hypostomal channel during salivation.

The ventral aspect of the hypostome in the Ixodidae is covered with rows of large, recurved teeth. These teeth engage host tissues and aid in anchoring ticks during feeding. Similar denticles on the hypostomes of mesostigmatid mites are either much reduced or absent.

The Dermanyssidae are primarily bird ectoparasites (Radovsky, 1985). The life cycles of these mites include the following stages: egg, non-feeding six-legged larva, eight-legged protonymph and deutonymph, and adult (male and female). Both nymphal stages and adult females feed rapidly and have similar mouthparts. Their long, slender chelicerae fit together to form a stylet-like tube that is used to pierce host tissues and suck blood. The chicken mite, *Dermanyssus gallinae* (De Geer), feeds on blood from chickens, turkeys and wild birds. The house mouse mite, *Liponyssoides sanguineus* (Hirst), is similar to *D. gallinae*, but infests small rodents. These and other dermanyssids are typical nest parasites, residing on the host only when feeding and capable of surviving long periods of time without a bloodmeal (Radovsky, 1994). These mites do this by taking relatively large bloodmeals. Engorged female *D. gallinae* contained an average of 204 µg of blood (Sikes and Chamberlain, 1954).

The Macronyssidae are ectoparasites of mammals, including bats, birds and reptiles (Radovsky, 1994). Their life cycles are the same as that of the Dermanyssidae, except that deutonymphs do not feed. In addition, protonymphs are fixed feeders, feeding in one place for up to 4 or 5 days if undisturbed. These nymphs spend all but the last few minutes of this time feeding on tissue serrous fluids. During the last minutes, they engorge on whole blood. Protonymph engorgement provides the nutrition for both nymphal stages. Adult females feed rapidly, engorging in minutes on blood pooling from a ruptured blood vessel.

Gorirossi (1950) observed adult tropical rat mites, *Ornithonyssus bacoti*, feeding on rat abdomens. Blood entered the gut soon after mites attached.

Mites completed feeding in 6 to 17 minutes. Chelicerae were the only mouthparts that apparently penetrated the skin. As in dermanyssids, protracted chelicerae united to form a tubular extension of the food canal. Skaliy and Hayes (1949) found that most female *O. bacoti* fed every 3 to 6 days and moved off the host after feeding. Females survived approximately 62 days. Northern fowl mites, *O. sylviarum* (Canestrini and Fanzago), and tropical fowl mites, *O. bursa* (Berlese), are important pests of domestic fowl, imbibing approximately 41 and 77 µg of blood per adult female, respectively, during each bloodmeal (Sikes and Chamberlain, 1954). DeLoach and DeVaney (1981) estimated that female *O. sylviarum* fed once during each 24 hour period.

Lavoipierre and Beck (1967) described feeding by *Chiroptonyssus robustipes* protonymphs and female adults on their bat hosts. Protonymphs fed by directing their chelicerae downwards between their legs and gouging out a small crater in the host's skin with their claw- or pincer-like cheliceral digits. Protonymphs fed on tissue serrous fluids exuding from these wounds for 10 to 12 hours. Tissue fluids exuding from these wounds were highly attractive to other protonymphs. After feeding extensively on tissue fluids, protonymphs then punctured a small blood vessel and fed avidly on the resulting blood pool for 15 to 20 minutes. Adult females also fed by directing their chelicerae downwards into unbroken skin. However, they probed repeatedly until a blood vessel was penetrated. Females then lacerated the vessel and fed to repletion from the resulting haemorrhage in 10 to 15 minutes.

Life histories, feeding mechanisms and behaviour of ticks in the families Ixodidae and Argasidae have been exhaustively studied and recently reviewed by Kemp *et al.* (1982) and Sonenshine (1991, 1994). Aspects of their biology will be covered here superficially. Ticks are obligate, blood-feeding ectoparasites of vertebrates. They reside on the host only when feeding. Developmental stages are eggs, six-legged larvae, eight-legged nymphs and eight-legged male and female adults. Bloodmeals are required for the development of larvae and nymphs and for reproduction by adults. Variable numbers of nymphal moults occur in argasids alone, with each nymphal instar requiring a bloodmeal to develop to the next stage.

Argasid nymphs and adults are rapid feeders, feeding to repletion in a few hours or less. *Ornithodoros* species required 33 to 162 minutes to complete feeding (Lavoipierre and Riek, 1955). These argasids fed by rapidly cutting into the host's skin with alternating motions of the chelicerae. Salivation began as soon as the blood pool formed. Extension of the haemorrhagic area was induced by action of the saliva on surrounding tissues and not from further mechanical trauma created by the chelicerae. The host range of argasids is large, encompassing most terrestrial vertebrates. However, due to their behaviour as nest parasites, an individual argasid generally will feed on the same host or host species throughout its life span of one or more years. In the laboratory, the life cycle of *Ornithodoros talaje* (Guerin-Meneville) was completed in a maximum of 849 days (Schumaker and Barros, 1995).

Larvae, nymphs and adults in the family Ixodidae require days to complete feeding. In general, larvae engorge in 4 days while nymphs and adults may feed for around a week. Each feeding stage of the ixodid life cycle may utilize a separate host, often of various species. However, in some species, subsequent life stages may remain and feed on the same host. In the Ixodidae, the chelicerae are used to create a shallow incision in the host's skin so that the hypostome may be inserted and anchored to host tissues. Although the cutting action of the cheliceral digits may be responsible for some damage to capillaries and other tissues at the site of attachment, salivary secretions are responsible for development of the extensive haemorrhage at the feeding lesion. Salivation by slow-feeding ixodids, such as *Dermacentor andersoni* Stiles, causes the progressive enlargement of and release of blood into the feeding lesion. This culminates approximately 2.5 hours after attachment with the sudden formation of a large haemorragic pool. Feeding from this source, ixodids may increase their engorged body weights to upwards of 100 times that of their unfed body weights.

## ACKNOWLEDGEMENTS

The author thanks Dr Richard C. Berberet and Dr Stephen K. Wikel for reading and editing this manuscript. Preparation of this chapter supported in part by Oklahoma Agricultural Experiment Station project OKL02174.

## REFERENCES

Anderson, J.R. (1987) Reproductive strategies and gonotrophic cycles of black flies. In: Kim, K.C. and Merritt, R.W. (eds), *Black Flies: Ecology, Population Management, and Annotated World List*. Pennsylvania State University, University Park, pp. 276–294.

Askew, R.R. (1971) *Parasitic Insects*. American Elsevier Publishing Co., New York.

Bequaert, J.C. (1952) The Hippoboscidae or louse-flies (Diptera) of mammals and birds. Part I. Structure, physiology and natural history. *Entomologica Americana* 32 (n.s.), 1–209.

Bosler, E.M. and Hansens, E.J. (1974) Natural feeding behavior of adult saltmarsh greenheads, and its relation to oogenesis. *Annals of the Entomological Society of America* 67, 321–324.

Braverman, Y. (1994) Nematocera (Ceratopogonidae, Psychodidae, Simuliidae and Culicidae) and control methods. *Revue Scientifique et Technique, Office International des Epizooties* 13, 1175–1199.

Braverman, Y., Linley, J.R., Marcus, R. and Frish, K. (1985) Seasonal survival and expectation of infective life of *Culicoides* spp. (Diptera: Ceratopogonidae) in Israel, with implications for blue tongue virus transmission and a comparison of the parous rate in *C. imicola* from Israel and Zimbabwe. *Journal of Medical Entomology* 22, 476–484.

Brinson, F.J., McKeever, S. and Hagan, D.V. (1993) Comparative study of mouthparts of the Phlebotomine sand flies *Lutzomyia longipalpis*, *L. shannoni*, and *Phylebotomus papatasi* (Diptera: Psychodidae). *Annals of the Entomological Society of America* 86, 470–483.

Burg, J.G., Knapp, F.W. and Silapanuntakul, S. (1994) Blood meal manipulation and *in vitro* colony maintenance of *Haematobia irritans* (Diptera: Muscidae). *Journal of Medical Entomology* 31, 868–874.

Busvine, J.R. (1948) The 'head' and 'body' races of *Pediculus humanus* L. *Parasitology* 39, 1–16.

Buxton, P.A. (1940) *The Louse*. Williams and Wilkins Co., Baltimore.

Buxton, P.A. (1955) *The Natural History of Tsetse Flies*. H.K. Lewis, London.

Carcavallo, R. and Martinez, A. (1972) Life cycles of some species of *Triatoma* (Hemiptera:Reduviidae). *Canadian Entomologist* 104, 699–704.

Chainey, J.E. (1993) Horse-flies, deer-flies and clegs (Tabanidae). In: Lane, R.P. and Crosskey, R.W. (eds), *Medical Insects and Arachnids*. Chapman & Hall, London, pp. 310–332.

Clay, T. (1949) Piercing-mouthparts in the biting lice (Mallophaga). *Nature* 164, 617.

Clements, A.N. (1992) *The Biology of Mosquitoes*, Volume 1. Chapman & Hall, London.

Cohen, S., Greenwood, M.T. and Fowler, J.A. (1991) The louse *Trinoton anserinum* (Amblycera: Phthiraptera), an intermediate host of *Sarconema eurycerca* (Filarioidea: Nematoda), a heartworm of swans. *Medical and Veterinary Entomology* 5, 101–110.

Cooksey, L.M. and Wright, R.E. (1987) Flight range and dispersal activity of the host-seeking horse fly, *Tabanus abactor* (Diptera: Tabanidae), in North Central Oklahoma. *Environmental Entomology* 16, 211–217.

Crosskey, R.W. (1990) *The Natural History of Blackflies*. John Wiley & Sons, Chichester.

Crosskey, R.W. (1993a) Introduction to the Diptera. In: Lane, R.P. and Crosskey, R.W. (eds), *Medical Insects and Arachnids*. Chapman & Hall, London, pp. 51–77.

Crosskey, R.W. (1993b) Stable-flies and horn-flies (bloodsucking Muscidae). In: Lane, R.P. and Crosskey, R.W. (eds), *Medical Insects and Arachnids*. Chapman & Hall (London), pp. 389–402.

Culpepper, G.H. (1948) Rearing and maintaining a laboratory colony of body lice on rabbits. *American Journal of Tropical Medicine* 28, 499–504.

Day, J.F., Edman, J.D. and Scott, T.W. (1994) Reproductive fitness and survivorship of *Aedes aegypti* (Diptera: Culicidae) maintained on blood, with field observations from Thailand. *Journal of Medical Entomology* 31, 611–617.

De Colmenares, M., Portus, M., Botet, J., Dobano, C., Gallego, M., Wolff, M. and Segui, G. (1995) Identification of blood meals of *Phlebotomus perniciosus* (Diptera: Psychodidae) in Spain by a competitive enzyme-linked immunosorbent assay biotin/avidin method. *Journal of Medical Entomology* 32, 229–233.

DeLoach, J.R. and DeVaney, J.A. (1981) Northern fowl mite, *Ornithonyssus sylviarum* (Acari: Macronyssidae), ingests large quantities of blood from White Leghorn hens. *Journal of Medical Entomology* 18, 374–377.

DeLoach, J.R. and Wright, F.C. (1981) Ingestion of rabbit erythrocytes containing $^{51}$Cr-labeled hemoglobin by *Psoroptes* spp. (Acari: Psoroptidae) that originated on cattle, mountain sheep, or rabbits. *Journal of Medical Entomology* 18, 345–348.

Deoras, P.S. and Prasad, R.S. (1967) A note on the feeding mechanism of two fleas. *Current Science* 36, 518–519.

Dickerson, G. and Lavoipierre, M.M.J. (1959a) Studies on the methods of feeding of blood-sucking arthropods II. The method of feeding adopted by the bed-bug (*Cimex lectularius*) when obtaining a blood-meal from the mammalian host. *Annals of Tropical Medicine and Parasitology* 53, 347–357.

Dickerson, G. and Lavoipierre, M.M.J. (1959b) Studies on the methods of feeding of blood-sucking arthropods III. The method by which *Haematopota pluvialis* (Diptera, Tabanidae) obtains its blood-meal from the mammalian host. *Annals of Tropical Medicine and Parasitology* 53, 465–472.

Dolling, W.R. (1991) *The Hemiptera*. Oxford University Press, New York.

Downes, J.A. (1958) The feeding habits of biting flies and their significance in classification. *Annual Review of Entomology* 3, 249–266.

Downes, J.A. (1971) The ecology of blood-sucking Diptera: an evolutionary perspective. In: Fallis, A.M. (ed.), *Ecology and Physiology of Parasites*. University of Toronto Press, Toronto, pp. 232–258.

Elnaiem, D.A., Morton, I., Brazil, R. and Ward, R.D. (1992) Field and laboratory evidence for multiple bloodfeeding by *Lutzomyia longipalpis* (Diptera: Psychodidae). *Medical and Veterinary Entomology* 6, 173–174.

El Sawaf, B.M., El Sattar, S.A., Shehata, M.G., Lane, R.P. and Morsy, T.A. (1994) Reduced longevity and fecundity in *Leishmania*-infected sand flies. *American Journal of Tropical Medicine and Hygiene* 51, 767–770.

Ferris, G.F. (1931) The louse of elephants. *Haematomyzus elephantis* Piaget (Mallophaga: Haematomyzidae). *Parasitology* 23, 112–127.

Ferris, G.F. (1951) *The Sucking Lice*. Memoirs of the Pacific Coast Entomological Society, San Francisco.

Ferro, C., Morrison, A.C., Torres, M., Pardo, R., Wilson, M.L. and Tesh, R.B. (1995) Age structure, blood-feeding behavior, and *Leishmania chagasi* infection in *Lutzomyia longipalpus* (Diptera: Psychodidae) as an endemic focus of visceral leishmaniasis in Columbia. *Journal of Medical Entomology* 32, 618–629.

Flemings, M.B. and Ludwig, D. (1964) Effects of temperature and parental age on the life cycle of the body louse, *Pediculus humanus humanus*. *Annals of the Entomological Society of America* 57, 560–563.

Friend, W.G. and Smith, J.J.B. (1971) Feeding in *Rhodnius prolixus*: mouthpart activity and salivation and their correlation with changes of electrical resistance. *Journal of Insect Physiology* 17, 233–243.

Gad, A.M., Riad, I.B. and Farid, H.A. (1995) Host-feeding patterns of *Culex pipiens* and *Cx. antennatus* (Diptera: Culicidae) from a village in Sharquya Governorate, Egypt. *Journal of Medical Entomology* 32, 573–577.

Gooding, R.H. (1963) Studies on the frequency of feeding on the biology of a rabbit adapted strain of *Pediculus humanus*. *Journal of Parasitology* 49, 516–521.

Gordon, R.M. and Lumsden, W.H.R. (1939) A study of the behaviour of the mouthparts of mosquitoes when taking up blood from living tissue; together with some observations on the ingestion of microfilariae. *Annals of Tropical Medicine and Parasitology* 33, 259–278.

Gordon, R.M., Crewe, W. and Willett, K.C. (1956) Studies on the deposition, migration and development to the blood forms of trypanosomes belonging to the *Trypanosoma brucei* group. I. An account of the process of feeding adopted by the tsetse fly when obtaining a blood meal from the mammalian host, with special reference to the ejection of saliva and the relationship of the feeding process to the deposition of the metacyclic typanosomes. *Annals of Tropical Medicine and Parasitology* 50, 426–437.

Gorirossi, F.E. (1950) The mouth-parts of the adult female tropical rat mite *Bdellonyssus bacoti* (Hirst, 1913) Fonseca, 1941 (=*Liponissus bacoti* [Hirst]), with observations on the feeding mechanism. *Journal of Parasitology* 36, 301–318.

Greenberg, B. (1973) *Flies and Disease*, Volume 2. Princeton University Press, Princeton.

Greenberg, B. and Povolny, D. (1971) Bionomics of flies. In: Greenberg, B. (ed.), *Flies and Disease*, Volume 1. Princeton University Press, Princeton, pp. 56–83.

Griffiths, R.B. and Gordon, R.M. (1952) An apparatus which enables the process of feeding by mosquitoes to be observed in tissues of a live rodent, together with an account of the ejection of saliva and its significance in malaria. *Annals of Tropical Medicine and Parasitology* 46, 311–319.

Guzman, H., Walters, L.L. and Tesh, R.B. (1994) Histologic detection of multiple blood meals in *Phlebotomus duboscqi* (Diptera: Psychodidae). *Journal of Medical Entomology* 31, 890–897.

Harris, R.L., Miller, J.A. and Frazar, E.D. (1974) Horn flies and stable flies: feeding activity. *Annals of the Entomological Society of America* 67, 891–894.

Hollander, A.L. and Wright, R.E. (1980) Impact of tabanids on cattle: blood meal size and preferred feeding sites. *Journal of Economic Entomology* 73, 431–433.

Jobling, B. (1926) A comparative study of the structure of the head and mouthparts in the Hippoboscidae. *Parasitology* 18, 319–349.

Jobling, B. (1928a) The structure of the head and mouth parts in *Culicoides pulicaris* L. (Diptera: Nematocera). *Bulletin of Entomological Research* 18, 211–236.

Jobling, B. (1928b) The structure of the head and mouth-parts in the Nycteribiidae (Diptera: Pupipara). *Parasitology* 20, 254–272.

Jobling, B. (1929) A comparative study of the structure of the head and mouth parts in the Streblidae (Diptera: Pupipara). *Parasitology* 21, 417–445.

Jobling, B. (1933) A revision of the structure of the head, mouth-part and salivary glands of *Glossina palpalis* Rob.-Desv. *Parasitology* 24, 449–490.

Jobling, B. (1976) On the fascicle of blood-sucking Diptera. In addition a description of the maxillary glands in *Phlebotomus papatasi*, together with the musculature of the labium and pulsatory organ of both the latter species and also of some other Diptera. *Journal of Natural History* 10, 457–461.

Johnson, P.T. and Hertig, M. (1961) The rearing of *Phlebotomus* sandflies (Diptera: Psychodidae) II. Development and behavior of Panamanian sandflies in laboratory culture. *Annals of the Entomological Society of America* 54, 764–776.

Jones, C.J., Hogsette, J.A., Patterson, R.S. and Milne, D.E. (1985) Effects of natural saccharide and pollen extract feeding on stable fly (Diptera: Muscidae) longevity. *Environmental Entomology* 14, 223–227.

Jones, C.J., Milne, D.E., Patterson, R.S., Schreiber, E.T. and Milio, J.A. (1992) Nectar feeding by *Stomoxys calcitrans* (Diptera: Muscidae): effects on reproduction and survival. *Environmental Entomology* 21, 141–147.

Jones, J.C. and Pilitt, D.R. (1973) Blood-feeding behaviour of adult *Aedes aegypti* mosquitoes. *Biological Bulletin* 145, 127–139.

Jordan, A.M. (1993) Tsetse-flies (Glossinidae). In: Lane, R.P. and Crosskey, R.W. (eds),

*Medical Insects and Arachnids*. Chapman & Hall, London, pp. 333–388.

Kemp, D.H., Stone, B.F. and Binnington, K.C. (1982) Tick attachment and feeding: role of the mouthparts, feeding apparatus, salivary gland secretions and the host response. In: Obenchain, F.D. and Galun, R. (eds), *Physiology of Ticks*. Pergamon Press, Oxford, pp. 119–168.

Klowden, M.J. and Briegel, H. (1994) Mosquito gonotrophic cycle and multiple feeding potential: contrasts between *Anopheles* and *Aedes* (Diptera: Culicidae). *Journal of Medical Entomology* 31, 618–622.

Klowden, M.J. and Lea, O.A. (1979) Effect of defensive host behavior on the blood meal size and feeding success of natural populations of mosquitoes (Diptera: Culicidae). *Journal of Medical Entomology* 15, 514–517.

Knaus, R.M., Foil, L.D., Issel, C.J. and Leprince, D.J. (1993) Insect blood meal studies using radiosodium $^{24}$Na and $^{22}$Na. *Journal of the American Mosquito Control Association* 9, 264–268.

Kniepert, F. (1980) Blood-feeding and nectar-feeding in adult Tabanidae (Diptera). *Oecologia* 46, 125–129.

Kramer, V.L., Carper, E.R., Beesley, C. and Reisen, W.K. (1995) Mark–release–recapture studies with *Aedes dorsalis* (Diptera: Culicidae) in coastal northern California. *Journal of Medical Entomology* 32, 375–380.

Lane, R.P. (1993) Sand flies (Phlebotominae). In: Lane, R.P. and Crosskey, R.W. (eds), *Medical Insects and Arachnids*. Chapman & Hall, London, pp. 78–119.

Langley, P.A. and Stafford, K. (1990) Feeding frequency in relation to reproduction in *Glossina morsitans morsitans* and *G. pallidipes*. *Physiological Entomology* 15, 415–421.

Lavoipierre, M.M.J. (1965) Feeding mechanism of blood-sucking arthropods. *Nature* 208, 302–303.

Lavoipierre, M.M.J. (1967) Feeding mechanism of *Haematopinus suis* on the transilluminated mouse ear. *Experimental Parasitology* 20, 303–311.

Lavoipierre, M.M.J. and Beck, A.J. (1967) Feeding mechanism of *Chiroptonyssus robustipes* on the transilluminated bat wing. *Experimental Parasitology* 20, 312–320.

Lavoipierre, M.M.J. and Hamachi, M. (1961) An apparatus for observations on the feeding mechanism of the flea. *Nature* 192, 998–999.

Lavoipierre, M.M.J. and Riek, R.F. (1955) Observations on the feeding habits of argasid ticks and on the effect of their bites on laboratory animals, together with a note on the production of coxal fluid by several of the species studied. *Annals of Tropical Medicine and Parasitology* 49, 96–113.

Lavoipierre, M.M.J., Dickerson, G. and Gordon, R.M. (1959) Studies on the methods of feeding of blood-sucking arthropods. I. The manner in which triatomine bugs obtain their blood-meal, as observed in the tissues of the living rodent, with some remarks on the effects of the bite on human volunteers. *Annals of Tropical Medicine and Parasitology* 53, 235–250.

Lavoipierre, M.M.J., Radovsky, F.J. and Budwiser, P.D. (1979) The feeding process of a tungid flea, *Tunga monositus* (Siphonaptera:Tungidae), and its relationship to the host inflammatory and repair response. *Journal of Medical Entomology* 15, 187–217.

Lee, R. (1974) Structure and function of the fascicular stylets, and the labral and cibarial sense organs of male and female *Aedes aegypti* (L.). *Quaestiones Entomologicae* 10, 187–215.

Lewis, D.J. (1975) Functional morphology of the mouth parts in New World phlebotomine sandflies (Diptera: Psychodidae). *Transactions of the Royal Entomological Society of London* 126, 497–532.

Linley, J.R. (1983) Autogeny in the Ceratopogonidae: literature and notes. *Florida Entomologist* 66, 228–234.

Lysyk, T.J. (1995) Temperature and population density effects on feeding activity of *Stomoxys calcitrans* (Diptera: Muscidae) on cattle. *Journal of Medical Entomology* 32, 508–514.

Maa, T.C. (1969) Notes on the Hippoboscidae (Diptera). *Pacific Insects Monograph* 20, 237–260.

Maa, T.C. (1971) Revision of the Australian batflies (Diptera: Streblidae and Nycteribiidae). *Pacific Insects Monograph* 28, 1–118.

Magnarelli, L.A. and Stoffolano, J.G., Jr (1980) Blood feeding, oogenesis, and oviposition by *Tabanus nigrovittatus* in the laboratory. *Annals of the Entomological Society of America* 73, 14–17.

Marshall, A.G. (1981) *The Ecology of Ectoparasitic Insects*. Academic Press, London.

McAlpine, J.F. (1981) *Manual of Nearctic Diptera*, Volume 1. Research Branch, Agriculture Canada, Monograph No. 27.

McCreadie, J.W., Colbo, M.H. and Hunter, F.F. (1994) Notes on sugar feeding and selected wild mammalian hosts of black flies (Diptera: Simuliidae) in Newfoundland. *Journal of Medical Entomology* 31, 566–570.

McKeever, S., Wright, M.D. and Hagan, D.V. (1988) Mouthparts of females of four *Culicoides* species (Diptera:Ceratopogonidae). *Annals of the Entomological Society of America* 81, 332–341.

Meola, R.W., Harris, R.L., Meola, S.M. and Oehler, D.D. (1977) Dietary-induced secretion of sex pheromone and development of sexual behavior in the stable fly. *Environmental Entomology* 6, 895–897.

Morrison, A.C., Ferro, C., Morales, A., Tesh, R.B. and Wilson, M.L. (1993a) Dispersal of the sand fly *Lutzomyia longipalpus* (Diptera: Psychodidae) at an endemic focus of visceral leishmaniasis in Columbia. *Journal of Medical Entomology* 30, 427–435.

Morrison, A.C., Ferro, C. and Tesh, R.G. (1993b) Host preferences of the sand fly *Lutzomyia longipalpus* at an endemic focus of American visceral leishmaniasis in Columbia. *American Journal of Tropical Medicine and Hygiene* 49, 68–75.

Mullens, B.A. and Schmidtmann, E.T. (1982) The gonotrophic cycle of *Culicoides variipennis* (Diptera: Ceratopogonidae) and its implications in age-grading field populations in New York state, USA. *Journal of Medical Entomology* 19, 340–349.

Nelson, W.A. and Petrunia, D.M. (1969) *Melophagus ovinus*: Feeding mechanism on transilluminated mouse ear. *Experimental Parasitology* 26, 308–313.

Nicholson, H.P. (1945) The morphology of the mouthparts of the non-biting blackfly, *Eusimulium dacotense* D. & S., as compared with those of the biting species, *Simulium venustum* Say (Diptera: Simuliidae). *Annals of the Entomological Society of America* 38, 281–297.

O'Rourke, F.J. (1956) Observations on pool and capillary feeding in *Aedes aegypti* (L.). *Nature* 177, 1087–1088.

Radovsky, F.J. (1985) Evolution of mammalian mesostigmate mites. In: Kim, K.C. (ed.), *Coevolution of Parasitic Arthropods and Mammals*. John Wiley & Sons, New York, pp. 441–504.

Radovsky, F.J. (1994) The evolution of parasitism and the distribution of some dermanyssoid mites (Mesostigmata) on vertebrate hosts. In: Houck, M.A. (ed.), *Mites: Ecological and Evolutionary Analyses of Life-History Patterns*. Chapman & Hall, New York, pp. 186–217.

Rafferty, D.E. and Gray, J.S. (1987) The feeding behavior of *Psoroptes* spp. mites on rabbits and sheep. *Journal of Parasitiology* 73, 901–906.

Robertson, L.D., Prior, S., Apperson, C.S. and Irby, W.S. (1993) Bionomics of *Anopheles quadrimaculatus* and *Culex erraticus* (Diptera: Culicidae) in the Falls Lake Basin, North Carolina: seasonal changes in abundance and gonotrophic status; and host-feeding patterns. *Journal of Medical Entomology* 30, 689–698.

Robinson, A. (1939) The mouthparts and their function in the female mosquito, *Anopheles maculipennis*. *Parasitology* 31, 212–242.

Rothschild, M. (1975) Recent advances in our knowledge of the order Siphonaptera. *Annual Review of Entomology* 20, 241–259.

Rothschild, M. and Clay, T. (1957) *Fleas, Flukes and Cuckoos: a Study of Bird Parasites*. Macmillan, New York.

Savage, H.M., Niebylski, M.L., Smith, G.C., Mitchell, C.J. and Craig, G.B. Jr (1993) Host-feeding patterns of *Aedes albopictus* (Diptera: Culicidae) at a temperate North American site. *Journal of Medical Entomology* 30, 27–34.

Schuh, R.T. and Slater, J.A. (1995) *True Bugs of the World (Hemiptera:Heteroptera): Classification and Natural History*. Cornell University Press, Ithaca.

Schuh, R.T. and Stys, P. (1991) Phylogenetic analysis of cimicomorphan family relationships (Heteroptera). *Journal of the New York Entomological Society* 99, 298–350.

Schumaker, T.T.S. and Barros, D.M. (1995) Life cycle of *Ornithodoros (Alectorobius) talaje* (Acari: Argasidae) in laboratory. *Journal of Medical Entomology* 32, 249–254.

Scott, T.W., Chow, E., Strickman, D., Kittayapong, P., Wirtz, R.A., Lorenz, L.H. and Edman, J.D. (1993a) Blood-feeding patterns of *Aedes aegypti* (Diptera: Culicidae) collected in a rural Thai village. *Journal of Medical Entomology* 30, 922–927.

Scott, T.W., Clark, G.G., Lorenz, L.H., Amerasinghe, P.H., Reiter, P. and Edman, J.D. (1993b) Detection of multiple blood feeding in *Aedes aegypti* (Diptera: Culicidae) during a single gonotrophic cycle using a histologic technique. *Journal of Medical Entomology* 30, 94–99.

Sikes, R.K. and Chamberlain, R.W. (1954). Laboratory observations on three species of bird mites. *Journal of Parasitology* 40, 691.

Skaliy, P. and Hayes, W.J., Jr (1949) The biology of *Liponyssus bacoti* (Hirst, 1913)(Acarina, Liponyssidae). *American Journal of Tropical Medicine* 29, 759–772.

Smith, J.J.B. (1985) Feeding mechanisms. In: Kerkut, G.A. and Gilbert, G.A. (eds), *Comprehensive Insect Physiology, Biochemistry and Pharmacology*, Volume 4. Pergamon Press, Oxford, pp. 33–85.

Snodgrass, R.E. (1935) *Principles of Insect Morphology*. McGraw-Hill, New York.

Sonenshine, D.E. (1991) *Biology of Ticks*, Volume 1. Oxford University Press, Oxford.

Sonenshine, D.E. (1994) *Biology of Ticks*, Volume 2. Oxford University Press, Oxford.

Stoffolano, J.G., Jr and Yin, L.R.S. (1983) Comparative study of the mouthparts and associated sensilla of adult male and female *Tabanus nigrovittatus* (Diptera: Tabanidae). *Journal of Medical Entomology* 20, 11–32.

Stojanovich, C.J. (1945) The head and mouthparts of the sucking lice (Insecta: Anoplura). *Microentomology* 10, 1–46.

Sutcliffe, J.F. and Deepan, P.D. (1988) Anatomy and function of the mouthparts of the biting midge, *Culicoides sanguisuga* (Diptera: Ceratopogonidae). *Journal of Morphology* 198, 353–365.

Sutcliffe, J.F. and McIver, S.B. (1984) Mechanics of blood-feeding in black flies (Diptera:Simuliidae). *Journal of Morphology* 180, 125–144.

Thornhill, A.R. and Hays, K.L. (1972) Dispersal and flight activities of some species of *Tabanus* (Diptera: Tabanidae). *Environmental Entomology* 1, 602–606.

Tobe, S.S. and Davey, K.G. (1972) Volume relationships during the pregnancy cycle of the tsetse fly *Glossina austeni*. *Canadian Journal of Zoology* 50, 999–1010.

Trpis, M., Hausermann, W. and Craig, G.B., Jr (1995) Estimates of population size, dispersal, and longevity of domestic *Aedes aegypti aegypti* (Diptera: Culicidae) by mark–release–recapture in the village of Shauri Moyo in eastern Kenya. *Journal of Medical Entomology* 32, 27–33.

Usinger, R.L. (1966) *Monograph of Cimicidae (Hemimptera-Heteroptera)*, Volume 7. Thomas Say Foundation, Entomological Society of America, College Park.

Wekesa, J.W., Yuval, B. and Washino, R.K. (1995) Multiple blood feeding in *Anopheles freeborni* (Diptera: Culicidae). *American Journal of Tropical Medicine and Hygiene* 52, 508–511.

Wenk, P. (1981) Bionomics of adult blackflies. In: Laird, M. (ed.), *Blackflies*. Academic Press, London, pp. 259–279.

West, L.S. (1951) *The Housefly. Its Natural History, Medical Importance and Control.* Comstock Publishing Company, Ithaca.

Wilson, B.H. (1967) Feeding, mating, and oviposition studies of horse flies *Tabanus lineola* and *T. fuscicostatus* (Diptera: Tabanidae). *Annals of the Entomological Society of America* 60, 1102–1106.

Wilson, F.H. (1933) A louse feeding on the blood of its host. *Science* 77, 490.

Xue, R.-D., Edman, J.D. and Scott, T.W. (1995) Age and body size effects on blood meal size and multiple blood feeding by *Aedes aegypti* (Diptera: Culicidae). *Journal of Medical Entomology* 32, 471–474.

Zarate, L.G. (1983) The biology and behavior of *Triatoma barberi* (Hemiptera: Reduviidae) in Mexico. *Journal of Medical Entomology* 20, 485–497.

Zarate, L.G., Lopez, G.M., Ozuna, M.C., Santiago, G.G. and Zarate, R.J. (1984) The biology and behavior of *Triatoma barberi* (Hemiptera: Reduviidae) in Mexico. *Journal of Medical Entomology* 21, 548–560.

Zeledon, R., Guardia, V.M., Zuniga, A. and Swartzwelder, J.C. (1970) Biology and ethology of *Triatoma dimidiata* (Latreille, 1811). *Journal of Medical Entomology* 7, 313–319.

Zumpt, F. (1973) *The Stomoxyine Biting Flies of the World*. Gustave Fischer Verlag, Stuttgart.

# 3

# Salivary Gland Physiology of Blood-feeding Arthropods

## John R. Sauer,[1] Alan S. Bowman,[1] Janis L. McSwain[1] and Richard C. Essenberg[2]

[1]*Department of Entomology, Oklahoma State University, Stillwater, Oklahoma 74078, USA;* [2]*Department of Biochemistry and Molecular Biology, Oklahoma State University, Stillwater, Oklahoma 74078, USA*

## INTRODUCTION

Salivary glands can be defined as organs that synthesize and secrete products that assist in the acquisition of food. Properties of factors in the saliva of arthropods that feed on blood which rarely, if ever, contain blood digesting enzymes (Lehane, 1991), are discussed in Chapter 4.

Blood is nutritious and when arthropods first became able to consume it, selection favoured the evolutionary development of adaptations leading to full haematophagy (Lehane, 1991). Mouthparts (Chapter 2) capable of penetrating host wounds and sores and then later host skin were important attributes favouring haematophagy. Important too, but not as well understood until recently, is the role of salivary secretions. Some early investigators questioned the importance of saliva in this adaptation to haematophagy because a few species were capable of taking blood even when rendered incapable of salivating (Hudson *et al.*, 1960; Hudson, 1964; Rossignol and Spielman, 1982). However, other studies indicate that one of the most important attributes of saliva is to facilitate probing and blood-finding to reduce the time an arthropod associates with the host (Mellink and Van Den Bovenkamp, 1981; Ribeiro, 1992). Saliva in longer-feeding ixodid ticks has a more essential role in blood-feeding and not unexpectedly an elaborate array of salivary products during tick feeding (Sauer *et al.*, 1995).

In this chapter we focus on the tissue where saliva originates by briefly reviewing salivary gland morphology, the relationship between feeding habits

---

© CAB INTERNATIONAL 1996. From Wikel, S.K. (ed.) *The Immunology of Host–Ectoparasitic Arthropod Relationships.* CAB INTERNATIONAL, Wallingford.

and salivary glands, control of arthropod salivary gland function and the ancillary but important role of the salivary glands in transmission and establishment of parasitic infection that may be aided, at least in some cases, by the immunomodulatory properties of the arthropod saliva.

## MORPHOLOGY OF ARTHROPOD SALIVARY GLANDS

The structures of insect salivary glands have been reviewed by Chapman (1985) and House and Ginsborg (1985), and tick salivary glands by Sauer *et al.* (1995). An excellent summary of mosquito salivary gland morphology is included in James (1994). Morphologically, the salivary glands are grouped into two types: tubular and racemose with ducts and acini (alveoli). The mosquito salivary gland is an example of the former and the tick of the latter. James (1994) refers to the apical cavity accumulating material from secretory cells of the mosquito salivary gland as an extracellular acinus and Novak *et al.* (1995) describe mosquito salivary glands as of the 'tri-lobed, acinar-type'. We prefer not to use acinar in this context because of the more common meaning of acinar as a spherical grouping of cells surrounding a lumen. Diagrammatic representations of the organization features and cells in a mosquito salivary gland and a tick salivary gland are shown in Figs 3.1 and 3.2. The cells of the salivary glands rest on a basement membrane with tracheoles embedded in this membrane (not shown in Figs 3.1. and 3.2). The structural features of cells appear to be associated with their functions. Functions may be carried out by separate cells, or single cells may have more than one function. The three main processes involved in the production of saliva include: (i) synthesis and secretion of enzymes and other proteins; (ii) secretion of fluid; and (iii) reabsorption of salts. Enzyme and protein secreting cells possess abundant rough endoplasmic reticulum, numerous Golgi bodies and vacuoles containing granules. Fluid secreting cells typically display an abundance of plasma membranes, especially deep invaginations of the apical plasma membrane with a branching system of channels and secretory canaliculi. Mitochondria are numerous and in close proximity to the plasma membrane. Reabsorptive cells are commonly cuboidal or slightly flattened and in a number of insects the basal plasma membrane is extensively infolded with associated mitochondria.

## FEEDING HABITS AND THE SALIVARY GLAND

There are a variety of feeding strategies among haematophagous arthropods. Duration of bloodmeal acquisition can be measured in seconds for mosquitoes to weeks for ixodid ticks. Host fluid can be obtained from the lumen of blood vessels (solenophagy) as for mosquitoes and reduviids, or host fluids can be

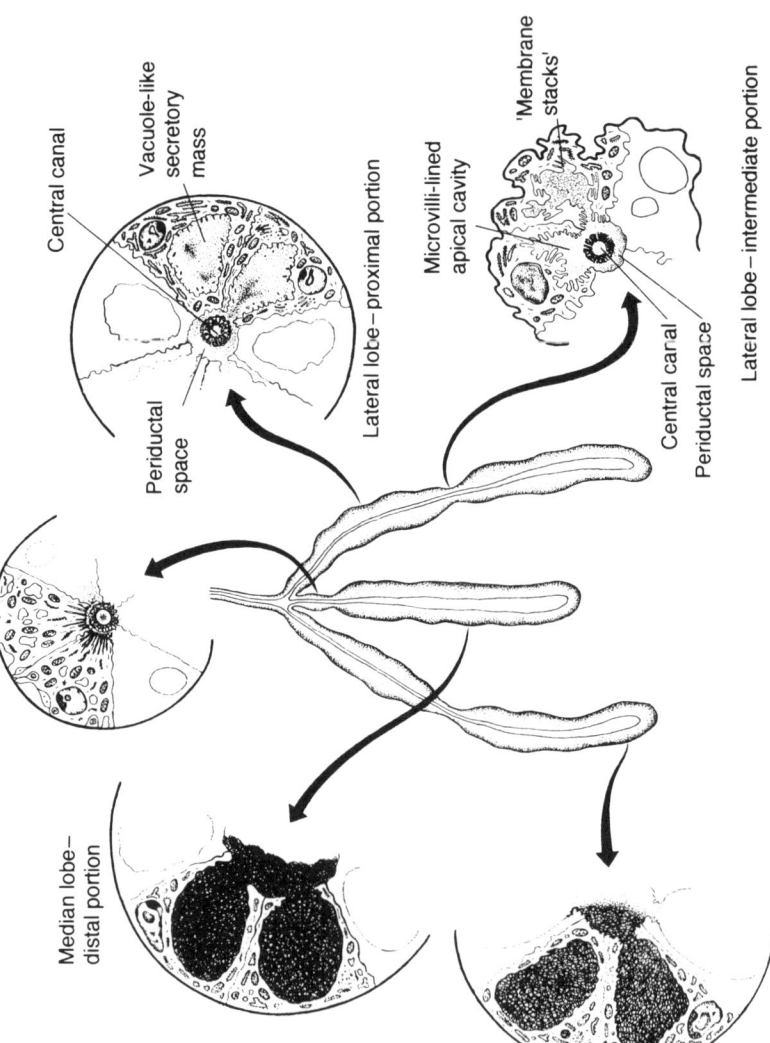

**Fig. 3.1.** Diagrammatic representation of the cellular architecture of the various portions of the trilobed salivary gland of female *Anopheles stephensi* Liston (adapted from Wright, 1969). The diagrams illustrate cross-sections of distal and proximal regions of the median and lateral lobes. The apical tips of the epithelial cells stop short of the cuticle-lined duct leaving a periductal space. The cross-sections show cells at the level of the opening of the secretory mass to the central canal or periductal space. Other cells are shown where the secretory mass appears vacuole-like because of the plane of sectioning. The cuticular-lined central canal is continuous through tortuous canals with the periductal space. The intermediate portion of the lateral lobe depicts a microvillus-lined apical cavity and a cavity bordered by 'membrane-stacks'. (Artwork by Kerry Stricker, Oklahoma State University.)

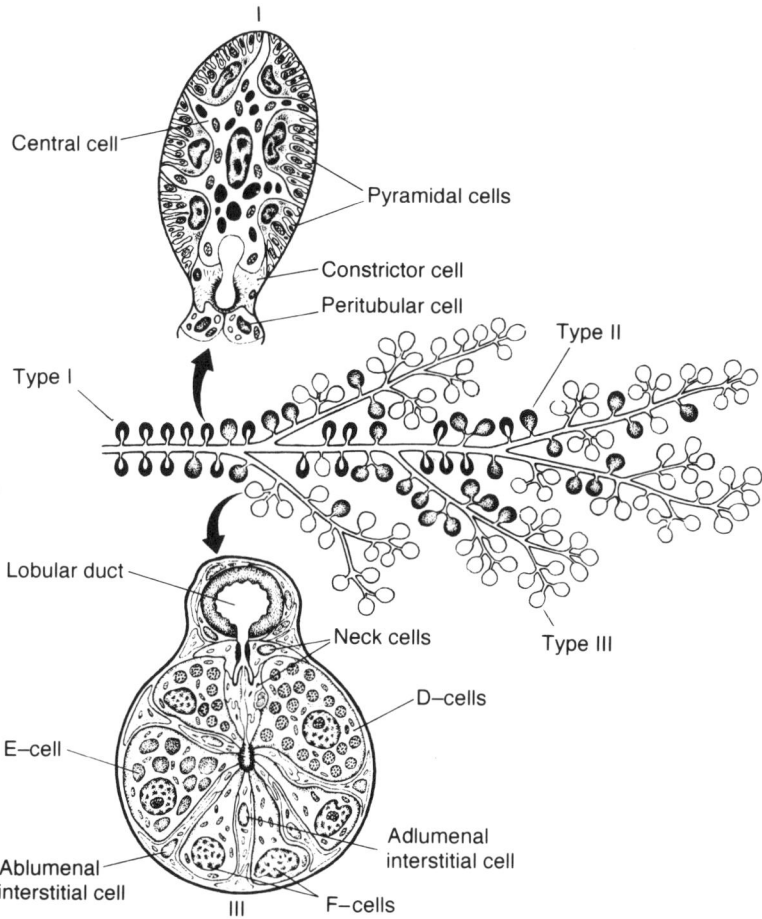

**Fig. 3.2.** Arrangement of the three alveoli types of the ixodid tick female salivary gland (adapted from Binnington, 1978). The alveolus types are distinguished by differential shading. Type I alveoli attach directly to the main salivary duct and to some principal branches of that duct. A diagrammatic section of a type I alveolus with cell types is shown above (adapted from Barker et al., 1984). Type II alveoli are more abundant in the proximal section and type III are more abundant in the distal section of the large branches. A schematic representation of the structure of the type III alveolus and cell types from an unfed tick is depicted below (from Fawcett et al., 1986 with permission). The types II and IV alveoli have a similar organization to type III alveoli but different cell types. Type IV alveoli are found only in males. Type II and especially type III undergo remarkable and complex cytological transformation to meet the tick's changing physiological requirements (including fluid transport) during tick feeding (see Fawcett et al., 1986). Nerve supply and tracheoles to the gland are not shown. (Artwork by Kerry Stricker, Oklahoma State University.)

imbibed from a haemorrhage created by physical and enzymatic rupturing of blood vessels (telmophagy) as for ticks and horse flies. Blood can form part of the diet in all life stages (e.g. ticks, reduviids, lice); only in the adult stage (e.g. dipterans, fleas); or, more rarely, in the larval and nymphal stages only (e.g. the argasid tick, *Otobius megnini*). Sexual dimorphism in the feeding habits of haematophagous arthropods also exists. While female ticks imbibe large volumes of blood over a lengthy period, male ixodid ticks are intermittent feeders with low blood intakes. Additionally, the salivary gland can serve some other functions as, for example, in ticks where it plays a vital role in maintaining water balance during both the feeding and non-feeding periods (Sauer *et al.* 1995). Because salivary glands and their secretions facilitate the acquisition of the bloodmeal, it is not surprising that for a given species the structure and physiology of the gland is related to the feeding strategy.

## Life Stage Differences and Developmental Changes

For many species, haematophagy is only present in one life stage with consequential effects on the salivary glands. In mosquitoes, although the imaginal buds that give rise to the adult salivary glands are associated with the larval glands (Jensen and Jones, 1957), the adult salivary glands are embryologically distinct from those of the larval stage (Trager, 1937) and remain undifferentiated throughout the larval stage. During metamorphosis the gland cells of the imaginal bud increase in number and change shape until they attain the final adult salivary gland form. With the change from the aquatic larval stage feeding on microorganisms to the adult form feeding on meals of sugar and blood, a commensurate change in salivary gland gene expression has been deduced by James and colleagues (James, 1994). In *Aedes aegypti*, the gene encoding α-glucosidase is expressed only in adult salivary glands (James *et al.*, 1989), when it would be beneficial during sugar feeding. Homologous α-glucosidase genes in the fruitfly *Drosophila melanogaster* are expressed in both the larval and adult stages (Synder and Davidson, 1983). However, unlike the fruitfly, the diet of mosquito larvae and adults is very different and this may be a reason for the non-overlap of gene expression from larval to adult salivary gland (James, 1994). Similarly, in *A. aegypti* the gene encoding apyrase, a potent platelet anti-aggregating factor (see Chapter 4), is only expressed in adult female salivary glands.

The argasid spinose ear tick (*Otobius megnini*) is unusual in that the larval and nymphal stages are haematophagous, but the adult form does not feed at all. Salivary glands of the larval and nymphal stages are relatively large and contain the type II acini associated with haematophagy. The non-feeding adult salivary gland degenerates to a small and simple organ consisting of type I acini alone which apparently function to absorb atmospheric water vapour and aid in water balance control (Stricker, 1993).

## Gender Differences

In several haematophagous species, the male does not take bloodmeals and this is reflected in the salivary gland morphology and secretions. Both male and female mosquitoes feed on sugar solutions from extrafloral nectaries and fruits (Foster, 1995), but only the female is haematophagous. The male salivary glands are much smaller than those of the female such that, for *A. aegypti*, the female salivary gland contains tenfold more protein than does the male gland (James, 1994). While female glands have a complex histology with many cell types, male glands seem to consist of a single cell-type resembling that of the proximal–lateral lobes of the female salivary gland (Fig. 3.1). James and colleagues (James, 1994) have identified an α-glucosidase gene (*MalI*) expressed in the salivary gland of both male and female *A. aegypti* and assumed to be involved in sugar feeding. *MalI* is expressed in all lobes of the male gland, but only in the proximal–lateral lobes of the female gland (Fig. 3.1). Female salivary glands synthesize and secrete many proteins for which functions in blood-feeding have been ascribed and are discussed in detail in Chapter 4. Female mosquitoes are able to salivate differentially depending upon the bloodmeal ingested. Only proteins produced in the proximal–lateral lobes are depleted when female *A. aegypti* feed on sugar, but when given a bloodmeal, protein in all lobes is decreased (Marinotti *et al.*, 1990).

Though both sexes of ixodid ticks are haematophagous, males feed intermittently while the female takes a prolonged and much larger bloodmeal. Consequently, growth and secretory capacity of male salivary glands are less than in the female (Till, 1961; Kaufman, 1976). However, only the male salivary glands possess type IV acini which are thought to secrete some factors used during the transfer of the spermatophore to the female, a process involving the secretion of copious volumes of saliva (Feldman-Muhsam *et al.*, 1970). Other blood-feeding species demonstrating sexual dimorphism include the biting midge (Perez de Leon *et al.*, 1994), tabanids (Elger *et al.*, 1980), and blackflies (Gosbee *et al.*, 1969).

## ALTERNATIVE FUNCTIONS OF THE SALIVARY GLANDS

In some instances the salivary gland performs alternative functions to simply producing saliva to aid in food acquisition. The ixodid tick salivary gland is multifunctional and the structure and physiology of this remarkable organ have evolved accordingly and were recently reviewed in detail (Sauer *et al.*, 1995). Types II and III acini of the salivary gland (Fig. 3.2) synthesize and secrete a cement that firmly secures the tick to the host via the mouthparts during the long feeding process. However, the major function of the ixodid tick salivary gland is osmoregulation. Blood is a relatively dilute food source. Thus, haematophagous arthropods need to concentrate the blood nutrients by reabsorbing water from the gut lumen and then eliminating the excess fluid in order to prevent haemolymph hypotonicity. Haematophagous insects utilize their Malphighian tubules for osmoregulation. In contrast, in ixodid ticks it is the salivary glands which are the organs of osmoregulation. Ticks spend the majority of the life cycle off-host during which time the salivary glands secrete, on to the mouthparts, a hygroscopic solution that absorbs water from unsaturated air and prevents the ticks from dehydrating (Sauer et al., 1995). Conversely, while on the host, the tick faces the opposite task of concentrating the bloodmeal and eliminating excess water and ions. This is achieved by continuously alternating between imbibing blood and secreting excess water and ions via the saliva back into the host. Additionally, there are sudden explosive secretions of saliva that are thought to clear the mouthparts and feeding lesion of any debris that may hinder the bloodmeal acquisition (Sweatman and Gregson, 1970).

## DYNAMIC EFFECTS OF THE SALIVARY GLAND–HOST RELATIONSHIP

A large part of this book discusses the immune response of vertebrate hosts to the salivary gland products of haematophagous arthropods. However, it should be noted that although the salivary glands' effect on the host is well documented, the host also has profound effects on the salivary gland. This reciprocal effect of the salivary gland on the host and vice versa is well illustrated in arachidonic acid metabolism in the tick salivary gland. Prostaglandins derived from arachidonic acid have been identified in the saliva of *Amblyomma americanum* (Ribeiro et al., 1992) and have been postulated to aid in tick feeding due to their vasodilatory, antihaemostatic, anti-inflammatory and, possibly, immunosuppressive effects (see Sauer et al., 1993). Ticks lack the ability to synthesize arachidonic acid (Bowman et al., 1995) and, thus, are dependent upon the host bloodmeal for their supply. Accordingly, the arachidonic acid content of the salivary gland of the unfed female *A. americanum* increases 40-fold during the feeding process (Shipley et al., 1993) and can be dramatically reduced in ticks fed on hosts whose blood lipids have been modified by dietary treatment (Madden et al., 1996).

## REGULATION OF SALIVARY SECRETION IN INSECTS AND TICKS

Surprisingly little research has been done on the control and physiology of salivary glands in blood-sucking insects; however, numerous studies have focused on control of salivation in non-parasitic cockroaches, migratory locusts and blowflies (House and Ginsborg, 1985). This information will be briefly reviewed with the assumption that it may parallel control strategies in blood-feeding insects. This hypothesis is supported by the correspondence noted in neural dopaminergic control of salivary secretion in cockroaches, locusts and ixodid ticks and some recent evidence for control by serotonin (5-hydroxytryptamine or 5-HT) in mosquitoes (Novak *et al.*, 1995). Additionally one of us observed that salivation could be induced in female horse flies and stable flies by injecting dopamine into their haemocoels (Bowman, unpublished).

## Insects

The salivary glands of the locust (*Schistocerca gregaria*) and cockroaches (*Periplaneta americana* and *Nauphoeta cinerea*) are innervated with neurones containing dopamine and 5-HT (House and Ginsborg, 1986). The neurotransmitters' respective receptors appear to be linked to adenylate cyclase, and when activated by dopamine or 5-HT induce increases in cyclic AMP (cAMP) (Lafon-Cazal and Bockaert, 1984; Ali *et al.*, 1993). Dopamine is also involved in the control of salivation in *Manduca sexta* (Robertson, 1974).

The salivary glands of *Schistocerca* are further influenced by the prothoracic posterior nerve containing biologically active neuropeptides (proctolin and YGGFMRFamide) but only if innervation from the salivary nerve from the suboesophageal ganglion is present (Baines *et al.*, 1989). It is hypothesized that neurosecretory-like factors released near the salivary glands influence the firing of the two suboesophageal salivary neurones that have their axons in the salivary nerve.

Intracellular recordings of the membrane potential across cockroach salivary gland acinar cells and secretion in response to a variety of dopamine $D_1$ and $D_2$ receptor agonists and selected $D_1$ and $D_2$ receptor antagonists indicated that the same receptors mediate both the hyperpolarizing and secretory responses and that they are similar to the mammalian $D_1$ receptor (Evans and Green, 1990, 1991).

Salivation in the cockroach appears to be induced by stimulation of sensilla on the mouthparts or antennae of the insect (Chapman, 1985). Saliva stored within the reservoirs of the salivary glands empties during stimulation apparently due to relaxation of muscles associated with the reservoir. The reservoirs are filled during intervals between feeding providing the insect has access to water.

### Blowflies

Blowfly salivary glands have provided an unprecedented model system for studying regulation of fluid secretion and signal transduction pathways in an insect tissue. Fluid secretion in the blowfly salivary gland is stimulated by

5-HT and cAMP (Berridge and Prince, 1971). The salivary glands of *Calliphora* are not innervated (Hansen-Bay, 1978a) suggesting that 5-HT acts as a hormone in this fly. In support of this hypothesis, haemolypmph from salivating flies stimulates fluid secretion when added to isolated glands. However, haemolymph from non-salivating flies is ineffective in stimulating salivary secretion by isolated glands.

Numerous studies have used blowfly salivary glands as a model system to study two messengers formed after receptor-stimulated phosphatidylinositol 4,5-bisphosphate hydrolysis (Berridge, 1987). Stimulation of a receptor causes breakdown of phosphoinositides to diacylglycerol and inositol trisphosphate (Fain and Berridge, 1979). Inositol trisphosphate mobilizes intracellular calcium and diacylglycerol activates protein kinase C, both important intracellular regulators in cells. In this system, 5-HT produces, through different receptors, an increase in levels of cAMP and $Ca^{2+}$ (via inositol trisphosphate) within gland cells (Berridge and Irving, 1984). The rise in levels of cAMP and $Ca^{2+}$ in the salivary gland increases the movement of potassium and chloride out of the cell and into the salivary duct. The gradient created is sufficient for the movement of water to occur. Second messenger levels also affect other functions of the glands such as enzyme release (Hansen-Bay, 1978).

## Mosquitoes

5-HT, pilocarpine, and the organophosphate insecticide malathion induce salivation in mosquitoes suggesting a role for nerves (Kerlin and Hughes, 1992). In support of this, 5-HT has been demonstrated in a dense plexus of axons surrounding the proximal medial lobe of the salivary gland of adult female *A. aegypti* (Novak et al., 1995). 5-HT-immunoreactive innervation (originating from the stomatogastric nervous system) is absent in male salivary glands, suggesting that 5-HT is involved in blood-feeding. Female mosquitoes treated with the 5-HT-depleting agent α-methyltryptophan (AMTP) and then allowed to feed on a rat exhibited a significantly longer probing period and a lower blood-feeding success rate than did control mosquitoes. AMTP-treated individuals secreted significantly less saliva and had lower concentrations of apyrase, an enzyme important in blood-feeding, than did control mosquitoes experimentally induced to salivate into mineral oil. 5-HT injected into both AMTP-treated and control mosquitoes elicited increases in the volume and/or its apyrase content. Female *A. aegypti* salivate differently depending on the type of meal being ingested (Marinotti et al., 1990).

## Ticks

We know more about control of salivary secretion in ixodid ticks than in any other blood-feeding arthropod. Secretion by tick salivary glands (Figure 3.2) is controlled by nerves (Coons and Roshdy, 1973; Roshdy and Coons, 1975) with no evidence of any humoral factors. The nerve(s) run(s) along salivary ducts, and enter at the base of all alveoli types (Fawcett *et al.*, 1986). Synapses are noted on glandular cells near the lumen (Fawcett *et al.*, 1986). Three types of vesicles are contained in the nerve endings: large vesicles that may contain peptides with a granular content of moderate density; small dense vesicles with a homogeneous content that may contain catecholamines; and numerous non-granular vesicles (Fawcett *et al.*, 1986) in individual axons (Coons and Roshdy, 1973). There is lack of agreement on the number of nerve branches originating from the fused central nervous system (i.e. synganglion) that affect the salivary glands (Sauer *et al.*, 1995).

The neurotransmitter dopamine stimulates fluid secretion when applied to isolated salivary glands from partially fed female adult ixodid ticks (Needham and Sauer, 1975; Kaufman, 1976) and when injected into the tick's haemocoel (McSwain *et al.*, 1992a). The presence of dopamine in salivary gland nerves has been established (Binnington and Stone, 1977). A dopamine $D_1$ receptor in the salivary gland of *A. americanum* linked to activation of adenylate cyclase has been demonstrated (Schmidt *et al.*, 1981). Exogenous cAMP stimulates fluid secretion by isolated salivary glands (Needham and Sauer, 1979). Dopamine stimulates an increase in salivary gland cAMP and phenylethylamines stimulate fluid secretion and adenylate cyclase activity in the order dopamine > norephinephrine = epinephrine > isoproterenol (Kaufman, 1977; Schmidt *et al.*, 1981). The concentration of dopamine required for half-maximal activation of adenylate cyclase is the same as the concentration required to stimulate fluid secretion half-maximally in isolated salivary glands (Sauer *et al.*, 1986). GTP and its non-hydrolysable analogue Gpp(NH)p can stimulate basal adenylate cyclase activity of a washed particulate fraction of tick salivary glands (Sauer *et al.*, 1986). These results demonstrate that dopamine released from a salivary nerve exerts its control through the action of a G protein-coupled receptor, which results in the activation of an adenylate cyclase and the formation of cAMP (Fig. 3.3). A dopamine $D_2$-like receptor may also exist in the salivary glands which inhibits adenylate cyclase, decreases cAMP and down-regulates fluid secretion at high concentrations of dopamine (Shipley *et al.*, 1996).

Although the salivary nerve controlling secretion is dopaminergic, pilocarpine, a cholinomimetic agent was the first pharmacological substance to successfully stimulate salivary secretion *in vivo* (Howell, 1966; Tatchell, 1967). Two sensory systems appear to impinge on the dopaminergic salivary nerve, with synaptic regions lying within the synganglion. One nerve is

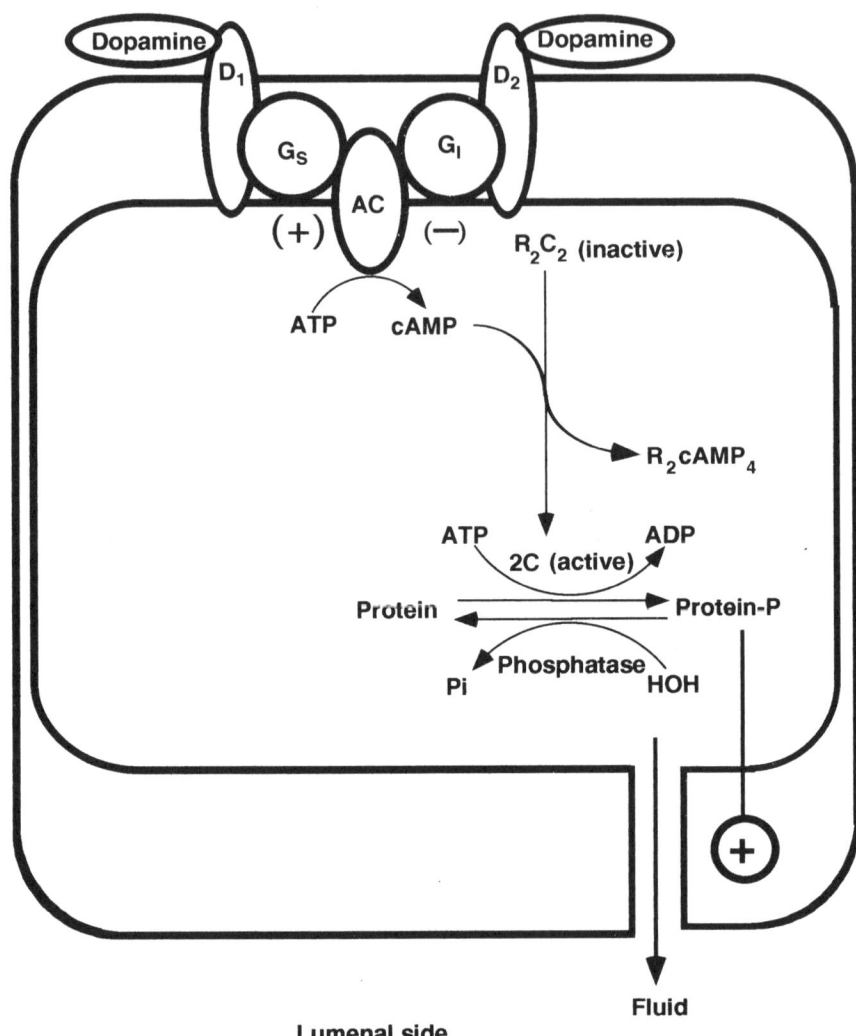

**Fig. 3.3.** Cellular control of salivary secretion in ixodid ticks. The salivary glands are innervated and dopamine is a neurotransmitter at the neuroeffector junction controlling secretion. Dopamine binds to a $D_1$-like receptor which acts through a G protein to activate adenylate cyclase. Adenylate cyclase converts ATP to cAMP which activates a protein kinase ($R_2C_2$) by binding to the regulatory subunits (R). The catalytic subunits (C) are released after cAMP binds to R and phosphorylate specific substrate proteins which effect secretion in some unknown way. Phosphorylated proteins are dephosphorylated (and presumably deactivated) by phosphatases. A dopamine $D_2$-like receptor may also

cholinergic and mimicked by pilocarpine while the other is activated by an increase in haemolymph volume and by stretch receptors in the abdominal body wall (Kaufman, 1989). The neurotransmitter at the synapse within the synganglion of the nerve mediating the stretch receptors' effect remains unstudied. The physiological condition that elicts activation of the cholinergic nerve is also unknown.

Other receptors appear to be involved in controlling secretion (not shown in Fig. 3.3.). Spiperone, pimozide, and haloperidol, which are usually considered antagonists of dopamine receptors in mammals, potentiate the effect of dopamine without stimulating secretion on their own (Wong and Kaufman, 1981). Sulpiride inhibits the potentiation by spiperone without reducing the effect of dopamine on secretion. Two models were postulated to explain the results: (i) spiperone allosterically modulates the dopamine receptor to increase the rate of secretion in response to dopamine; or (ii) spiperone inhibits an inhibitory pathway that responds to dopamine (Lindsay and Kaufman, 1986). Gamma-aminobutyric acid (GABA) is an inhibitory neurotransmitter in the brain of vertebrates and the central nervous system of insects (Harrow *et al*, 1991), but when applied to isolated salivary gland of *A. hebraeum*, it potentiates the effect of dopamine on fluid secretion in the same manner as spiperone (Lindsay and Kaufman, 1986). The GABA antagonists picrotoxin and bicuculline blocked both GABA- and spiperone-induced potentiation of salivary secretion. Sulpiride also blocked GABA-induced potentiation. These findings suggest that sulpiride and spiperone may interact with a GABA receptor in the salivary gland, an unexpected result in that both spiperone and sulpiride are dopamine antagonists in mammals. However, the results could also be explained by spiperone inhibiting the postulated dopamine $D_2$-like inhibitory receptor (Shipley *et al*., 1995).

Ergot alkaloids stimulate still another receptor in the salivary glands (Kaufman and Wong, 1983). The receptor is not likely a dopamine receptor because sulpiride inhibits the ability of ergot alkaloids to stimulate salivation but does not attenuate the salivary gland's response to dopamine. Ergot alkaloids have affinity for 5-HT receptors, but they probably do not act via a

---

**Fig. 3.3.** *(continued)* exist in the salivary glands which inhibits adenylate cyclase, decreases cAMP and down-regulates fluid secretion at high concentrations of dopamine (Shipley *et al*., 1996).

Although the mechanism and cellular site of fluid secretion are unproven, recent studies have shown that a vacuolar (V)-ATPase coupled with a cation/H$^+$ exchanger drive fluid transport in arthropod tissues including insect Malpighian tubules (Pannabecker, 1995). A clone has been isolated from a tick salivary gland cDNA library that shows significant similarity to the 44 kDa subunit of V-ATPase and bafilomycin A1, which specifically inhibits V-ATPase, was shown to inhibit fluid secretion by tick salivary glands (Sauer *et al*., 1995). The link between control by cAMP-dependent protein phosphorylation and the proposed mechanism of secretion is unknown.

5-HT receptor in tick salivary glands because 5-HT is an ineffective agonist of secretion (Needham and Sauer, 1975; Kaufman, 1977).

A factor in the tick synganglion sensitive to heat and trypsin induces metabolism of phosphatidylinositol 4,5-bisphosphate to inositol trisphosphate ($IP_3$) through the apparent activity of phospholipase C (PLC), indicating a receptor for a neuropeptide and a phosphoinositide signal-transduction pathway in the salivary gland of ticks as in blowflies (McSwain et al., 1989a,b). $IP_3$ mobilizes intracellular calcium in the tick salivary glands (Roddy et al., 1990). Another product of PLC-catalysed phosphoinositide turnover, diacylglycerol, activates salivary gland protein kinase C in the presence of calcium and phosphatidylserine (McSwain et al., 1992b). Protein kinase C activity in tick salivary glands is further indicated by the stimulation of phosphate incorporation into endogenous proteins in response to specific activators of protein kinase C 12-O-tetradecanoyl phorbol-13-acetate (TPA) and 1-oleoyl-2-acetyl-sn-glycerol (OAG) (McSwain et al., 1989b, 1992a). TPA and OAG do not stimulate fluid secretion by isolated salivary glands and only marginally affect the ability of dopamine to stimulate secretion and to increase salivary gland cAMP (McSwain et al., 1992b). The function of protein kinase C and phosphoinositide turnover in tick salivary gland fluid secretion remains elusive at this time.

Despite the above information about control of fluid secretion, nothing is known about the mechanism of secretion except that the adluminal cell in type III acini functions as a myoepithelial cell by causing an expanded and fluid-filled acinus to contract and eject fluid from the lumen into the salivary ducts (Coons et al, 1994, Lamoreaux et al., 1994).

## ROLE OF THE SALIVARY GLAND IN PATHOGEN TRANSMISSION

### Arthropod-borne Pathogens

Blood-sucking arthropods are important vectors of disease caused by a variety of pathogens, including viruses, bacteria, rickettsia, and protozoa (Kaufman, 1989). Based on feeding strategies, two vector strategies can be distinguished (Nuttal et al., 1994). Strategy one is exemplified by ixodid ticks which feed on a single host and remain attached for a considerable period. Thus infection by pathogens must persist through the moult (trans-stadial) and accompanying tissue remodelling. In some instances, pathogens are transmitted through eggs to the next generation (transovarial). In strategy two, most other vectors, even some adult male ixodid ticks (Kocan et al., 1992a,b, 1993), ingest several bloodmeals from several hosts, and feed for relatively short periods on each host. The pathogen does not require survival through a moult, but it may require some growth, development and migration in the vector during the blood-feeding stage.

In both types of vector strategies, the initial contact with the pathogen occurs when the bloodmeal is taken into the gut. The gut cells become infected, providing a reservoir for subsequent spread to other tissues. In most instances, the pathogen eventually escapes from the gut and may infect other tissues in the animal. Some vectors transmit pathogens via coxal fluid (Burgdorfer and Hayes, 1990), faeces (Brener, 1973) or regurgitation (Brown, 1988; Connat, 1991), but the most common mode of transmission is through the salivary glands. Studies with arboviruses have distinguished several barriers to the transmission of pathogens, including a salivary gland infection barrier and a salivary gland escape barrier (Hardy, 1988).

## Effects of Growth Stage on Pathogens

In trans-stadial or transovarial transmission in ixodid ticks, the pathogen changes its site of growth depending on the physiological state of the tick vector. For many pathogens, the infection locus and/or the state of the pathogen changes in response to the tick's development or activities, such as feeding. A number of pathogens require that the vector feed on the host before they are detected in the salivary glands or transmitted to the host (Stiller *et al.*, 1989; Booth *et al.*, 1991). This dependence on the tick's condition is most obvious in the piroplasms (Mehlhorn and Schein, 1984). They have a rather complex life cycle with several morphologically distinct stages that infect different tissues, and these changes are tied to changes in the tick, so, for example, the salivary gland stage is not found until after moulting, and it does not become the infective form until feeding starts. This behaviour implies that the tick signals the pathogen in some way, but little information is yet available as to what this signal may be. Some pathogens appear not to depend so much on the tick and produce a generalized infection after escape from the gut (Burgdorfer, 1977; Burgdorfer and Hayes, 1990). In this case, the amount of infectious pathogen in the salivary gland increases, but only to the extent that the salivary glands have enlarged. For vectors that feed on multiple hosts, the unpredictable, short feeding makes it less likely for this sort of change to be important, though the first feeding may prepare the vector for the growth of the pathogen. In addition, the cells of the various tissues may have characteristics that direct that pathogen's development to a certain form.

## Tissue and Cell Specificity of the Pathogen

Many pathogens show specificity for the tissues of the vector they infect. In some cases, this specificity extends to specific cell types within the tissue; for example, piroplasms infect only cell types d and e of the type III alveolus of their ixodid tick vectors (Fawcett et al., 1982). There is some evidence for arboviruses in mosquitoes that the virus must replicate in the cells of the salivary gland that are involved in blood-feeding, as opposed to sugar feeding (Hardy, 1988). In a study of *Trypanosoma rangeli* in *Rhodnius prolixus*, Marinkelle et al. (1986) found that forms of the pathogen in the salivary gland differed from those in other tissues by reacting with a specific lectin, suggesting the presence of $\alpha$-D-galactose on the surface. These sorts of interactions imply the presence of specific receptors in these cell types, and possibly specific signalling molecules.

## Salivary Secretions and Pathogen Transmission

Blood-sucking arthropods, especially ixodid ticks, secrete a variety of substances to allow their feeding to proceed, including anticoagulants, vasodilators and platelet aggregation inhibitors. In addition, saliva of some vectors leads to immunosupression in the host (Chapter 4). These factors may provide a better environment for the pathogen to gain an advantage in establishing an infection in the new host. A rather interesting case is Thogotovirus and tick-borne encephalitis virus, in which a factor in tick saliva increases the infection rate of other ticks feeding on the same host (Jones et al., 1987, 1989; Labuda et al., 1993). Another example may be in *Leishmania* and sandflies, where a factor in saliva inhibits the growth of the pathogen, possibly directing it to differentiate into the form that is infective in the host (Charlab and Ribeiro, 1993).

## Effects of the Pathogen on the Vector

The examples where the salivary gland is the pathway of transmission involve an infection of the vector by the pathogen. Studies indicate that the vector is not greatly affected by infection with the pathogen (e.g. Youdeowei, 1976; Paulson et al., 1992), which is to be expected, because the vector must be healthy enough to feed several times in order to transmit the pathogen. However, various morphological changes are observed in infected salivary glands. There is also some evidence for the pathogen altering the vector's physiology in subtle ways to improve its chances of transmission. Malaria sporozoites affected mosquito salivary glands so that less apyrase was produced (Rossignol et al., 1984). Infected mosquitoes spent more time probing a host, apparently because apyrase assists in locating blood by inhibiting blood clotting, allowing more opportunity for the pathogen to infect. The volume of saliva, which provides the vehicle to move the pathogen, was unaffected.

## Observations on Types of Studies

The vast majority of the studies in this area are morphological, with some support from studies of infectivity of various preparations. This type of study can provide a great deal of interesting information as to what happens and when it happens, but leaves the question of how it happens largely unanswered. More recently, immunological and hybridization techniques to detect and measure specific proteins and nucleic acids and their locations in tissues and cells have been applied. These are more sensitive, and allow detection of pathogens in conditions where microscopic detection is very difficult. However, physiological and biochemical studies looking at the details of the interaction between the pathogen and the vector are largely lacking. In part, this is due to the difficulty in finding a good experimental system, but some recent advances in methodology, such as DNA hybridization and immunological techniques may now allow these studies to be done.

## SUMMARY

The salivary glands are essential in the ability of blood-feeding arthropods to feed. The size and development of the glands reflect the arthropod's feeding habits. In many species, males do not feed on blood in contrast to females; thus, female salivary glands are often larger and more complex (Perez de Leon et al., 1994; Elger et al., 1980; Gosbee et al., 1969) and there is some evidence that control of secretion in some cases may be different in the sexes (Novak et al., 1995). Although the salivary glands are important sources of profeeding molecules, ixodid tick salivary glands are also essential organs of osmoregulation. Cells in ixodid females are transformed into a fluid-transporting epithelium as the salivary glands gain increasing ability to secrete fluid back to the host during tick feeding (Fawcett et al., 1986). The salivary glands of both argasid and ixodid ticks secrete a hygroscopic solution on the mouthparts that absorb water from unsaturated air and prevents the ticks from dehydrating.

The salivary glands of most arthropods studied are controlled by nerves; the blowfly is a notable exception where they are controlled by a neurohormone, probably serotonin. There is good evidence that serotonin and/or dopamine serve as neurotransmitters to control secretion in locusts, cockroaches, ticks, mosquitoes and possibly other arthropods. Surprisingly little research has been done on the physiology and control of salivary secretions in blood-feeding insects. Much more is known about control of salivary secretion in ixodid ticks but the mechanism of secretion remains poorly understood. Blood-sucking arthropods are important vectors of disease caused by a variety of pathogens and most are transmitted through the salivary glands. However, we know very little about how the physiology of the arthropod affects salivary gland pathogen transmission. It seems probable that the saliva of some vectors provides an enhanced environment for establishment of the pathogen in the host.

The salivary glands are the origin for most of the factors that stimulate or modulate immune reactions in response to arthropod feeding. Although we are beginning to learn much about the nature of salivary secretory products, increased effort directed at studying the control, development, metabolic pathways, and mechanisms of salivary secretion in blood-feeding arthropods should yield valuable information.

## ACKNOWLEDGEMENTS

Approved for publication by the director, Oklahoma Agricultural Experimental Station. Research from our laboratories reported in this chapter was supported in part by Grants AI-26158 and AI-31460 from the National Institutes of Health. We thank LaFonda Barrera for typing and Drs Doug Bergman and Bob Barker for critically reviewing this manuscript.

## REFERENCES

Ali, D.W., Orchard, I. and Lange, A.B. (1993) The aminergic control of locust (*Locusta migratoria*) salivary glands: Evidence for dopaminergic and serotonergic innervation. *Journal of Insect Physiology* 39, 623–632.

Baines, R.A., Tyrer, N.M. and Mason, J.C. (1989) The innervation of locust salivary glands. I. Innervation and analysis of transmitters. *Journal of Comparative Physiology* A 165, 395–405.

Barker, D.M., Ownby, C.L., Krolak, J.M., Claypool, P.L. and Sauer, J.R. (1984) The effects of attachment, feeding and mating on the morphology of the type I alveolus of salivary glands of the lone star tick, *Amblyomma americanum* (L.). *Journal of Parasitology* 70, 99–113.

Berridge, M.J. (1987) Inositol trisphosphate and diacylglycerol: two interacting second messengers. *Annual Review of Biochemistry* 56, 159–193.

Berridge, M.J. and Irvine, R.F. (1984) Inositol trisphosphate, a novel second messenger in cellular signal transduction. *Nature* 312, 315–321.

Berridge, M.J. and Prince, W.T. (1971) The electrical response of isolated salivary glands during stimulation with 5-hydroxytryptamine and cyclic AMP. *Philosphical Transactions of the Royal Society of London* B 262, 111–120.

Binnington, K.C. (1978) Sequential changes in salivary gland structure during attachment and feeding of the cattle tick, *Boophilus microplus*. *International Journal for Parasitology* 8, 97–115.

Binnington, K.C. and Stone, B.F. (1977) Distribution of catecholamines in the cattle tick *Boophilus microplus*. *Comparative Biochemistry and Physiology* 58C, 21–28.

Booth, T.F., Steele G.M., Marriott, A.C. and Nuttall, P.A. (1991) Dissemination, replication, and *trans*-stadial persistence of Dugbe virus (nairovirus, bunyaviridae) in the tick vector *Amblyomma variegatum*. *American Journal of Tropical Medicine and Hygiene* 45, 146–157.

Bowman, A.S., Sauer, J.R., Neese, P.A. and Dillwith, J.W. (1995) Origin of arachidonic acid in the salivary glands of the lone star tick, *Amblyomma americanum*. *Insect Biochemistry and Molecular Biology* 25, 225–233.

Brener, Z. (1973) Biology of *Trypanosoma cruzi*. *Annual Review of Microbiology* 27, 347–382.

Brown, S. (1988) Evidence for regurgitation by *Amblyomma americanum*. *Veterinary Parasitology* 28, 335–342.

Burgdorfer, W. (1977) Tick-borne diseases in the United States: Rocky Mountain spotted fever and Colorado tick fever. *Acta Tropica* 34, 103–126.

Burgdorfer, W. and Hayes, S.F. (1990) Vector–spirochete relationships in Louse-borne and tick-borne borrelioses with emphasis on Lyme disease. In: Harris, K.F. (ed.), *Advances in Disease Vector Research*, Volume 6. Springer Verlag, New York, pp. 127–150.

Chapman, R.F. (1985) Coordination of digestion. In: Kerkut, G.A. and Gilbert, L.I. (eds), *Comprehensive Insect Physiology, Biochemistry and Pharmacology: Digestion, Nutrition, Excretion*. Pergamon Press, New York, pp. 213–240.

Charlab, R. and Ribeiro, J.M.C. (1993) Cytostatic effect of *Lutzomyia longipalpis* salivary gland homogenates on *Leishmania* parasites. *American Journal of Tropical Medicine and Hygiene* 48, 831–838.

Connat, J.L. (1991) Demonstration of regurgitation of gut content during blood meals of the tick *Ornithodorus moubata*. Possible role in the transmission of pathogenic agents. *Parasitology Research* 77, 452–454.

Coons, L.B. and Roshdy, M.A. (1973) Fine structure of the salivary glands of unfed male *Dermacentor variabilis* (Say) (Ixodoidea: Ixodidae). *Journal of Parasitology* 59, 900–912.

Coons L.B., Lessman, C.A., Ward, M.W., Berg, R.H. and Lamoreaux, W.J. (1994) Evidence of a myoepithelial cell in tick salivary glands. *International Journal for Parasitology* 24, 551–562.

Elger, M., Hentschel, H. and Peohling, H.-M. (1980) The salivary gland of the cleg *Haematopota pluvialis* L. (Diptera: Tabanidae): sex differences in the protein patterns and the fine structure of the distal glandular portion. *European Journal of Cell Biology* 20, 209–216.

Evans, A.M. and Green, K.L. (1990) Characterization of the dopamine receptor mediating the hyperpolarization of the cockroach salivary gland acinar cells *in vitro*. *British Journal of Pharmacology* 101, 103–108.

Evans, A.M. and Green, K.L. (1991) Effects of selective D-1 and D-2 dopamine agonists on cockroach salivary gland acinar cells *in vitro*. *British Journal of Pharmacology* 104, 787–792.

Fain, J.N. and Berridge, M.J. (1979) Relationship between hormonal activation of phosphatidylinositol hydrolysis, fluid secretion and calcium flux in the blowfly salivary gland. *Biochemical Journal* 178, 45–58.

Fawcett, D.W., Buscher, G. and Doxsey, S. (1982) Salivary gland of the tick vector of East Coast fever. IV. Cell type selectivity and host cell responses to *Theileria parva*. *Tissue and Cell* 14, 397–414.

Fawcett, D.W., Binnington, K. and Voight, W.P. (1986) The cell biology of the ixodid tick salivary gland. In: Sauer, J.R. and Hair, J.A. (eds), *Morphology, Physiology, and Behavioral Biology of Ticks*. Ellis Horwood, Chichester, pp. 22–45.

Feldman-Muhsam, B., Borut S. and Saliternik-Givant S. (1970) Salivary secretion of the male tick during copulation. *Journal of Insect Physiology* 16, 1945–1949.

Foster, W.A. (1995) Mosquito sugar feeding and reproductive energetics. *Annual Review of Entomology* 40, 443–474.

Gosbee, J., Allen, J.R. and West, A.S. (1969) The salivary glands of adult blackflies. *Canadian Journal of Zoology* 47, 1341–1344.

Hansen-Bay, C.M. (1978a) Control of salivation in the blowfly *Calliphora*. *Journal of Experimental Biology* 75, 189–201.

Hansen-Bay, C.M. (1978b) The secretion and action of the digestive enzymes' of the salivary glands of the blowfly, *Calliphora*. *Journal of Insect Physiology* 24, 141–149.

Hardy, J.L. (1988) Susceptibility and resistance of vector mosquitoes. In Monath, T.P. (ed.), *The Arboviruses: Epidemiology and Ecology*. CRC Press, Inc., Boca Raton, FL, pp. 87–126.

Harrow, I.D., Gration, K.F. and Evans, N.A. (1991) Neurobiology of arthropod parasites. *Parasitology* 102, S59–S69.

House, C.R. and Ginsborg, B.L. (1985) Salivary gland. In: Kerkut, G.A. and Gilbert, L.I. (eds), *Comprehensive Insect Physiology, Biochemistry and Pharmacology*, Volume II. Pergamon Press, New York, pp. 195–224.

Howell, C.J. (1966) Collection of salivary gland secretion from the argasid, *Ornithodoros savignyi* Audouin (1827) by the use of a pharmacological stimulant. *Journal of the South African Veterinary Medicine Association* 37, 236–239.

Hudson, A. (1964) Some functions of the salivary glands of mosquitoes and other blood-feeding insects. *Canadian Journal of Zoology* 42, 113–120.

Hudson, A., Bowman, A.L. and Orr, C.W.M. (1960) Effects of absence of saliva on blood feeding by mosquitoes. *Science* 131, 1730–1731.

James, A.A. (1994) Molecular and biochemical analysis of the salivary glands of vector mosquitoes. *Bulletin Institute Pasteur* 92, 133–150.

James, A.A., Blackmer, K. and Racioppi, J.V. (1989) A salivary gland-specific, maltase-like gene of the vector mosquito, *Aedes aegypti*. *Gene* 75, 73–83.

Jensen, D.V. and Jones, J.C. (1957) The development of the salivary glands in *Anopheles albimanus* Wiedemann (Diptera, Culicidae). *Annals of the Entomological Society of America* 50, 464–469.

Jones, L.D., Davies, C.R., Steele, G.M. and Nuttall, P.A. (1987) A novel mode of arbovirus transmission involving a nonviraemic host. *Science* 237, 775–777.

Jones, L.D., Hodgson, E. and Nuttall, P.A. (1989) Enhancement of virus transmission by tick salivary glands. *Journal of General Virology* 70, 1895–1898.

Kaufman, W.R. (1976) The influence of various factors on fluid secretion by *in vitro* salivary glands of ixodid ticks. *Journal of Experimental Biology* 64, 727–742.

Kaufman, W.R. (1977) The influence of adrenergic agonists and their antagonists on isolated salivary glands of ixodid ticks. *European Journal of Pharmacology* 45, 61–68.

Kaufman, W.R. (1989) Tick host interaction: a synthesis of current concepts. *Parasitology Today* 5, 47–56.
Kaufman, W.R. and Wong, D.L.P. (1983) Evidence for multiple receptors mediating fluid secretion in salivary glands of ticks. *European Journal of Pharmacology* 87, 43–52.
Kerlin, R.L. and Hughes, S. (1992) Enzymes in saliva from four parasitic arthropods. *Medical and Veterinary Entomology* 6, 121–126.
Kocan, K.M., Goff, W., Stiller, D., Claypool, P., Edwards, W., Ewing, S., Hair, J. and Barron, S. (1992a) Persistence of *Anaplasma marginale* (Rickettsiales: *Anaplasma taceae*) in male *Dermacentor andersoni* (Acari: Ixodidae) transferred successively from infected to susceptible calves. *Journal of Medical Entomology* 29, 657–668.
Kocan, K.M., Stiller, D., Goff, W.L., Claypool, P.L., Edwards, W., Ewing, S., McGuire, T., Hair, J.A. and Barron, S. (1992b) Development of *Anaplasma marginale* in male *Dermacentor andersoni* transferred from infected to susceptible cattle. *American Journal of Veterinary Research* 53, 499–507.
Kocan, K.M., Goff, W.L., Stiller, D., Edwards, W., Ewing, S., Claypool, P.L., McGuire, T., Hair, J.A. and Barron, S. (1993) Development of *Anaplasma marginale* in salivary glands of male *Dermacentor andersoni*. *American Journal of Veterinary Research* 54, 107–112.
Labuda, M., Jones, L.D., Williams, T. and Nuttall, P.A. (1993) Enhancement of tick-borne encephalitis virus transmission by tick salivary gland extracts. *Medical and Veterinary Entomology* 7, 193–196.
Lafon-Cazal, M. and Bockaert, J. (1984) Pharmacological characterization of dopamine-sensitive adenylate cyclase in the salivary glands of *Locusta migratoria* L. *Insect Biochemistry* 14, 541–545.
Lamoreaux, W.J., Needham, G.R. and Coons, L.B. (1994) Fluid secretion by isolated tick salivary glands depends on an intact cytoskeleton. *International Journal for Parasitology* 24, 563–567.
Lehane, M.J. (1991) *Biology of Blood-sucking Insects*. Harper Collins, London. 288pp.
Lindsay, P.J. and Kaufman, W.R. (1986) Potentiation of salivary fluid secretion in ixodid ticks: a new receptor system for $\gamma$-aminobutyric acid. *Canadian Journal of Physiology and Pharmacology* 64, 1119–1126.
Madden, R.D., Sauer, J.R., Dillwith, J.W. and Bowman, A.S. (1996) Alteration of arachidonate levels in tick salivary glands by dietary modification of host blood lipids. *Archives of Insect Biochemistry and Physiology* 31, 53–72.
Marinkelle, C.J., Vallejo, G.A., Schottelius, J., Guhl, F. and De Sanchez, N. (1986) The affinity of the lectins *Ricinus communis* and *Glycine maxima* to carbohydrates on the cell surface of various forms of *Trypanosoma cruzi* and *Trypanosoma rangeli*, and the application of these lectins for the identification of *T. cruzi* in the feces of *Rhodnius prolixus*. *Acta Tropica* 43, 215–223.
Marinotti, O., James, A.A. and Ribeiro, J.M.C. (1990) Diet and salivation in female *Aedes aegypti* mosquitoes. *Journal of Insect Physiology* 36, 545–548.
McSwain, J.L., Essenberg, R.C. and Sauer, J.R. (1989a) Second messenger molecules and regulation of ixodid tick salivary gland function: a role for protein kinase C? In: Borovsky, D. and Spielman, A. (eds), *Host Regulated Developmental Mechanisms in Vector Arthropods*, Volume 2. Univ. Fla - IFAS, Vero Beach, Fla., pp. 213–220.
McSwain, J.L., Tucker, J.S., Essenberg, R.C. and Sauer, J.R. (1989b) Brain factor

induced formation of inositol phosphates in tick salivary glands. *Insect Biochemistry* 19, 343–349.

McSwain, J.L., Essenberg, R.C. and Sauer, J.R. (1992a) Oral secretion elicited by effectors of signal transduction pathways in the salivary glands of *Amblyomma americanum* (Acari: Ixodidae). *Journal of Medical Entomology* 29, 41–48.

McSwain, J.L., Masaracchia, R.A., Essenberg, R.C., Tucker, J.S. and Sauer, J.R. (1992b) *Amblyomma americanum* (L): protein kinase C-independent fluid secretion by isolated salivary glands. *Experimental Parasitology* 74, 324–331.

Mehlhorn, H. and Schein, E. (1984) The piroplasms: life cycle and sexual stages. *Advances in Parasitology* 23, 37–103.

Mellink, J.J. and Van Den Bovenkamp, W. (1981) Functional aspects of mosquito salivation in blood feeding of *Aedes aegypti*. *Mosquito News* 41, 115–119.

Needham, G.R. and Sauer, J.R. (1975) Control of fluid secretion by isolated salivary glands of the lone star tick. *Journal of Insect Physiology* 21, 1893–1898.

Needham, G.R. and Sauer, J.R. (1979) Involvement of calcium and cyclic AMP in controlling ixodid tick salivary fluid secretion. *Journal of Parasitology* 65, 531–542.

Novak, M.G., Ribeiro, J.M.C. and Hildebrand, J.G. (1995) 5-Hydroxytryptamine in the salivary glands of adult female *Aedes aegypti* and its role in regulation of salivation. *Journal of Experimental Biology* 198, 167–174.

Nuttall, P.A., Jones, L.D., Labuda, M. and Kaufman, W.R. (1994) Adaptations of arboviruses to ticks. *Journal of Medical Entomology* 31, 1–9.

Pannabecker, T. (1995) Physiology of the Malpighian tubule. *Annual Review of Entomology* 40, 493–510.

Paulson, S.L., Poirier, S.J., Grimstad, P.R. and Craig, G.B., Jr (1992) Vector competence of *Aedes hendersoni* (Diptera: Culicidae) for LaCrosse virus: Lack of impaired function in virus-infected salivary glands and enhanced virus transmission by sporozoite-infected mosquitoes. *Journal of Medical Entomology* 29, 483–488.

Perez De Leon, A.A., Lloyd, J.E. and Tabachnick, W.J. (1994) Sexual dimorphism and developmental change of the salivary glands in adult *Culicoides variipennis* (Diptera: Ceratopogonidae). *Journal of Medical Entomology* 31, 898–902.

Ribeiro, J.M.C. (1992) Characterization of a vasodilator from the salivary glands of the yellow fever mosquito *Aedes aegypti*. *Journal of Experimental Biology* 165, 61–71.

Ribeiro, J.M.C., Evans, P.M., McSwain, J.L. and Sauer, J.R. (1992) *Amblyomma americanum*: characterization of salivary prostaglandins $E_2$ and $F_{2\alpha}$ by RP-HPLC/bioassay and gas chromatography-mass spectrometry. *Experimental Parasitology* 74, 112–116.

Robertson, H.A. (1974) The innervation of the salivary gland of the moth *Manduca sexta*. *Cell and Tissue Research* 148, 237–245.

Roddy, C.W., McSwain J.L., Kocan K.M., Essenberg, R.C. and Sauer, J.R. (1990) The role of inositol 1,4,5-trisphosphate in mobilizing calcium from intracellular stores in the salivary glands of *Amblyomma americanum* (L). *Insect Biochemistry* 20, 83–89.

Roshdy, M.A. and Coons, L.B. (1975) The subgenus *Persicargas* (Ixodoidea: Argasidae: *Argas*.) 23. Fine structure of the salivary glands of unfed *A (P) aboreus* Kaiser, Hoogstraal and Kohls. *Journal of Parasitology* 61, 743–752.

Rossignol, P.A. and Spielman, A. (1982) Fluid transport across the ducts of the salivary glands of a mosquito. *Journal of Insect Physiology* 28, 579–583.

Rossignol, P.A., Ribeiro, J.M.C. and Spielman, A. (1984) Increased intradermal probing time in sporozoite-infected mosquitoes. *American Journal of Tropical Medicine and Hygiene* 33, 17–20.
Sauer, J.R., Mane, S.D., Schmidt, S.P. and Essenberg, R.C. (1986) Molecular basis for salivary secretion in ixodid ticks. In: Sauer, J.R. and Hair, J.A. (eds), *Morphology, Physiology and Behavioral Biology of Ticks*. Ellis Harwood, Chichester, pp. 55–74.
Sauer, J.R., Bowman, A.S., Shipley, M.M., Gengler, C.L., Surdick, M.R., Luo, C., Essenberg, R.C. and Dillwith, J.W. (1993) Arachidonate metabolism in tick salivary glands. In: Stanley-Samuelson, D.W. and Nelson D.R. (eds), *Insect Lipids: Chemistry, Biochemistry, and Biology*. Univ. Nebraska Press, Lincoln, NE, pp. 99–138.
Sauer, J.R., McSwain, J.L., Bowman, A.S. and Essenberg, R.C. (1995) Tick salivary gland physiology. *Annual Review of Entomology* 40, 245–267.
Schmidt, S.P., Essenberg, R.C. and Sauer, J.R. (1981) Evidence for a $D_1$ dopamine receptor in the salivary glands of *Amblyomma americanum* (L). *Journal of Cyclic Nucleotide Research* 7, 375–384.
Shipley, M.M., Dillwith, J.W., Bowman, A.S., Essenberg, R.C. and Sauer, J.R. (1993) Changes in lipids of the salivary glands of the lone star tick, *Amblyomma americanum* during feeding. *Journal of Parasitology* 79, 834–842.
Shipley, M.M., Sauer, J.R., McSwain, J.L., Essenberg, R.C., Forest, D.A., Hickey, R.D. and Barker, R.W. (1996) Indication of dopamine $D_1$ and $D_2$ receptors in the salivary glands of *Dermacentor variabilis*. In: Horn, D.J., Needham, G.R. and Welborn, W.C. (eds), *Proceedings of the IX International Congress of Acarology*. Ohio Biological Survey (in press).
Snyder, M. and Davidson, N. (1983) Two gene families clustered in a small region of the *Drosophila* genome. *Journal of Molecular Biology* 166, 101–118.
Stiller, D., Kocan, K.M., Edwards, W., Ewing, S.A., Hair, J.A. and Barron, S.J. (1989) Detection of colonies of *Anaplasma marginale* Theiler in salivary glands of three *Dermacentor* spp. infected as nymphs or adults. *American Journal of Veterinary Research* 50, 1381–1386.
Stricker, K. (1993) Internal anatomy and salivary gland ultrastructure of the spinose ear tick (*Otobius megnini*: Duges) with notes on water vapor uptake and salivary gland degeneration. PhD thesis. Oklahoma State University, Stillwater, Oklahoma.
Sweatman, G.K. and Gregson, J.D. (1970) Feeding electrograms of *Hyalomma aegyptium* ticks at different temperatures. *Journal of Medical Entomology* 7, 575–584.
Tatchell, R.J. (1967) A modified method for obtaining tick oral secretion. *Journal of Parasitology* 53, 1106–1107.
Till, W.M. (1961) A contribution to the anatomy and histology of the brown ear tick *Rhipicephalus appendiculatus* Neumann. *Memoirs of the Entomological Society of South Africa* 6, 3–124.
Trager, W. (1937) Cell size in relation to the growth and metamorphosis of the mosquito, *Aedes aegypti*. *Journal of Experimental Zoology* 76, 467–489.
Wong, D.L.P. and Kaufman, W. (1981) Potentiation by spiperone and other butyrophenones of fluid secretion by isolated salivary glands of ixodid ticks. *European Journal of Pharmacology* 73, 163–173.

Wright, K.A. (1969) The anatomy of salivary glands of *Anopheles stephensi* Liston. *Canadian Journal of Zoology* 47, 579–587.

Youdeowei, A. (1976) Salivary secretion in wild *Glossina pallidipes* Austen (Diptera, Glossinidae). *Acta Tropica* 33, 369–375.

# 4

# Pharmacology of Haematophagous Arthropod Saliva

## Donald E. Champagne[1] and Jesus G. Valenzuela[2]

[1]*Department of Veterinary Science and Center for Insect Science, University of Arizona, Tucson, Arizona 85721, USA;* [2]*Department of Biochemistry and Center for Insect Science, University of Arizona, Tucson, Arizona 85721, USA*

Vertebrates in general and mammals in particular defend against blood loss with a variety of interacting haemostatic mechanisms. Circulating platelets are activated by exposure to injury-associated agonists, aggregate into a platelet plug, and degranulate to release ADP (which recruits further platelets to the developing plug) and the potent vasoconstrictors serotonin and thromboxane $A_2$ (Ribeiro, 1987). In addition, non-vascular cells and activated platelets present tissue factor, which initiates the extrinsic coagulation pathway leading to stabilization of the plug into a fibrin clot. These responses can curtail blood loss from capillaries, arterioles, and venules on a time scale of a few seconds. As blood-feeding arthropods feed for periods ranging from minutes to weeks, it is clear that they must have some capacity to inhibit these haemostatic responses. Antihaemostatic factors are contained in saliva, which is injected into the vertebrate host during probing (where they may assist in the location of a suitable vessel) as well as during blood-feeding. As the ability to feed on blood has evolved independently many times among arthropods, it may not be surprising that a varied array of substances have been employed in antihaemostatic roles. The diversity of compounds produced by a single species is also impressive, and reflects the need to counteract the redundancy built into the vertebrate haemostatic response. Perhaps most intriguing are those single compounds, such as the protein nitrophorin 2 from *Rhodnius prolixus*, which affect multiple sites in haemostasis.

In the following review we discuss in turn salivary factors which inhibit platelet aggregation, act as vasodilators, and inhibit the coagulation pathway.

© CAB INTERNATIONAL 1996. From Wikel, S.K. (ed.) *The Immunology of Host–Ectoparasitic Arthropod Relationships.* CAB INTERNATIONAL, Wallingford.

Attention will be given to appropriate bioassay techniques, illustrated by examples from the arthropod saliva literature, to assist those wishing to enter this fascinating and diverse area of research. Readers can also consult other recent reviews (Law et al., 1992; Champagne, 1994; Ribeiro, 1995) for other perspectives on the subject.

## PLATELET AGGREGATION INHIBITORS

Haemostasis is initiated when platelets respond to subendothelial collagen exposed at the injury site, ADP released from damaged cells, and thrombin produced following activation of the coagulation pathway (Marcus and Safier, 1993). Detection of these agonists is followed by changes in shape and modulation of cell surface receptors, particularly integrins, leading to increased adhesion and platelet aggregation (Shattil et al., 1994). At the same time activated platelets degranulate, releasing ADP (which recruits more platelets to the developing plug) and serotonin and thromboxane $A_2$, which cause the vessel to constrict around the platelet plug, restricting blood flow. The study of platelet function was greatly assisted by the development of spectrophotometric techniques which measure the increase in light transmission through a suspension of platelets as they aggregate in response to agonists (Born, 1962; O'Brian, 1962; Born and Cross, 1963). This technique has been developed in a variety of commercially available platelet aggregometers (Zucker, 1989). However, these devices are limited in the number of samples which can be examined simultaneously. Recently a technique has been described in which platelet aggregation assays are performed in a 96-well microtitre plate (Fratantoni and Poindexter, 1990; Bednar et al., 1995). This technique can facilitate isolation and characterization of arthropod-derived platelet aggregation inhibitors, as it permits the simultaneous assay of numerous samples following HPLC fractionation of salivary gland extracts.

Salivary antiaggregation factors fall into several classes which reflect the varied endogenous signals which can trigger platelet aggregation.

### Apyrase

The term apyrase has been applied to enzymes that hydrolyse both ATP and ADP, but not AMP (that is, ATP diphosphohydrolase, EC 3.6.1.5) (Meyerhof, 1945). The activity of this enzyme is conveniently monitored by assaying the release of inorganic phosphate from ATP and ADP according to the method of Fiske and Subbarow (1925). This assay is easily scaled down to a volume of 100 µl, which allows measurement in microtitre plate assays. Initially it is necessary to determine the optimal pH and divalent cation ($Ca^{2+}$ or $Mg^{2+}$, usually at 5 mM) for the particular apyrase under consideration. Chromatographic purification protocols involve the use of buffered mobile phases (e.g. Champagne et al.,1995b), in which case the buffer concentration of the reaction medium should be increased to 150 mM. The saliva of blood-feeding insects lacks detectable protease activity (Gooding, 1972), and inclusion of protease inhibitors in salivary homogenates is usually not necessary.

The importance of ADP in mediating platelet aggregation is reflected in the nearly universal occurrence of apyrase activity in the saliva of blood-feeding arthropods (Table 4.1). The only species which appears to lack apyrase is the tick *Ambylomma americanum*, in which the function of platelet aggregation inhibition is carried out by high concentrations of prostaglandins (Ribeiro et al., 1992), and the deer fly *Chrysops* which produces a disintegrin-like molecule (Grevelink et al., 1993) discussed below. In addition, those species that feed primarily on birds, such as *Culex* mosquitoes, tend to have lower apyrase titres which reflect the absence of platelets and poor thrombocyte function of their hosts. The importance of this enzyme in blood-feeding is further illustrated by studies which show an inverse correlation between apyrase titres and duration of probing in surgically altered or *Plasmodium*-infected mosquitoes (Ribeiro et al., 1984a; Rossignol et al., 1984). The same correlation is seen when different species of mosquitoes (Ribeiro et al., 1985b) or black flies are compared. Highest apyrase titres are typically seen in anthropophilic species.

Comparison of apyrases reveal interspecific differences in pH optimum, divalent cation requirement ($Ca^{2+}$ or $Mg^{2+}$), and substrate preference (ATP/ADP ratio) (Table 4.1). An apyrase has also been described from *Glossina* (Mant and Parker, 1981), but its biochemical characteristics have not been determined and it is not included in Table 4.1. The only arthropod apyrase to be characterized at the molecular level is the enzyme from *Aedes aegypti* (Champagne et al., 1995b). This 68 kDa protein is homologous to a family of vertebrate 5'-nucleotidases, which hydrolyse the ester linkages of nucleoside monophosphates such as AMP. It has been hypothesized that this apyrase evolved following gene duplication of an ancestoral broad-substrate enzyme, with subsequent selection for activity towards either the pyrophosphate (apyrase) or the ester bond (5'-nucleotidase). Vertebrate ecto-ATPases with apyrase activity have been purified from a variety of tissues, and a liver ecto-ATPase was cloned and sequenced and found to be related to intercellular adhesion molecules (Lin and Guidotti, 1989). It remains to be determined if the arthropod salivary apyrases are derived from a common progenitor molecule during the evolution of blood-feeding, or if they are derived from a variety of unrelated origins.

**Table 4.1.** Characteristics of salivary apyrase.

| Species | Divalent Ca$^{2+}$ | Cation Mg$^{2+}$ | ATP/ADP ratio | Optimal pH | Activity (mUnits)[a] | Reference |
|---|---|---|---|---|---|---|
| Diptera: Culicidae | | | | | | |
| Aedes aegypti | 1.0 | 0 | 1.4 | 9.0 | 157.7 | 1, 2 |
| Anopheles gambiae | 1.0 | 0.30 | | 8.5 | 30[b] | 3 |
| An. merus | 1.0 | 0.61 | | 8.5 | 25[b] | 3 |
| An. quadriannulatus | 1.0 | 0.39 | | 8.5 | 17[b] | 3 |
| An. arabiensis | 1.0 | 0.73 | | 8.0 | 32[b] | 3 |
| An. melas | 1.0 | 0.63 | | 8.0 | 30[b] | 3 |
| An. albimanus | 1.0 | 0.80 | | 8.0 | 25[b] | 3 |
| An. freeborni | 1.0 | 0.81[c] | 1.13 | 9.0 | 25.7[d] | 4 |
| An. stevensi | 1.0 | 0.74[c] | 1.30 | 9.0 | 10.6[d] | 4 |
| An. sp nr salbaii | 1.0 | 0.57[c] | 1.13 | 9.0 | 5.3[d] | d |
| Diptera: Phlebotomidae | | | | | | |
| Lutzomyia longipalpis | 1.0 | 0 | 0.59 | 7.0–8.0 | 39 | 5 |
| Phlebotomus perniciosus | 1.0 | 0 | 1.35 | 7.0–8.5 | 54 | 6 |
| P. papatasi | 1.0 | 0 | 0.80 | 6.5–8.5 | 85 | 6 |
| P. argentipes | 1.0 | 0.14 | 0.97 | 8.0–9.0 | 29 | 6 |
| P. colabaensis | 1.0 | ND | ND | 1.1 | 6 | 6 |
| Diptera: Simulidae | | | | | | |
| Simulium vittatum | 1.0 | 0.37 | 0.61 | 8.0 | 8.9 | 7 |
| Diptera: Culicoideae | | | | | | |
| Culicoides varipennis | 1.0 | 0.9 | 1.57 | 8.5 | 7.3 | 8 |

| | | | | | |
|---|---|---|---|---|---|
| **Siphonaptera** | | | | | |
| *Oropsylla bacchi* | 1.0 | 0.81 | 1.07 | 7.5 | 21.5 | 9 |
| *Orchopea howardi* | 1.0 | 0.96 | 1.0 | 8.0 | 17.5 | 9 |
| *Xenopsylla cheopis* | 1.0 | 0.82 | 1.3 | 7.5 | 14.5 | 9 |
| **Heteroptera: Reduviidae** | | | | | |
| *Rhodnius prolixus* | 1.0 | 0 | 1.7 | 7.5–8.5 | 37.3 | 10 |
| **Acarida** | | | | | |
| *Ornithodoros moubata* | 0.94 | 1.0 | 1.3 | 7.0–8.0 | (0.97)[e] | 11 |
| *Ixodes dammini* | [f] | | 1.1 | 8.5–9.0 | (3.0)[e] | 12 |

References: (1) Ribeiro et al., 1984b; (2) Marinotti et al., 1990; (3) Cupp et al., 1994a; (4) Ribeiro et al., 1985b; (5) Ribeiro et al., 1986; (6) Ribeiro et al., 1989 (7) Cupp et al., 1993; (8) Perez de Leon and Tabachnick, 1996; (9) Ribeiro et al., 1990b; (10) Ribeiro and Garcia, 1980; Sarkis et al., 1986; (11) Ribeiro et al., 1991; (12) Ribeiro et al., 1985a.

Notes: [a] ADPase activity in mUnits per salivary gland pair; [b] interpolated from reference 3 figure 1; [c] calculated from reference 4 table 1; [d] calculated from reference 4 table 1, ADP as substrate; [e] activity in mUnits ml$^{-1}$ saliva, ATP as substrate; [f] Mg$^{2+}$>Ca$^{2+}$.

## Disintegrin-like Peptides

Activated platelets express suface adhesion receptor proteins, collectively known as integrins. The $\alpha_2\beta_1$ integrin mediates the response to collagen, and the $\alpha_{IIb}\beta_3$ integrin binds fibrinogen and von Willebrant factor by recognizing repetitive RGD domains (Shattil et al., 1994). Snake venoms and leach saliva contain peptides with RGD domains, termed disintegrins, that are able to inhibit platelet aggregation. A disintegrin-like molecule, named disagregin, was isolated from saliva of the tick *Ornithodoros moubata* (Karczewski et al., 1994). This 60-residue-long peptide was shown to inhibit the aggregation response to a number of agonists including fibrinogen, collagen, platelet activating factor, and thrombin. Another peptide, moubatin, which is more specific for collagen-mediated aggregation has also been isolated from this tick (Waxman and Connolly, 1993). The triatomine bug *Triatoma pallidipennis* produces pallidipin, which also inhibits the response to collagen (Noeske-Jungblutt et al., 1994); an homologous protein has been cloned from *Rhodnius prolixus* (Champagne and Riberio, unpublished data). Curiously none of these peptides contains the RGD domain typical of disintegrins. Recently it has been shown that the initial binding of fibrinogen to the $\alpha_{IIb}\beta_3$ receptor is dependent on the carboxy terminus of the ligand rather than either of the two RGD domains (Farrell and Thiagarajan, 1994; Shattil et al., 1994), and it seems likely that the arthropods are exploiting one of these alternative domains. A disintegrin-like factor is also present in the saliva of deer flies (Grevelink et al., 1993), and *Aedes aegypti* saliva inhibits collagen-mediated platelet aggregation via an unknown mechanism (Riberio et al., 1984a). It is possible that such peptides are a typical component of the antihaemostatic armament of blood-feeding species in general.

## Second Messenger Activators

Platelet aggregation must be self-limiting if the thrombus is to be limited to the vicinity of an injury. The plug is limited by the prostaglandins $PGI_2$ and $PGE_2$, and by endothelium-derived relaxing factor (EDRF), released from the endothelium (Moncada et al., 1976; Furchgott and Zawadzki, 1980; Furlong et al., 1987; Radomski et al., 1987). $PGI_2$ and $PGE_2$ act through a G-protein receptor, which activates adenylate kinase leading to increased cyclic AMP levels. EDRF (now recognized as the gas nitric oxide [NO]) activates guanylate cyclase, which increases cyclic GMP titres (Moncada et al., 1991). NO also has fibrinolytic activity, which helps limit the growth of the developing clot (Korbut et al., 1990). Arthropods have exploited these feedback controls to their own advantage. The saliva of the ticks *Amblyomma americanum*, *Boophilus microplus* and *Ixodes scapularis*, contains $PGE_2$ at mean concentrations of 469 ng ml$^{-1}$, 153 ng ml$^{-1}$, and 97 ng ml$^{-1}$ respectively (Dickinson et al., 1976; Higgs et al., 1976; Ribeiro et al., 1985b, 1992). The concentration of salivary $PGE_2$ greatly exceeds the levels found in vertebrate inflammatory exudates, 0.5–20 ng ml$^{-1}$ (Moncada et al., 1978). In addition, saliva of Amblyomma americanum contains $PGF_{2\alpha}$ (Ribeiro et al., 1992).

Prostaglandins actually fill several roles in tick feeding: besides their role in platelet aggregation, prostaglandins act as vasodilators (discussed below), and $PGE_2$ inhibits production of cytokines from Th1 cells, as well as activation of macrophages, neutrophils, and mast cells (Bach, 1982). A similar multifunctional role is associated with the production of NO in salivary glands of Rhodnius prolixus (Ribeiro et al., 1990a, 1993) and Cimex lectularis (Valenzuela et al., 1995). NO-mediated production of cGMP inhibits platelet aggregation (Moncada et al., 1991), but the main role of NO is associated with vasodilation (discussed below).

## VASODILATORS

Vasodilators serve several functions in relation to blood-feeding. Blood vessels occupy less than 5% of the volume of the skin (Ribeiro, 1987), and vasodilators may enhance the probability of locating a suitable vessel (and decrease searching time) by increasing the diameter of the target vessels. Once feeding is initiated, they counter the vasoconstrictive activity of serotonin and thromboxane $A_2$ released from platelets, and enhance blood flow to the feeding site.

Several techniques are available for evaluating salivary vasodilators. Intradermal injection of salivary homogenates or suitably prepared chromatographic fractions into naive shaved rabbits can reveal persistent vasodilators by the presence of erythema (e.g. Lerner et al., 1991; Cupp et al., 1994b). Tension measurements can be most useful in characterizing a vasodilator (Moncada et al., 1978; Karaki, 1987). In this technique isolated organs (such as rabbit aorta rings) are suspended in an appropriate buffer (such as Krebs-Ringer solution) and their tension is measured upon exposure to a potential vasodilator and compared with isometrically maintained tissue. Aortic rings must be pre-constricted with noradrenaline to observe vasodilation. Some vasodilators (e.g. tachykinins) act on the vascular endothelium to promote synthesis of EDRF, which leads to smooth muscle relaxation. Other substances (e.g. NO) act directly on the muscle. These may be distinguished by measuring the response of intact rings compared to those in which the endothelium has been removed by gently rubbing the inside of the rings with a wooden stick (e.g. Ribeiro et al., 1990a; Ribeiro, 1992). It is important to note that not all vasodilators relax rabbit aortic rings. For example, $PGE_2$ actually contracts the aorta, despite the fact that it relaxes superficial vessels efficiently (Moncada et al., 1978). In this respect the response of different tissues can be informative. Tachykinins are endothelium-dependent vasodilators in aortic rings, but they contract the guinea pig ileum (e.g. Ribeiro, 1992; Champagne and Ribeiro, 1994). Prostaglandins can have very specific effects on preparations from rabbit aortas, rat stomach strips, and guinea pig ileums, among others, which allow characterization of the molecule involved (Moncada et al., 1978). Further, tension measurements allow characterization of vasodilators in relation to agonists and antagonists. For example Ribeiro et al. (1990a) identified the Rhodnius prolixus vasodilator as a nitrovasodilator based in part on the agonistic activity of superoxide dismutase and the antagonistic activity of methylene blue and hydroquinone. Since salivary vasodilators act on the peripheral circulation, quantitative methods which directly measure effects in skin are also desirable. One such approach is photoelectric plethysmography, which measures blood volume in superficial vessels (e.g. Pappas et al., 1986). Alternatively, blood flow may be measured by laser Doppler methods (e.g. Grevelink et al., 1995).

Vasodilators have been characterized at the molecular level more than other antihaemostatic compounds, and here the diversity of solutions arthropods have developed to solve the problem posed by haemostasis is particularily apparent.

## Peptide Vasodilators

Several insect species have evolved peptidic vasodilators; the independent evolution of these substances is reflected in the diversity of peptides employed. The yellow fever mosquito *Aedes aegypti* produces two tachykinins, named sialokinins, which mimic the endogenous tachykinins used to regulate vascular tone (Ribeiro, 1992; Champagne and Ribeiro, 1994). It is not clear if these peptides are homologous with the vertebrate neuromodulators or if they have arisen through convergent evolution, as the *Aedes* cDNA sequence is very different from the corresponding vertebrate preprotachykinin (outside of the region coding for the active peptide) (Champagne, unpublished data). Tachykinins are of interest in that they are also involved in modulating macrophage activation and primary antibody presentation (Brunelleschi et al., 1990; Ealezos et al., 1991), with possible consequences for parasites co-inoculated with saliva. This class of peptides may be characteristic of *Aedes* in general, as the vasodilator from *Ae. triseriatus* is also a tachykinin (Ribeiro et al., 1994).

The vasodilator from the sandfly *Lutzomyia longipalpus* is the most potent such peptide known, with an erythema-inducing activity 500-fold greater than human calcitonin gene-related peptide (CGRP) (Ribeiro et al., 1989b). Although early experiments suggested some similarity to CGRP (Ribeiro et al., 1989b), when the peptide (named maxadilan) was purified, cloned and sequenced, no marked identity with CGRP was found (Lerner et al., 1991; Lerner and Shoemaker, 1992). Further, maxadilan affects a different range of vessels than CGRP (skin is the only tissue that responds to both peptides), and it is now clear that maxadilan is acting through a different receptor than CGRP (Grevelink et al., 1995). Interest in maxadilan was increased by the discovery that *Lutzomyia longipalpus* saliva enhances *Leishmania* infection in CBA mice (Titus and Ribeiro, 1988; Theodos et al., 1991). Recently '*Lutzomyia longipalpus*' has been found to be a sibling species complex (Lanzaro et al., 1993), and the composition of the saliva, and its ability to enhance *Leishmania* infections varies between sibling species (Warburg et al., 1994). The contribution of maxadilan to the *Leishmania*-enhancing activity of the saliva is an active area of research.

Blackfly (*Simulium vittatum*) saliva also contains a potent erythema-inducing factor (Cupp et al., 1994b). This peptide has been cloned, sequenced, and expressed in a baculovirus system; it has no obvious sequence identity with maxadilan or CGRP (Cupp, M.S., Champagne, D.E. and Cupp, E.W., unpublished results).

## Catechol Oxidase/Peroxidase

A very different strategy is followed by *Anopheles albimanus* and *An. gambiae* (Ribeiro and Nussenzveig, 1993a; Ribeiro et al., 1994). Rather than produce vasodilatory substances directly, these mosquitoes secrete a peroxidase that generates $H_2O_2$, which is used to oxidize host-produced vasoconstrictors including serotonin and catecholamines. This produces a slow relaxation of the vessel, which may help account for the relatively long time required for these insects to take a bloodmeal.

## Prostaglandins

As noted above, ticks produce high concentrations of the prostaglandins $PGE_2$, $PGF_{2\alpha}$, and $PGI_2$. These substances are produced during feeding and are not stored in the salivary glands, so they may be absent from salivary gland homogenates. They can be detected in saliva produced in response to treatment with the secretagogue pilocarpine (Tatchell, 1967). It is interesting that $PGF_{2\alpha}$, produced by the cattle tick *Boophilus microplus*, is the most vasodilatory prostaglandin when tested on cattle skin (Kemp et al., 1983), which indicates how well adapted these blood-feeders are to their host species.

## Nitric Oxide

The vasodilatory activity of *Rhodnius prolixus* saliva was shown to have properties of nitrovasodilators, substances which release the gas NO in the blood (Ribeiro et al., 1990a). The NO is associated with abundant haem proteins, which reversibly bind NO in a concentration and pH-dependent manner (Ribeiro et al., 1993). Reversibility of NO binding is an unusual property, and in this case it is due to the presence of $Fe^{3+}$ haem rather than the $Fe^{2+}$ haem typical of haemoglobin and other haem proteins. Four NO-carrying proteins, named nitrophorin 1 to 4 (NP1–4), are present in *Rhodnius* saliva (Champagne et al., 1995a). The cDNA sequence of all four has been obtained (Champagne et al., 1995a; Champagne, unpublished results); NP 1 and 4 are 90% identical in sequence, and NP 2 and 3 are 80% identical, but the NP1+4 group has only 42% identity with the NP2+3 group. The reason for such a diversity of NO-transport proteins is unknown. The nitrophorins are stored in the salivary gland lumen; NO is synthesized by an intracellular nitric oxide synthetase (NOS) associate with vesicles, and NO appears to be directed towards the nitrophorins through abundant microvilli which project into the lumen cavity (Nussenzveig and Ribeiro, 1995). The NOS itself is a highly regulated enzyme with cofactor requirements similar to the vertebrate constitutive brain NOS (Ribeiro and Nussenzveig, 1993b).

Remarkably, the *Rhodnius* nitrophorins have multiple roles in inhibiting host haemostasis. These proteins have a high affinity for histamine, which is released from mast cells of allergic hosts (Ribeiro and Walker, 1994). Indeed histamine can displace NO from nitrophorins under conditions where NO would not readily be released. Further, nitrophorin 2 is identical with the anticlotting factor prolixin S, and accounts for all the anti-factor VIII activity of the saliva (Ribeiro et al., 1996). This single group of proteins accounts for the vasodilatory, anticlotting, and antihistaminic activity of the saliva, and they also contribute to the platelet antiaggregating activity through the release of NO (as discussed above).

Another blood-feeding hemipteran, the bedbug *Cimex lectularis*, also has a vasodilator with properties of nitrovasodilators (Valenzuela et al., 1995). The saliva of this insect yielded a single nitrophorin, which was also an $Fe^{3+}$ haem protein. However, the spectroscopic properties of this protein, and its greater size (32 kDa compared with 19–20 kDa for the *Rhodnius*) suggest that the *Cimex* nitrophorin might be different from the *Rhodnius* nitrophorins. Blood-feeding evolved independently in the Triatominae and the Cimicidae, and it is possible that, although both insects have arrived at the use of NO as a vasodilator, different molecules have been employed to accomplish this task.

## COAGULATION

Stabilization of the haemostatic plug in a blood clot is the final phase of the haemostatic response. The coagulation cascade involves a number of essential proteolytic enzymes and factors (Fig. 4.1a). Factors VII, IX, X, XI and

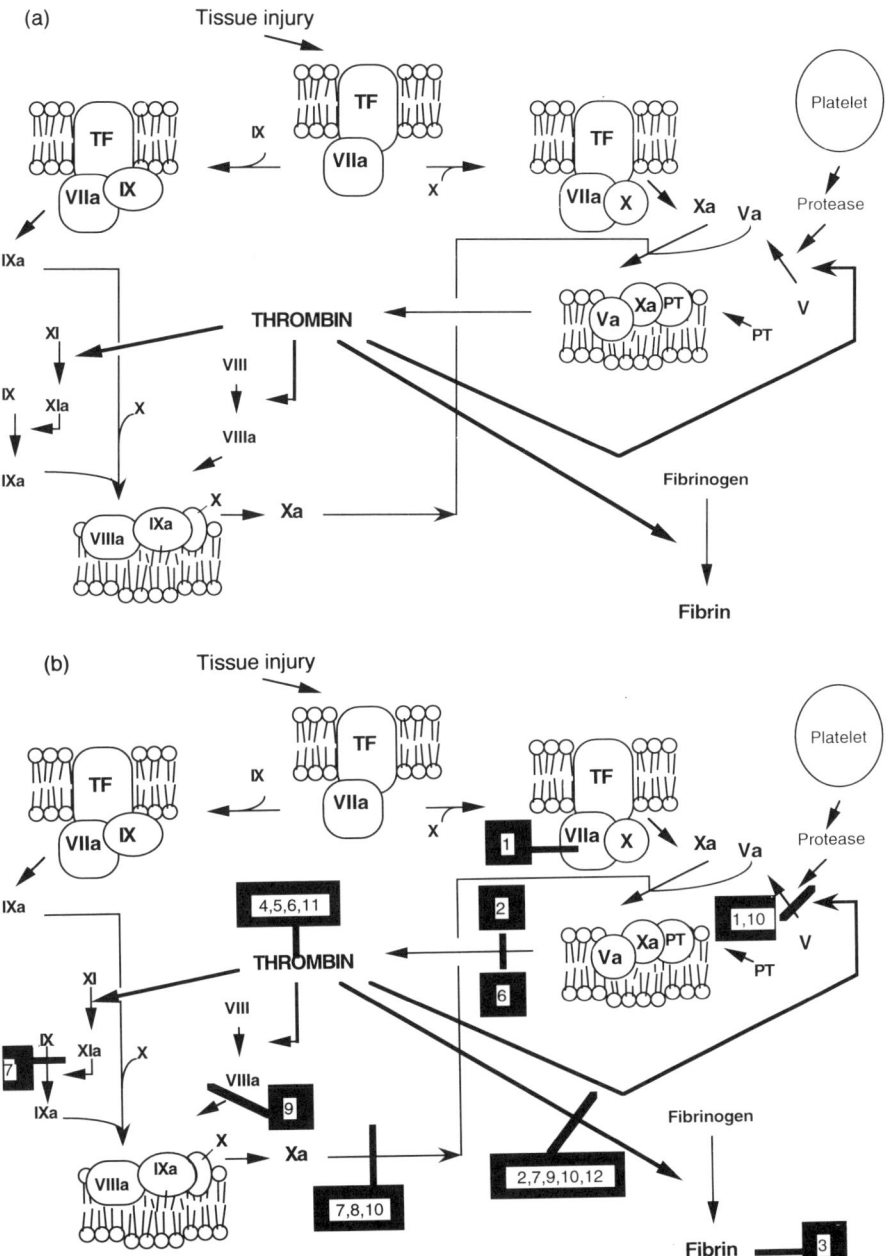

**Fig. 4.1.** (a) Blood coagulation cascade integrating the intrinsic and the extrinsic pathways. TF, tissue factor; PT, prothrombin. (b) Numbered black boxes represent the mode of action of the reported blood-clotting inhibitors from blood-sucking arthropods: (1) *Dermacentor andersoni*; (2) *Eutriatoma maculata*; (3) *Glossina austeni*; (4) *Glossina morsitans*; (5) *Ixodes holocyclus*; (6) *Ixodes ricinus*; (7) *Ornithodorus moubata*; (8) *Rhipicephalus appendiculatus*; (9) *Rhodnius prolixus*; (10) *Similium vittatum*; (11) *Tabanus bovinus*; (12) *Triatoma infestans*.

prothrombin are relatives of the proteases trypsin and chymotrypsin, and are normally found in the inactive form in the blood circulation (Furie et al., 1982). These proteases are activated by the cleavage of one or two peptide bonds. Two essential cofactors, factors V and VIII, are also activated by peptide bond cleavage (Furie and Furie, 1992). In addition there is tissue factor (factor III), the only transmembrane protein in the coagulation cascade, which is normally expressed in the surface of non-vascular cells (Spicer et al., 1987). The endpoint of the coagulation cascade is the formation of the fibrin clot resulting from the action of thrombin on fibrinogen. It has been proposed for many years that two different pathways are involved in the formation of thrombin: the intrinsic and the extrinsic pathways (Davie and Ratnoff, 1964; Macfarlane, 1964).

Data accumulated over the years from *in vitro* and patient bleeding disorder studies (Ratnoff and Colopy, 1955; Hathaway et al., 1965; Roberts and Lozier, 1991) have resulted in a better understanding of the blood coagulation pathway and a more physiological and integrated mechanism has been proposed (Furie and Furie, 1992; Esmon, 1993). Firstly, as factor XII has been shown not to be essential in patients lacking this factor and since the importance of factor XII was described only in *in vitro* studies, it has been concluded that this factor is not physiologically relevant in the initiation of the coagulation cascade. The question then arises as to how coagulation is initiated and how the two pathways are connected. The following proposed mechanism (Fig. 4.1) integrates the two pathways and gives a better explanation for the relevance of specific factors, cofactors and their regulation.

The initiation of the blood coagulation cascade can be envisioned as a slow initiation mechanism which is then augmented by the products or factors activated in the process; in other words the blood clotting cascade possesses a self-enhancing mechanism in which the main purpose is the massive production of thrombin. The process is initiated by the limited amount of thrombin produced when factor VIIa, which is initially activated by an unknown enzyme, binds to tissue factor in the membrane of non-vascular cells (Rao and Rapaport, 1988). By this stage in the process of haemostasis, platelets are already activated; one of the results of this activation is the change in the composition of the platelet membrane, which now consists mainly of negatively charged phospholipids (Davie et al., 1991). In the environment of negatively charged membrane and in the presence of the factor VIIa–tissue factor complex, factor X binds and is activated to factor Xa (Rao and Rapaport, 1988). The next step is the formation of the complex factor Xa–factor Va in the membrane. However, this complex is formed only when factor V is active. The initial activation of factor V to Va has been proposed to be by a protease released by platelets (Kane et al., 1982). The activity of factor Va is not very high by this mechanism, but presumably it is sufficient to form an active factor Xa–factor Va complex. This membrane-associated complex binds prothrombin, which is activated to thrombin and released to the circulation (Mann et al., 1988; Furie and Furie, 1992).

The amount of thrombin initially formed by this process is limited and is mainly due to the activation of factors V and X. Although factor IX is activated to IXa by the factor VIIa–tissue factor complex (Rao and Rapaport, 1988), it does not play any role in the initial production of thrombin because active factor VIIIa is absent in this initial process and these two factors (VIIIa and IXa) need to form a complex to activate factor X (Gilbert et al., 1990). Initiation of coagulation would therefore appear to be a function of the extrinsic pathway. The amount of thrombin and the rate at which it is initially produced probably would not be sufficient to induce massive conversion of fibrinogen to fibrin. However, at this point the integration of the extrinsic and intrinsic pathways takes place, and this integration results in the amplification of the clotting cascade and a large increase in the production of thrombin. This amplification event is led by low concentrations of thrombin, suggesting that the initial concentration of thrombin is more regulatory than enzymatic. This is based on the following observations: (i) thrombin activates factor XI to XIa (Naito and Fujikawa, 1991); (ii) thrombin activates factor VIII to VIIIa (Tiarks, 1992); and (iii) thrombin activates factor V to Va (Willems, 1992). Following activation of these enzymes by thrombin, the cascade is greatly amplified: factor XIa activates factor IX to IXa, which in the presence of thrombin-activated factor VIIIa cleaves prothrombin to thrombin. In addition, factor Xa activates factor VII to VIIa. Therefore, thrombin is not only important as the final enzyme in the coagulation cascade, but it is necessary to activate other enzymes which will accelerate the cascade to produce more thrombin.

## Inhibition of the Blood Coagulation Cascade

The cross-interaction of all the factors of the blood coagulation cascade to accomplish the formation of the fibrin clot is impressive. However, blood-sucking arthropods have been no less ingenious in strategies they have developed to circumvent this process. These strategies are summarized on Fig. 4.1b. While many species produce anticoagulation factors, some produce a remarkable variety of these substances that interfere whith several points in the clotting process. These latter species mainly feed from large haemorrhagic pools, in which blood is mixed with non-vascular tissues. Exposure to such a procoagulant environment may necessitate a high level of redundancy in anticoagulant factors (Ribeiro, 1995).

The coagulation cascade is initiated by the formation of the factor VIIa–tissue factor complex. The tick *Dermatocentor andersoni* has been reported to inhibit factors VII and V (Gordon and Allen, 1991), which inhibits this initial step of coagulation. Should the haemostatic system bypass the first level of inhibition the second step of the cascade, formation of a factor Xa–Va complex, would also be inhibited by the anti-factor V activity. The blackfly *Simulium vittatum* has also been reported to have anti-factor V, anti-Xa, and antithrombin activity (Abebe *et al.*, 1996).

The formation of thrombin has been reported to be inhibited by the triatomine *Eutriatoma maculatus* (Hellmann and Hawkins, 1966) and by the tick *Ixodes ricinus* (Markwardt *et al.*, 1964). Antithrombin activity has been reported in the tsetse fly *Glossina morsitans* (Parker and Mant, 1979), the horse fly *Tabanus bovinus* (Markwardt and Schultz, 1960a), and the ticks *Ixodes holocyclus* (Anastopoulos *et al.* 1991) and *Ixodes ricinus* (Hoffmann *et al.*, 1991). Antithrombin activity inhibiting fibrin formation has been reported in the tick *Ornithodoros moubata* (Hellmann and Hawkins, 1967), salivary glands of *Simulium vittatum* (Abebe and Cupp, unpublished data), and the triatomines *Eutriatoma maculatus* (Hellmann and Hawkins, 1966) and *Triatoma infestans* (Markwardt and Schultz, 1960b). The kazal-type thrombin inhibitor from *Rhodnius prolixus* (Friedrich *et al.*, 1993) is the same as prolixin G, and is of gut origin (Ribeiro *et al.*, 1996; Hellmann and Hawkins, 1965). *Glossina austeni* has been reported to have fibrinolytic activity (Hawkins, 1966).

Inhibitors of other coagulation factors has also been reported. For example the tick *Ornithodoros moubata* has anti-factor IX activity (Hellmann and Hawkins, 1967). The triatomine *Rhodnius prolixus* posseses an anti-factor VIII in the salivary glands; originally described as Prolixin S (Hellmann and Hawkins, 1965), this factor is identical to the vasodilator nitrophorin 2 (Ribeiro *et al.*, 1995). Anti-factor Xa activity has been reported in *Ornithodoros moubata* (Waxman *et al.*, 1990), *Rhipicephalus appendiculatus* (Limo *et al.*, 1991), *Simulium vittatum* (Jacobs *et al.*, 1990), the biting midge *Culicoides varipennis* (Perez de Leon, unpublished results) and the mosquito *Aedes aegypti* (Stark and James, unpublished results).

It is interesting to note the number of species with antithrombin activity. Most likely this reflects the central role of thrombin in accelerating the coagulation cascade, as well as its importance in converting fibrinogen to fibrin. Further, thrombin can act directly to stimulate platelet aggregation (Coughlin *et al.*, 1992). At least two functional sites are present on the thrombin molecule, only one of which is involved in fibrinogen recognition, and several sites were suggested as targets for antithrombic drugs (Stubbs and Bode, 1994). The diversity of antithrombins suggests that the arthropods may be ahead of humans in this area. The common occurrence of anti-Xa activities represents another strategy of attacking a point central to several steps in the cascade.

The assay chosen by the investigator to determine thrombin activity is important, firstly, in determining if the thrombin inhibitor is participating in the inihibition of the blood coagulation cascade, and secondly, for assessing the type of thrombin inhibitor. The two most common tests are the thrombin–fibrinogen reaction (Hellmann and Hawkins, 1966) and the chromogenic substrate assay (Greissbach *et al.*, 1985). Although the latter can represent true antithrombin activity in the coagulation cascade, it may also represent inhibition of other thrombin activities such as activation of platelet aggregation. For example *Cimex lectularis* possesses an antithrombin activity which is not related to the anticlotting activity (Valenzuela *et al*, submitted). Therefore it is important to ensure that any antithrombin activity revealed with the chromogenic substrate assay is paralleled by anticoagulant activity. Recently, Rupin *et al.* (1995) have described a method of thrombin-induced fibrin–clot microassay. This methodology not only evaluates the anticlotting activity of a thrombin inhibitor, it also evaluates the kinetics of an antithrombin component.

As it is not unusual for an insect to produce more than one anticoagulant molecule, parallel tests need to be performed with a variety of anti-factor activities to evaluate if any one can account for all of the anticlotting activity or if there is more than one factor involved. A convenient technique is the fractionation of the salivary homogenate by HPLC or FPLC, followed by testing for anticlotting activity using the recalcification time of citrated plasma. This method was used to characterize the anticoagulant component from salivary glands of *Rhodnius prolixus* (Ribeiro *et al.*, 1996) and *Cimex lectularis* (Valenzuela *et al.*, submitted).

## CONCLUDING COMMENTS

Of the more than 14,000 species of blood-feeding arthropods, fewer than 20 (primarily those responsible for human disease) have been studied for antihaemostatic compounds (Ribeiro, 1996). The diversity of antihaemostatic compounds found within that limited sample suggests that a wealth of novel pharmacologically active substances awaits discovery. These compounds may yield new drugs relevant to control of, for instance, cardiovascular disease. Further, it is evident that vectors are active participants in the infection process, with a role much greater than mere 'flying syringes' (e.g. Titus and Ribeiro, 1988; Warburg *et al.*, 1994). Elucidation of the effects of vector saliva on host physiology at the site of parasite introduction, and identification of the specific molecules participating in the interaction, can be expected to enhance our understanding of the process of parasite transmission. Eventually this may suggest novel strategies for interfering with transmission, including immune neutralization of specific salivary components (Ribeiro, 1996). Finally, the question arises as to the evolutionary source of the antihaemostatic compounds. Assuming that salivary gland factors are derived by modification and/or overexpression from components of other systems (such as the nervous system), study of these factors may lead to an improved undertanding of the function of those systems. In time this may point to novel avenues for vector control.

## ACKNOWLEDGEMENTS

We are grateful to Drs J.M.C. Ribeiro and J.A. Guimaraes for critical comments on the manuscript. DEC's research is supported by NIH grant AI-35591.

## REFERENCES

Abebe, M., Cupp, M.S., Champagne, D. and Cupp, E.W. (1995) Simulidin: a black fly (*Simulium vittatum*) salivary gland protein with anti-thrombin activity. *Journal of Insect Physiology* 41, 1001–1006.

Abebe, M., Ribeiro, J.M.C., Cupp, M.S. and Cupp, E.W. (199b) A novel anticoagulant from the salivary glands of *Simulium vittatum* (Diptera: Simuliidae) inhibits activity of coagulation factor V. *Journal of Medical Entomology*, 33, 173–176.

Anastopoulos, P., Thurn, M.J. and Broady, K.W. (1991) Anticoagulant in the tick *Ixodes holocyclus*. *Australian Veterinary Journal* 68, 366–367.

Bach, M.K. (1982) Mediators of anaphylaxis and inflammation. *Annual Review of Microbiology* 36, 371–413.

Bednar, B., Condra, C., Gould, R.J. and Connolly, T.M. (1995) Platelet aggregation monitored in a 96 well microplate reader is useful for evaluation of platelet agonists and antagonists. *Thrombosis Research* 77, 453–463.

Born, G.V.R. (1962) Aggregation of blood platelets by adenosine diphosphate and its reversal. *Nature* 194, 927–929.

Born, G.V.R. and Cross, M.J. (1963) The aggregation of blood platelets. *Journal of Physiology (London)* 168, 178–195.

Brunelleschi, S., Vanni, L., Giotti, A., Maggi, C.A. and Fantozzi, R. (1990) Tachykinins activate guinea pig alveolar macrophages: involvement of NK2 and NK1 receptors. *British Journal of Pharmacology* 10, 417–420.

Champagne, D.E. (1994) The role of salivary vasodilators in bloodfeeding and parasite transmission. *Parasitology Today* 10, 430–433.

Champagne, D.E. and Ribeiro, J.M.C. (1994) Sialokinin I and II: Vasodilatory tachykinins from the yellow fever mosquito *Aedes aegypti*. *Proceedings of the National Academy of Science U.S.A.* 91, 138–142.

Champagne, D.E., Nussenzveig, R.H. and Ribeiro, J.M.C. (1995a) Purification, partial characterization, and cloning of nitric oxide-carrying heme proteins (Nitrophorins) from salivary glands of the blood-sucking insect *Rhodnius prolixus*. *Journal of Biological Chemistry* 270, 8691–8695.

Champagne, D.E., Smartt, C.T., James, A.A. and Ribeiro, J.M.C. (1995b) The salivary gland-specific apyrase of the mosquito *Aedes aegypti* is a member of the 5'-nucleotidase family. *Proceedings of the National Academy of Science USA* 92, 694–698.

Coughlin, S.R., Vu, T.K.H., Hung, D.T. and Wheaton, V.I. (1992) *Journal of Clinical Investigation* 89, 351–353.

Cupp, M.S., Cupp, E.W. and Ramberg, F.B. (1993) Salivary gland apyrase in black flies (*Simulium vittatum*). *Journal of Insect Physiology* 39, 817–821.

Cupp, E.W., Cupp, M.S. and Ramberg, F. (1994a) Salivary apyrase in African and New World vectors of *Plasmodium* species and its relationship to malaria transmission.

*The American Journal of Tropical Medicine and Hygiene* 50, 235–240.
Cupp, M.S., Ribeiro, J.M.C. and Cupp, E.W. (1994b) Vasodilative activity in black fly salivary glands. *The American Journal of Tropical Medicine and Hygiene* 50, 241–246.
Davie, E.W. and Ratnoff, O.D. (1964) Waterfall sequence for intrinsic blood clotting. *Science* 145, 1310–1312.
Davie, E.W., Fujikawa, K. and Kisiel, W. (1991) The coagulation cascade: initiation, maintenance and regulation. *Biochemistry* 30, 10363–10370.
Dickinson, R.G., O'Hagan, J.E., Shotz, M., Binnington, K.C. and Hegarty, M.P. (1976) Prostaglandin in saliva of the cattle tick *Boophilus microplus*. *Australian Journal of Experimental Biology and Medical Sciences* 54, 475–486.
Ealezos, A., Andrews, P.V., Boyd, R.L. and Helme, R.D. (1991) Tachykinin-mediated modulation of the primary antibody response in rats: evidence for mediation by an NK2 receptor. *Journal of Neuroimmunology* 32, 11–18.
Esmon, C.T. (1993) Cell mediated events that control blood coagulation and vascular injury. *Annual Review of Cell Biology* 9, 1–26.
Farrell, D.H. and Thiagarajan, A.P. (1994) Binding of recombinant fibrinogen mutants to platelets. *Journal of Biological Chemistry* 269, 226–231.
Fiske, C.H. and Subbarow, Y. (1925) The colorimetric determination of phosphorus. *The Journal of Biological Chemistry* 6, 375–400.
Fratantoni, J.C. and Poindexter, B.J. (1990) Measuring platelet aggregation with a microplate reader. *Journal of Clinical Pathology* 94, 613–617.
Friedrich, R., Kroger, B., Biajolan, S., Lemarie, H.G., Hoffken, H.W., Reuschenbach, P., Otte, M. and Dodt, J.A. (1993) A Kazal-type inhibitor with thrombin specificity from *Rhodnius prolixus*. *Journal of Biological Chemistry* 268, 16216–16222.
Furchgott, R.F. and Zawadzki, J.V. (1980) The obligatory role of endothelial cells in the relaxation of arterial smooth muscle by acetylcholine. *Nature* 299, 373–376.
Furie, B. and Furie B.C. (1992) Molecular and cellular biology of blood coagulation. *New England Journal of Medicine* 326, 800–806.
Furie, B., Bing, D.H., Feldman, R.J., Robinson, D.J., Burnier, J.P. and Furie, B.C. (1982) Computer-generated models of blood coagulation factor Xa, factor IXa, and thrombin based upon structural homology with other serine proteases. *Journal of Biological Chemistry* 257, 875–3882.
Furlong, B., Henderson, A.H., Lewis, M.J. and Smith, J.A. (1987) Endothelium-derived relaxing factor inhibits *in vitro* platelet aggregation. *British Journal of Pharmacology* 90, 687–692.
Gilbert, G.E., Furie, B.C. and Furie, B. (1990) Binding of human factor VIII to phospholipid vesicles. *Journal of Biological Chemistry* 265, 815–822.
Gooding, R.H. (1972) Digestive processes of haematophagous insects I: A literature review. *Quaestiones Entomoloicae* 8, 5–60.
Gordon, J.R. and Allen, J.R. (1991) Factors V and VII anticoagulants activities in the salivary glands of feeding *Dermacentor andersoni*. *Journal of Parasitology* 77, 167–170.
Greissbach, U., Strurzebecher, J. and Markwardt, F. (1985) Assay of huridin in plasma using a chromogenic substrate. *Thrombosis Research* 37, 347–350.
Grevelink, S.A., Youssef, D.E., Loscalzo, J. and Lerner, E.A. (1993) Salivary-gland extracts from the deerfly contain a potent inhibitor of platelet aggregation. *Proceedings of the National Academy of Science USA* 90, 9155–9158.

Grevelink, S.A., Osborne, J., Loscalzo, J. and Lerner, E.A. (1995) Vasorelaxant and second messenger effects of maxidilan. *The Journal of Pharmacology and Experimental Therapeutics* 272, 33–37.

Hathaway, W.E., Belhasen, L.P. and Hathaway, H.S. (1965) Evidence for a new plasma thromboplastin factor. I. Case report, coagulation studies and physicochemical properties. *Blood*. 26,521–532.

Hawkins, R.I. (1966) Factors affecting blood clotting from salivary glands and crop of *Glossina austeni*. *Nature* 207, 738–739.

Hellmann, K. and Hawkins, R.I. (1965) Prolixin-S and Prolixin-G: Two anticoagulants from *Rhodnius prolixus* Stahl. *Nature (London)* 207, 265–267.

Hellmann, K. and Hawkins, R.I. (1966) An antithrombin (Maculatin) and a plasminogen activator extractable from the blood-sucking Hemipteran *Eutriatoma maculatus*. *British Journal of Haematology* 12, 376–384.

Hellmann, K. and Hawkins, R.I. (1967) The action of tick extract on blood coagulation and fibrinolysis. *Thrombosis et Diathesis Haemorrhagica* 18, 617–625.

Higgs, G.A., Vane, J.R., Hart, R.J., Porter, C. and Wilson, R.G. (1976) Prostaglandins in the saliva of the cattle tick, *Boophilus microplus* (Canestrini) (Acarina, Ixodidae). *Bulletin of Entomological Research* 66, 665–670.

Hoffmann, A., Walsmann, P., Riesener, G., Paintz, M. and Markwardt, F. (1991) Isolation and characterization of a thrombin inhibitor from the tick *Ixodes ricinus*. *Pharmazie* 46, 209–212.

Jacobs, J.W., Cupp, E.W., Sardana, M. and Friedman, P.A. (1990) Isolation and characterization of a coagulation factor Xa inhibitor from black fly salivary glands. *Thrombosis and Haemostasis* 64, 235–238.

Kane, W.H., Mruk, J.S. and Majerus, P.W. (1982) Activation of coagulation factor V by a platelet protease. *Journal of Clinical Investigation* 70, 1092–1100.

Karaki, H. (1987) Use of tension measurements to delineate the mode of action of vasodilators. *Journal of Pharmacological Methods* 18, 1–21.

Karczewski, J., Endris, R. and Connolly, T.M. (1994) Disagregin is a fibrinogen receptor antagonist lacking the Arg–Gly–Asp sequence from the tick, *Ornithodoros moubata*. *Journal of Biological Chemistry* 269, 6702–6708.

Kemp, D.H. *et al.* (1983) Comparison of cutaneous hyperemia in cattle elicited by larvae of *Boophilus microplus* and by PGs and other mediators. *Experientia* 39, 725

Korbut, R., Lidbury, P.S. and Vane, J.R. (1990) Prolongation of fibrinolytic activity of tissue plasminogen activator by nitrovasodilators. *Lancet* 335, 699.

Lanzaro, G.C., Ostrovska, K., Herrero, M.V., Lawyer, P.G. and Warburg, A. (1993). *Lutzomyia longipalpis* is a species complex: Genetic divergence and interspecific hybrid sterility among three populations. *American Journal of Tropical Medicine and Hygiene* 48, 839–847.

Law, J., Ribeiro, J.M.C. and Wells, M. (1992) Biochemical insights derived from diversity in insects. *Annual Review of Biochemistry* 61, 87–112.

Lerner, E.A. and Shoemaker, C.B. (1992) Maxadilan: Cloning and functional expression of the gene encoding this potent vasodilator. *The Journal of Biological Chemistry* 267, 1062–1066.

Lerner, E.A., Ribeiro, J.M.C., Nelson, R.J. and Lerner, M.R. (1991) Isolation of maxidilan, a potent vasodilatory peptide from the salivary glands of the sand fly *Lutzomyia longipalpis*. *Journal of Biological Chemistry* 266, 11234–11236.

Limo, M.K., Voigt, W.P., Tumbo-Oeri, A.G., Njogu, R.M. and Ole-MoiYoi, O.K. (1991) Purification and characterization of an anticoagulant from the salivary glands of the ixodid tick, Rhipicephalus appendiculatus. Experimental Parasitology 72, 418–429.
Lin, S.H. and Guidotti, G. (1989) Cloning and expression of a cDNA coding for a rat liver plasma membrane ecto-ATPase. The primary structure of the ecto-ATPase is similar to that of the human biliary glycoprotein I. The Journal of Biological Chemistry 264, 14408–14414.
Macfarlane, R.G. (1964) An enzymatic cascade in the blood mechanism and its function as a biochemical amplifier. Nature 202, 498–499.
Mann, K.G., Jenny, R.J. and Krishnaswamy, K. (1988) Cofactor proteins in the assembly and expression of blood clotting enzyme complexes. Annual Review of Biochemistry 57, 915–956.
Mant, M.J. and Parker, K.R. (1981) Two platelet aggregation inhibitors in tsetse (Glossina) saliva with studies of roles of thrombin and citrate in in vitro platelet aggregation. British Journal of Haematology 48, 601–608.
Marcus, A.J. and Safier, L.B. (1993) Thromboregulation: multicellular modulation of platelet reactivity in hemostasis and thrombosis. Federation of the American Society for Experimental Biology Journal 7, 516–522.
Marinotti, O., James A.A. and Ribeiro, J.M.C. (1990) Diet and salivation in female Aedes aegypti mosquitoes. Journal of Insect Physiology 36, 545–548.
Markwardt, F. and Schultz, E. (1960a) Uber den mechanismus der blutgerinnungshemmenden wirkung des Tabanins. Naunyn-Schmiedebergs Arch. Exp. Path. Pharmak 238, 320–328.
Markwardt, F. and Schultz, E. (1960b) Uber einen hemmstoff des gerinnungsfermentes thrombin aus blutsaugenden raubwanzen (reduviiden). Naturwissenschaften 47, 43.
Markwardt, F., Hoffmann, A. and Landmann, H. (1964) Versuche zur autoproprothrombin C wirkung. Thrombs. Diathes, Haemorrh (Stuttgart) 11, 230–233.
Meyerhoff, O. (1945) The origin of the reaction of Harden and Young in cell-free alcoholic fermentation. The Journal of Biological Chemistry 157, 105–119.
Moncada, S., Gryglewski, D., Bunting, S. and Vane, G.R. (1976) An enzyme isolated from arteries transforms prostaglandin endoperoxides to an unstable substance that inhibits platelet aggregation. Nature 263, 663–665.
Moncada, S., Ferreira, S.H. and Vane, J.R. (1978) Bioassay of prostaglandins and biologically active substances derived from arachidonic acid. Advances in Prostaglandin and Thromboxane Research 5, 211–236.
Moncada, S., Palmer, R.M.J. and Higgs, E.A. (1991) Nitric oxide: Physiology, pathophysiology, and pharmacology. Pharmacological Review 43, 109–142.
Naito, K. and Fujikawa, K. (1991) Activation of human blood coagulation factor XI independent of factor XII: factor XI is activated by thrombin and factor XIa in the presence of negatively charged surfaces. Journal of Biological Chemistry 266, 7353–7358.
Noeske-Jungblutt, C., Kratzschmar, J., Haendler, B., Alagon, A., Possani, L., Verhallen, P. and Schleuning, W.-D. (1994) An inhibitor of collagen-mediated platelet aggregation from the saliva of Triatoma pallidipennis. Journal of Biological Chemistry 269, 5050–5053.
Nussenzveig, R.H. and Ribeiro, J.M.C. (1995) Nitric oxide loading of the salivary nitric-

oxide-carrying hemoproteins (nitrophorins) in the blood- sucking bug *Rhodnius prolixus*. *The Journal of Experimental Biology* 198, 1093–1098.

O'Brian, J.R. (1962) Platelet aggregation.II. Some results from a new method of study. *Journal of Clinical Pathology* 15, 452–464.

Pappas, L.G., Pappas, C.D. and Grossman, G.L. (1986) Hemodynamics of human skin during mosquito (Diptera:Culicidae) blood feeding. *Journal of Medical Entomology* 23, 581–587.

Parker, K.R. and Mant, M.J. (1979) Effects of tsetse salivary gland homogenate on coagulation and fibrinolysis. *Thrombosis and Haemostasis (Stuttgart)* 42, 743–751.

Perez de Leon, A.A. and Tabachnick, W.J. (1996) Apyrase activity and ADP- induced platelet aggregation inhibition by the salivary gland proteins of *Culicoides varipennis*, the North American vector of bluetongue viruses. *Veterinary Parasitology*, 61, 327–338.

Radomski, M.W., Palmer, R.M.J. and Moncada, S. (1987) Comparative pharmacology of endothelium-derived relaxing factor, nitric oxide and prostacyclin in platelets. *British Journal of Pharmacology* 92, 181–187.

Rao, L.V. and Rapaport, S.I. (1988) Activation of factor VII bound to tissue factor: a key early step in the tissue factor pathway of blood coagulation. *Proceedings of the National Academy of Science USA* 85, 6687–6691.

Ratnoff, O.D. and Colopy, J.E. (1955) A familial hemorragic trait associated with a deficiency of a clot-promoting fraction of plasma. *Journal of Clinical Investigation* 34, 602 613.

Ribeiro, J.M.C. (1987) Role of saliva in blood-feeding by arthropods. *Annual Review of Entomology* 32, 463–478.

Ribeiro, J.M.C. (1992) Characterization of a vasodilator from the salivary glands of the yellow fever mosquito *Aedes aegypti*. *Journal of Experimental Biology* 165, 61–71.

Ribeiro, J.M.C. (1995) Blood feeding arthropods: live syringes or invertebrate pharmacologists? *Infectious Agents and Disease* 4, 143–152.

Ribeiro, J.M.C. and Garcia, E.S. (1980) The salivary and crop apyrase activity of *Rhodnius prolixus*. *Journal of Insect Physiology* 26, 303–307.

Ribeiro, J.M.C. and Nussenzveig, R.H. (1993a) The salivary catechol oxidase/peroxidase activities of the mosquito *Anopheles albimanus*. *Journal of Experimental Biology* 179, 273–287.

Ribeiro, J.M.C. and Nussenzveig, R.H. (1993b) Nitric oxide synthase activity from a hematophagous insect salivary gland. *FEBS Letters* 330, 165–168.

Ribeiro, J.M.C. and Walker, F.A. (1994) High-affinity histamine binding and antihistaminic activity of the salivary NO–heme carrying proteins (nitrophorins) in the blood-sucking bug *Rhodnius prolixus*. *Journal of Experimental Medicine* 180, 2251–2257.

Ribeiro, J.M.C., Rossignol, P.A. and Spielman, A. (1984a) Role of mosquito saliva in blood vessel location. *Journal of Experimental Biology* 108, 1–7.

Ribeiro, J.M.C., Sarkis, J.J.F., Rossignol, P.A. and Spielman, A. (1984b) Salivary apyrase of *Aedes aegypti*: characterization and secretory fate. *Comparative Biochemistry and Physiology* 79B, 81–86.

Ribeiro, J.M.C., Makoul, G.T., Levine, J., Robinson, D.R. and Spielman, A. (1985a) Antihemostatic, antiinflammatory, and immunosuppressive properties of the saliva of a tick, *Ixodes dammini*. *Journal of Experimental Medicine* 161, 332–344.

Ribeiro, J.M.C., Rossignol, P.A. and Spielman, A. (1985b) Salivary gland apyrase determines probing time in anopheline mosquitoes. *Journal of Insect Physiology* 31, 689–692.

Ribeiro, J.M.C., Rossignol, P.A. and Spielman, A. (1986) Blood-finding strategy of a capillary-feeding sandfly, *Lutzomyia longipalpis*. *Comparative Biochemistry and Physiology* 83A, 683–686.

Ribeiro, J.M.C., Modi, G.B. and Tesh, R.B. (1989a) Salivary apyrase activity of some old world phlebotomine sand flies. *Insect Biochemistry* 19, 409–412.

Ribeiro, J.M.C., Vachereau, A., Modi, G.B. and Tesh, R.B. (1989b) A novel vasodilatory peptide from salivary glands of the sand fly *Lutzomyia longipalpis*. *Science* 243, 212–214.

Ribeiro, J.M.C., Marinotti, O. and Gonzales R. (1990a) A salivary vasodilator in the blood-sucking bug, *Rhodnius prolixus*. *British Journal of Pharmacology* 101, 932–936.

Ribeiro, J.M.C., Vaughan, J.A. and Azad, A.F. (1990b) Characterization of the salivary apyrase of three rodent flea species. *Comparative Biochemistry and Physiology* 95B, 215–218.

Ribeiro, J.M.C., Endris, T.M. and Endris, R. (1991) Saliva of the soft tick, *Ornithodoros moubata*, contains anti-platelet and apyrase activities. *Comparative Biochemistry and Physiology* 100A, 109–112.

Ribeiro, J.M.C., Evans, P.M., MacSwain, J.L. and Sauer, J. (1992) *Amblyomma americanum*: Characterization of salivary prostaglandins $E_2$ and $F_2$ by RP-HPLC/bioassay and gas chromatography-mass spectrometry. *Experimental Parasitology* 74, 112–116.

Ribeiro, J.M.C., Hazzard, J.M.H., Nussenzveig, R.H., Champagne, D.E. and Walker, F.A. (1993) Reversible binding of nitric oxide by a salivary heme protein from a bloodsucking insect. *Science* 260, 539–541.

Ribeiro, J.M.C., Nussenzveig, R.H. and Tortorella, G. (1994) Salivary vasodilators of *Aedes triseriatus* and *Anopheles gambiae* (Diptera: Culicidae). *Journal of Medical Entomology* 31, 747–753.

Ribeiro, J.M.C., Schneider, M. and Guimaraea, J.A. (1996) Purification and characterization of prolixin S (nitrophorin 2), the anticoagulant of the blood-sucking bug *Rhodnius prolixus*. *Biochemical Journal*, in press.

Roberts, H.R. and Lozier, J.N. (1991) Other clotting factor deficiencies. In: Hoffman, R., Benz, E.J., Shattil, S.J., Furie, B. and Cohen, H.J. (eds), *Hematology: Basic Principles and Practice*. Churchill Livingstone, New York, pp. 1332–1342.

Rossignol, P.A., Ribeiro, J.M.C. and Spielman, A. (1984) Increased intradermal probing time in sporozoite-infected mosquitoes. *American Journal of Tropical Medicine and Hygiene* 33, 17–20.

Rupin, A., Mennecier, P., De Nanteuil, G., Laubie, M. and Verbeuren, T.J. (1995) A screening procedure to evaluate the anticoagulant activity and kinetic behaviour of direct thrombin inhibitors. *Thrombosis Research* 78, 217–225.

Sarkis, J.J.F., Guimaraes, J.A. and Ribeiro, J.M.C. (1986) Salivary apyrase of *Rhodnius prolixus*-kinetics and purification. *Biochemical Journal* 233, 885–891.

Shattil, S.J., Ginsberg, M.H. and Brugge, J.S. (1994) Adhesive signalling in platelets. *Current Opinion in Cell Biology* 6, 695–704.

Spicer, E.K., Horton, R. and Bloem, L. (1987) Isolation of cDNA clones coding for human tissue factor: primary structure of the protein and cDNA. *Proceedings of the*

National Academy of Science USA 84, 5148–5152.

Stubbs, M.T. and Bode, W. (1994) Coagulation factors and their inhibitors. Current Opinion in Structural Biology 4, 823–892.

Tatchell, R.J. (1967) A modified method for obtaining tick oral secretion. Journal of Parasitology 53, 1106–1107.

Theodos, C.M., Ribeiro, J.M.C. and Titus, R.G. (1991) Analysis of enhancing effect of sand fly saliva on Leishmania infection in mice. Infection and Immunity 59, 1592–1598.

Tiarks, C. (1992) Characterization of a factor VIII immunogenic site using factor VIII synthetic peptide 1687–1695 and rabbit antipeptide antibodies. Thrombosis Research 65, 301–310.

Titus, R.G. and Ribeiro, J.M.C. (1988) Salivary gland lysates of the sand fly Lutzomyia longipalpis enhance Leishmania infectivity. Science 239, 212–214.

Valenzuela, J.G., Walker F.A. and Ribeiro, J.M.C. (1995) A salivary vasodilator from the salivary glands of the bed bug, Cimex lectularis. Journal of Experimental Biology, 198, 1519–1526.

Warburg, A., Saraiva, E., Lanzaro, G.C., Titus, R.G. and Neva, F. (1994) Saliva of Lutzomyia longipalpis sibling species differs in its composition and capacity to enhance leishmaniasis. Philosophical Transactions of the Royal Society of London B 345, 223–230.

Waxman, L. and Connolly, T.M. (1993) Isolation of an inhibitor selective for collagen-stimulated platelet aggregation from the soft tick Ornithodorus moubata. Journal of Biological Chemistry 268, 5445–5449.

Waxman, L., Smith, D.E., Arcuri, K.E. and Vlasuk, G.P. (1990) Tick anticoagulant peptide (TAP) is a novel inhibitor of blood coagulation factor Xa. Science 248, 593–596.

Willems, G. (1992) Linearized model for the initiation of factor Va, and thrombin generation. Hemostasis 21, 248–253.

Zucker, M.B. (1989) Platelet aggregation measured by photometric method. Methods in Enzymology 169, 117–133.

# 5
# Arthropod Modulation of Host Immune Responses

## Stephen K. Wikel, Rangappa N. Ramachandra and Douglas K. Bergman

*Department of Entomology, Oklahoma State University, Stillwater, Oklahoma 74078, USA*

## INTRODUCTION

Haematophagous arthropods are faced with a complex array of host responses that can inhibit bloodmeal acquisition. Haemostatic mechanisms reduce availability of blood for ingestion. Host immune defences (antibodies and specifically sensitized cells) reactive with tissues and saliva can disrupt bloodmeal acquisition, impair physiological responses and/or kill the arthropod (Wikel, 1982a, 1996). The development of arthropod countermeasures to inhibit host haemostasis, inflammation and immune defences has been reported for both fast and slow blood-feeders (Ribeiro, 1987a, 1989; Titus and Ribeiro, 1990; Champagne 1994). Pharmacological properties of arthropod saliva facilitate both bloodmeal acquisition and pathogen transmission. This chapter focuses upon arthropod countermeasures to host immune defences.

Antihaemostatic and vasoactive activities of arthropod saliva are reviewed in this book by Champagne and Titus and elsewhere (Ribeiro, 1987a, 1989; Champagne, 1994). Blood coagulation and immune systems are interconnected. A close association exists between the intrinsic clotting (contact activation) system of plasma and elements of inflammatory/immune responses (Porteu and Nathan, 1990; Kozin and Cochrane, 1992). Activated Hageman factor (XII) initiates conversion of factor XI to XIa and conversion of prekallikrein to kallikrein. Kallikrein acts upon plasminogen to form plasmin and upon kininogens to form bradykinin. In turn, bradykinin

---

© CAB INTERNATIONAL 1996. From Wikel, S.K. (ed.) *The Immunology of Host–Ectoparasitic Arthropod Relationships.* CAB INTERNATIONAL, Wallingford.

stimulates contraction of smooth muscle, increases vascular permeability, causes leukocyte margination and induces pain (Kozin and Cochrane, 1992). Kallikrein is responsible for generation of the anaphylatoxin C5a from complement (Wiggins et al., 1981), contributing to a variety of inflammatory and immune responses.

Duration of feeding and bloodmeal volume vary greatly among haematophagous arthropods (Ribeiro, 1987a). Both fast and slow blood-feeders must deal with the rapid onset of host haemostatic mechanisms. The triatomine bug, *Rhodnius prolixus* may engorge for a period of a few minutes to an hour with its mouthparts in a venule or arteriole of the dermis (Lavoipierre et al., 1959; Friend and Smith, 1971). Mosquito probing and blood ingestion occur in less than ten minutes (Ribeiro, 1987a). These rapid rates of engorgement are in sharp contrast to ixodid ticks which require days of continuous feeding to obtain a bloodmeal (Sonenshine, 1991). In addition to host haemostasis, variations in the duration of engorgement pose potentially different problems in dealing with host inflammatory responses and immune defences.

The complexity of interactions at the ectoparasitic arthropod–host interface is increased when arthropod-borne pathogens are added to the equation. Host immune responses to infestation are often the same ones stimulated by arthropod-transmitted disease-causing agents (Playfair, 1993; Wikel, 1996). Arthropod factors influence both infection and the resulting immune responses. Saliva of the sandfly, *Lutzomyia longipalpis*, enhanced infectivity of *Leishmania major*, and may be essential for initiation of vertebrate host infection (Titus and Ribeiro, 1990). However, the saliva of sibling species of *L. longipalpis* differed in its pharmacological properties and ability to promote *Leishmania* infections (Warburg et al., 1994). Costa Rican *L. longipalpis* saliva contained low vasodilatory activity and very little of an erythema-inducing peptide, maxadilan, but it strongly enhanced cutaneous proliferation of *Leishmania*. Sandflies that vector visceral leishmaniasis had greater amounts of erythema-inducing peptide, but they were less potent in enhancing cutaneous disease.

The importance of the arthropod vector is further illustrated by differences in immune responses elicited by a tick-borne pathogen following arthropod transmission versus needle inoculation. Antibody responses of hamsters following tick transmission of *Borrelia burgdorferi* differed from that induced by needle inoculation of cultured spirochaetes (Roehrig et al., 1992). Tick-transmitted infection resulted in a lack of production of antibodies to outer surface proteins A and B, similar to that observed in natural infections of humans. Mice infected with *B. burgdorferi* by *Ixodes ricinus* infestation were more infective for a subsequent tick challenge than mice infected by subcutaneous inoculation (Gern et al., 1993). These studies clearly show that ectoparasitic arthropods contribute to the success of infestation and pathogen transmission. They are not merely walking or flying hypodermic needles and syringes. Therefore, ectoparasite modulation of host immune responses should be a central consideration in the analysis of arthropod–host–pathogen interactions.

## HOST IMMUNITY TO ECTOPARASITIC ARTHROPODS

A variety of host immune responses are stimulated by infestation with ectoparasitic arthropods (Willadsen, 1980; Wikel, 1982a, 1996; Brossard *et al.*, 1991). Response patterns and their contributions to the pathophysiology of infestation depend upon the host–arthropod association, history of prior exposure, intensity of infestation, host physiology, and immune response capabilities of the infested animal. An understanding of immune responses to infestation is essential to an understanding of arthropod strategies for modulation of host immune defences. Therefore, a brief overview of immunity to arthropod infestation is presented here prior to an in-depth examination of ectoparasite modulation of host immune defences. Chapters describing immunity to a specific arthropod should be consulted for a detailed analysis of host immune responses to infestation.

There is limited information is regarding host immune reactivity to many ectoparasitic arthropods. Mice developed immunological based resistance to the sucking louse, *Polyplax serrata*, which was characterized by louse antigen induced immediate and delayed skin test hypersensitivity and *in vitro* proliferation of lymph node cells (Ratzlaff and Wikel, 1990). Infestation with triatomine bugs stimulated antibody responses (Pinnas *et al.*, 1986). Mosquito bites elicited salivary-gland-specific IgE and IgG antibody production (Shen *et al.*, 1989; Konishi, 1990) and delayed cutaneous hypersensitivity (Wilson and Clements, 1965), indicating stimulation of both B- and T-lymphocyte effector pathways. Mice bitten by the blackfly, *Simulium vittatum*, developed antibodies cross-reactive with other *Simulium* species (Cross *et al.*, 1993b). Antibodies that reacted with pharmacologically active molecules in saliva did not develop. *Sarcoptes scabiei* induced antibodies to a variety of mite proteins/polypeptides (Arlian, 1989). Histopathological studies of *S. scabiei* lesions revealed a predominance of T lymphocytes over B lymphocytes infiltrating lesions (Arlian, 1989). Infestation with the mange mite, *Psoroptes cuniculi*, induced an intense antibody response reactive with a variety of mite proteins/polypeptides (Wikel, 1989). However, questions remain as to how host immune responses affect the ectoparasitic arthropod, the nature of immunogens involved and the impact of host immunity upon pathogen transmission.

Immunity to ticks has been the most extensively studied host–arthropod association (Willadsen, 1980; Wikel, 1982a, 1996; Brossard *et al.*, 1991). Tick feeding stimulated host immune regulatory and effector responses involving antigen-presenting cells, cytokines, T lymphocytes, complement, homocytotropic antibodies and circulating immunoglobulins. Acquired resistance to ticks is expressed as reduced engorgement weight, prolonged feeding, impaired production of ova, fewer viable ova, prevention of moulting and death of the feeding tick.

Cutaneous reactions at tick attachment sites on cattle and laboratory animals expressing acquired resistance were characterized by infiltrates of basophils and eosinophils (Allen, 1973; Allen et al., 1977; Brossard and Fivaz, 1982). The basophil infiltrate was a cutaneous basophil hypersensitivity response (Allen, 1973), which is a form of delayed type hypersensitivity (Dvorak et al., 1970) mediated by $T_H 1$ lymphocytes (Mossman and Coffman, 1989).

Infestation induced formation of tick salivary gland reactive homocytotropic antibodies, those immunoglobulins which bound to Fc receptors on basophils and mast cells (Brossard and Girardin, 1979). Tick feeding over a period of days resulted in essentially continuous introduction of saliva into the bite-site with the potential for homocytotropic antibody-mediated degranulation of mast cells and infiltrating basophils. Basophils degranulated more readily during infestations of previously exposed animals (Brossard et al., 1982). In addition, histamine, leukotrienes, prostaglandins, eosinophil major basic protein, enzymes and other bioactive molecules derived from basophils and eosinophils likely contribute to the expression of acquired resistance to ticks (Wikel, 1996). However, additional research is needed to determine the importance of these molecules in acquired resistance.

Langerhans cells are the primary antigen presenting cells of the skin (Salmon et al., 1994). Langerhans cells trap tick salivary gland antigens in the skin and function as antigen-presenting cells (Nithiuthai and Allen, 1984a, 1985). Depletion of Langerhans cells by ultraviolet radiation prior to an initial tick infestation limited to the treated site reduced the acquisition of resistance, and initial treatment prior to a second infestation reduced the expression of resistance (Nithiuthai and Allen, 1984b). The production of cytokines by Langerhans cells during primary and secondary tick infestations has not been determined. Langerhans cells are macrophage-like cells important in the acquisition and expression of anti-tick immunity.

Skin and lymph nodes of BALB/c mice infested three successive times with nymphal *I. ricinus* were examined by *in situ* hybridization for interleukin-2 (IL-2), interleukin-4 (IL-4) and interferon-gamma (IFN-$\gamma$) mRNAs (Mbow et al., 1994). Lymph nodes obtained during primary and repeated infestations had stronger signals for IL-2 and IFN-$\gamma$ with consistently faint positive reactivity for IL-4. These findings indicate the predominance of a $T_H 1$-lymphocyte response to infestation.

In addition to homocytotropic antibody production, circulating immunoglobulins reactive with salivary gland proteins/polypeptides were induced by infestation (Brossard and Girardin, 1979; Willadsen, 1980; Wikel, 1982a, 1996). Tick salivary gland proteins changed during feeding (McSwain et al., 1982), a finding that indicated a potentially varying population of immunogens presented to the host immune system. Tick reactive antibody was involved in expression of acquired resistance (Willadsen, 1980; Wikel, 1982a, 1996; Brown, 1985; Whelen and Wikel, 1993). However, immunogens and specific roles of antibodies in acquisition and expression of resistance remain to be fully characterized.

Complement is a group of serum proteins involved in innate and acquired immune defences (Goldstein, 1988). Antigen–antibody complexes of an appropriate immunoglobulin class (IgM or IgG) induce a series of molecular interactions referred to as the classical pathway of complement activation. The alternative pathway of complement activation does not require antigen–antibody complexes. Expression of acquired resistance was reduced in the absence of an intact alternative pathway of complement activation (Wikel, 1979). Tick salivary gland immunogens, IgG and complement were localized at the dermal–epidermal junctions of bite sites on resistant animals (Allen *et al.*, 1979). A role has yet to be established in acquired resistance to ticks for the classical pathway of complement activation. Complement-derived anaphylatoxins C3a and C5a cause smooth muscle contraction and the release of histamine and other mediators from mast cells and basophils (Goldstein, 1988). C5a is chemotactic for neutrophils, monocytes/macrophages and basophils (Ward *et al.*, 1975; Goldstein, 1988). Anaphylatoxins can contribute to the resistance response by: (i) stimulating the release of biologically active molecules which cause the accumulation of cells; (ii) increasing vascular permeability allowing the egress of immunoglobulins and other molecules into the bite site; and (iii) direct action of basophil and mast cell mediators on the tick.

Lymphocytes collected from tick infested laboratory animals (Wikel *et al.*, 1978; Schorderet and Brossard, 1993) and cattle (Wikel and Osburn, 1982; George *et al.*, 1985) proliferated *in vitro* with the addition of tick salivary gland extracts and proliferative responses were generally more intense with repeated exposures. Specific lymphocyte populations reacting were not identified.

Both fast and slow blood-feeding arthropods are affected by host immune responses to infestation. One can visualize how a tick feeding over a period of days can be the target of immune defences. How can an arthropod feeding for minutes be influenced by host immunity? The number of fast feeding arthropods attacking a host over time can be very high. Each repeated feeding potentially stimulates the host immune system. Subsequent infestations are thus susceptible to antibodies and other immune effectors stimulated by previous exposures. Rapid blood feeders can be viewed as a 'long-term' population stimulating and interacting with host immune defences.

This general overview of host immune responses to ectoparasites provides the framework for understanding why certain immune regulatory and effector pathways are modulated by infesting arthropods. As new information emerges about host immune responses to infestation, the analysis of arthropod countermeasures and development of novel control methods for arthropods and vector-borne pathogens will be advanced.

# HOST IMMUNE FUNCTION AND THE SURVIVAL OF PARASITES

Disease causing agents have developed strategies for coping effectively with the complex challenges presented by host defences. General survival schemes include: (ii) rapid replication and/or mutation; and (ii) inhibition of the development and expression of host immunity (Dessaint and Capron, 1993). A vast amount of attention has been focused upon endoparasitic organisms, and many of the findings pertaining to modulation of host defences may apply to ectoparasitic arthropod–host associations. Many parasitic infections persist for months to years in the presence of a host immune response. Host defences against certain bacterial and viral pathogens result in sterilizing immunity, but most parasitic infections induce non-sterilizing immunity that provides a balance between parasite and host (Dessaint and Capron, 1993). With non-sterilizing immunity, parasite numbers and host pathology might be reduced while host defences are weakened, adaptations beneficial to both parties. The parasite must not damage the host to the point where it cannot survive. This dynamic equilibrium represents a balanced host–parasite relationship. Ecto-parasitic arthropods appear to have achieved similar equilibria with the defences of their hosts (Wikel et al., 1994). Exposure to fast and/or slow feeding arthropods over months and years can be compared to chronic infections with endoparasites.

Multiple mechanisms have emerged to facilitate the survival of endoparasites in the immunocompetent host (Mims, 1987; Sher, 1988; Dessaint and Capron, 1993; Marrack and Kappler, 1994). Immunomodulation strategies utilized include: (i) induction of tolerance to parasite antigens; (ii) residence of parasites in cysts or tissues that make them inaccessible to immune attack; (iii) variation of surface antigens; (iv) suppression of immune regulatory and effector pathways by release of immunomodulatory molecules; (v) molecular mimicry of host antigens; (vi) rapid replacement of parasite immunogens; (vii) masking of the parasite surface with host molecules; and, (viii) diversion of the host immune response to antigens not essential for parasite survival. Both non-specific and specific host defences need to be suppressed. Non-specific defences such as natural killer (NK) cells (Trinchieri, 1989) and complement (Goldstein, 1988) are rapidly inducible, while specific antibody and T-lymphocyte responses require days to weeks to develop (Goodman, 1991). Particular attention has been focused in recent years upon the modulation of host cytokine networks by infectious agents (Marrack and Kappler, 1994).

Parasites elicit host antibody and cell-mediated immune responses reactive with a variety of immunogens (Sher and Scott, 1993). These complex and intense responses often do not eliminate the parasite. Although host immunity can reduce the parasite burden (Sher and Scott, 1993), many of the responses are directed or 'diverted' towards antigens that have little to do with parasite survival (Sher, 1988; Dessaint and Capron, 1993). Attention should be focused upon those immune response pathways suppressed by parasites, since their full expression would likely be deleterious to the parasite (Sher, 1988; Wikel et al., 1994). Blockage of parasite-induced host immunosuppression should be considered as an approach to vaccination-induced protection against parasites. Identification of parasite-derived molecules responsible for the modulation of host responses and development of strategies for their use in vaccines is a fertile area of investigation.

## ARTHROPOD MODULATION OF HOST IMMUNE RESPONSES

Blood-feeding arthropods modulate the host immune responses that impair their ability to obtain a bloodmeal, which in turn could facilitate transmission of arthropod-borne disease-causing agents (Wikel et al., 1994; Wikel, 1996). Common elements of the host immune system are involved in defences against both arthropods and vector-borne pathogens (Playfair, 1993; Wikel, 1996). Both fast and slow blood-feeders can modulate host immunity. The most extensive body of information on this topic involves tick–host associations (Wikel et al., 1994). However, ticks and other blood-feeders will be considered.

Blackfly feeding elicits host antibodies reactive with salivary gland components (Cross et al., 1993b). When BALB/c mice were immunized with *S. vittatum* salivary gland extract or allowed to be bitten by flies (Cross et al., 1993b), antibodies induced were of the IgM, IgG and IgE isotypes. Immunoblot analyses revealed antibodies reactive with proteins/polypeptides in the 26 to 67 kilodalton (kDa) molecular mass range. Sera collected from mice bitten by *S. vittatum* contained fewer salivary-gland-reactive antibodies than those animals immunized with salivary gland extracts and variable levels of immunoblot cross-reactivity were detected for the *Simulium* sp. salivary gland extracts tested. Further, mice did not produce antibodies reactive with recognized pharmacologically active components in blackfly saliva (Cross et al., 1993b). The absence of host antibodies to pharmacologically active molecules in fly saliva would facilitate bloodmeal acquisition. Do the antibodies reactive with blackfly saliva alter feeding, fecundity or other physiological processes?

In addition, *Simulium vittatum* salivary gland extract has been shown to be immunosuppressive (Cross et al., 1993a). Influence of salivary gland extract on the afferent limb of the immune response was tested *in vivo* and *in vitro* by assessing expression of major histocompatibility complex (MHC) class II by lymphomyeloid and epidermal cells. In addition, the influence of salivary gland extract on *in vitro* responsiveness of mononuclear splenocytes to the T-lymphocyte mitogen concanavalin A (Con A) and the B-lymphocyte mitogen *Salmonella typhosa* lipopolysaccharide (StLPS) was also tested. Mice used for both MHC class II and mitogen responsiveness studies were inoculated with blackfly salivary gland extract or placebo prior to collection of cells, and harvested cells were then cultured *in vitro* in the presence of the same salivary gland extracts or placebo.

The effect of salivary gland extract on the effector arm of the immune response was evaluated by immunization of mice with sheep red blood cells (SRBC) co-administered with blackfly salivary gland extract or placebo protein. *In vivo* inoculation of blackfly salivary gland extract reduced the percentage of MHC class II positive cells in the spleen, but had no effect on lymph node and epidermal cells (Cross *et al.*, 1993a). This observation suggests that the reduced capacity of spleen cells to present antigen modulates the development of an immune response. Splenic mononuclear cells were not altered in their ability to respond to Con A or StLPS by *in vivo* inoculation of blackfly salivary gland extract (Cross *et al.*, 1993a). However, *in vitro* exposure of the same cell population to 5 µg ml$^{-1}$ of salivary gland extract reduced Con A responsiveness by 50 to 60% and reactivity to StLPS by 49 to 58%. Salivary gland extracts were not cytotoxic so the reduction was not due to the death of cells. Therefore, blackfly salivary-gland-derived molecules are suppressive *in vitro* for both B and T lymphocytes, which could modulate host ability to mount antibody and cell-mediated immune responses. Inoculation of blackfly salivary gland extract with SRBC consistently enhanced anti-SRBC antibody responses (Cross *et al.*, 1993a).

The saliva of the sand fly *L. longipalpis* has an established role in enhancing infectivity of *Leishmania major* (Titus and Ribeiro, 1990). Salivary gland extract of *L. longipalpis* inhibited macrophage presentation of *L. major* antigens to parasite-specific T lymphocytes (Theodos and Titus, 1993). Adherent peritoneal cells (macrophages) were collected from *L. major* resistant C3H/H and CBA/T6J mice and susceptible BALB/c mice. Syngeneic antigen-presenting cells were cultured with stationary phase *L. major* promastigotes and parasite specific T lymphocytes in the presence or absence of sand fly salivary gland extract. Ability of parasite-infected macrophages to present antigen to *L. major* specific T cells was assessed by lymphocyte incorporation of radiolabel. Presence of sand fly salivary gland extract decreased T-lymphocyte proliferative response for adherent cells from all mouse strains tested. The mechanism by which sand fly saliva modulates adherent cell (macrophage) function is unknown. Impaired macrophage function at the levels of either antigen processing and presentation or cytokine elaboration could impact antibody and cell-mediated immune defences.

Salivary gland extracts prepared from female *Aedes aegypti* inhibited release of the pro-inflammatory cytokine TNF from rat mast cells, but not antigen-induced release of histamine (Bissonnette *et al.*, 1993). Inhibitory activity was detected in an ultrafiltration retentate of 10 kDa and greater, and the ability to inhibit TNF release was destroyed by boiling for ten minutes or heating to 56°C for 30 minutes. The inhibitor of TNF release from mast cells was found only in female mosquitoes, suggesting a specific role in blood-feeding. Diminished TNF levels would decrease the pro-inflammatory effects of this cytokine at the feeding site (Bissonnette et al., 1993).

*Sarcoptes scabiei* is an important ectoparasitic mite of humans and other animal species. Patients infested with *S. scabiei* had lower levels of circulating IgA than uninfested individuals (Hancock et al., 1974). This observation was confirmed by the finding that patients had significantly lower serum IgA concentrations during active infestation than six to nine months later after successful treatment (Falk, 1980, 1981). However, a role was not established for IgA in limiting infestation or contributing to the pathophysiology of disease.

Infestation of canines with the follicle mite, *Demodex canis*, resulted in reduced *in vitro* responsiveness of T lymphocytes to the mitogen phytohae-magglutinin (PHA) and the B- and T-lymphocyte polyclonal activator pokeweed mitogen (PWM) (Scott et al., 1974). Lymphocytes obtained from dogs with chronic generalized demodectic mange were 30% less responsive to PHA than cells from normal controls, and the magnitude of suppression correlated directly with intensity of infestation (Corbett et al., 1975). Sera collected from infested dogs significantly reduced *in vitro* proliferation of normal canine lymphocytes cultured in the presence of PHA. Therefore, *D. canis* infestation appeared to modulate T-lymphocyte proliferative capacity, as measured by altered PHA responsiveness. However, the specific targets in terms of T-lymphocyte subpopulations, B lymphocytes and other cells involved in immune regulatory and effector functions remain to be defined. A role has not yet been established for T lymphocytes in the control of mite infestation. Suppressed T-lymphocyte functions can affect both antibody production and cell-mediated immunity.

## Tick Modulation of Host Immune Function

The days required for an ixodid tick to obtain a bloodmeal provide ample opportunity for interaction with host immune defences. Ticks developed countermeasures to many of the pathways involved in resistance to infestation. Tick modulation targeted both 'pre-programmed' and specific immune defences of the host (Wikel et al., 1994; Wikel, 1996). This strategy facilitates bloodmeal acquisition in the presence of both the rapidly activated, 'pre-programmed' defences, such as the alternative pathway of complement and natural killer cells, and the specific antibody and cell-mediated responses which, because of clonal expansion, require days to develop.

The alternative pathway of complement activation and natural killer cells can be immediately activated by a variety of different molecules, providing a potentially broad spectrum of protection. Tick salivary-gland-derived factors modulate complement activation, natural killer function, antibody production, T-lymphocyte proliferative responses, and cytokine elaboration by antigen-presenting cells and T lymphocytes. In this section, each of the reported host immunosuppressive properties associated with ticks is described and a model is developed to show the interrelationships of recognized host defences and tick immunomodulatory factors.

The conspecificity of *Ixodes dammini* and *Ixodes scapularis* has been established (Oliver et al., 1993). Each paper cited in this chapter will refer to the species designation given the manuscript being discussed. The alternative pathway of complement activation is a natural defence mechanism activated by a variety of bacteria, viruses and other factors in the absence of immunoglobulin (Muller-Eberhard, 1988). The specific nature of activating molecules is unknown; however, recognition and binding of C3b is essential for pathway initiation (Muller-Eberhard, 1988). The alternative pathway of complement was shown to be involved in the expression of acquired resistance to ticks (Wikel, 1979). *Ixodes dammini* saliva inhibited activation of the alternative complement pathway (Ribeiro, 1987b) and antagonized complement generated anaphylatoxin (Ribeiro and Spielman, 1986). Anti-complement activity of *I. dammini* saliva inhibited alternative pathway deposition of C3b and C5b on an activating surface, and release of the anaphylatoxin C3a (Ribeiro, 1987b). This saliva factor blocked activation of the alternative complement pathway of human, rat, mouse, hamster and guinea pig sera. Fractionation of *I. dammini* saliva by gel permeation chromatography resulted in identification of a single peak of inhibitory activity with a molecular weight of 49 kDa (Ribeiro, 1987b). The inhibitor prevented C3 hydrolysis after binding to an activator surface, but the mechanism of action was unknown. The effect of this inactivator on the classical pathway of complement activation was not described.

In addition to blocking activation of the alternative complement pathway, *I. dammini* saliva contained an anaphylatoxin inhibitor, which was likely a carboxypeptidase-N-like enzyme (Ribeiro and Spielman, 1986). Tick saliva blocked both anaphylatoxin-induced contraction of guinea pig ileum and increased vascular permeability. In addition, bradykinin activity was inhibited by incubation with *I. dammini* saliva.

The inhibition of both the alternative pathway activation and anaphylatoxins would reduce inflammatory and immune responses. The biological activities of C3b-derived molecules and the anaphylatoxins C3a and C5a were described by Goldstein (1988). Blocking the activity of C3b could prevent the subsequent release of active fragments of complement components involved in the assembly of the alternative pathway, such as C5a and the membrane attack complex, C5b–C9, responsible for formation of transmembrane pores. Inactivation of anaphylatoxin activity inhibits the recognized biological activities of these molecules which include: (i) smooth muscle contraction; (ii) histamine release from mast cells and basophils; (iii) increased permeability of small blood vessels; and (iv) C5a-mediated chemotaxis of neutrophils, eosinophils, basophils and monocytes. These activities could contribute to lesion formation at the tick bite-site and potentially modify pathogen transmission and development. By an undescribed mechanism, histamine inhibited the ability of *Dermacentor andersoni* to salivate and engorge (Paine *et al.*, 1983). Reduction in the amount of anaphylatoxin would reduce degranulation of resident cutaneous mast cells and the large number of infiltrating basophils observed at tick attachment sites on hosts expressing acquired resistance. Inhibition of these selected aspects of complement activation provides a more favourable environment for the tick to feed.

Homocytotropic and circulating antibodies reactive with tick salivary-gland-derived molecules were induced during infestation, and immunoglobulins were reported to contribute to the expression of resistance (Brossard and Girardin, 1979; Willadsen, 1980, Wikel, 1982a, 1996; Brown, 1985). The inhibition of host antibody responses provides a survival advantage by reducing direct damage caused by antibody binding to tick antigens and the subsequent activation of the classical complement pathway or possibly by antibody-dependent cellular cytotoxicity.

Guinea pigs infested twice with *D. andersoni* larvae were tested for their ability to develop a primary, IgM, antibody response to sheep red blood cells (SRBC), a thymic-dependent antigen (Wikel, 1985). Antibody production was determined on: the fifth day of a first infestation, the second day of a second exposure, the fifth day of a second exposure, and four days after termination of the second infestation. Generation of SRBC-specific immunoglobulins was assessed by determination of the number of antibody-producing spleen cells by using the highly sensitive direct haemolytic plaque-forming cell assay. SRBC-specific antibodies were significantly reduced for tick infested animals, when compared with normal controls. The SRBC antibody responses of tick exposed animals returned to normal by four days after termination of the second infestation. Maximum suppression of host antibody responsiveness to SRBC was 41.6% on day two of the second infestation. This significant reduction in antibody production did not leave the host without immunoglobulin defences. Tick-induced host immunosuppression appears to achieve a balance between reducing immunological responses that impair ixodid feeding and survival of the host in an environment filled with challenges requiring some level of host immunocompetence. In addition, suppression of immunoglobulin responses could facilitate establishment of tick-borne pathogens within the host.

Infestation of rabbits with 20 pairs of adult *Rhipicephalus appendiculatus* suppressed their ability to generate antibodies to bovine serum albumin (BSA), administered during the peak period of tick feeding (Fivaz, 1989). Rabbits were immunized with 10 mg of BSA in Freund's complete adjuvant by intramuscular injection at the time when female *R. appendiculatus* were at least three-quarters engorged. Tick infestation was terminated at approximately day 17 post-immunization. Mean anti-BSA titre of sera from infested rabbits was 1:64, while mean titre of uninfested control rabbit sera was 1:2000. Immunosuppression was attributed to lymphocytotoxicity of salivary gland extracts.

This study (Fivaz, 1989) was the only report of tick-induced immunosuppression due to lymphocytotoxicity of tick salivary gland extracts. Contrariwise, *D. andersoni* salivary gland extract suppression of host immune function was not due to cytotoxicity of tick salivary gland extracts for macrophages and lymphocytes obtained from laboratory animals or bovines (Wikel and Osburn, 1982; Ramachandra and Wikel, 1992, 1995). The mechanism of salivary-gland-extract-mediated immunosuppression is unknown at this time.

The mitogen Con A was a substitute for simulation of T lymphocytes by a specific immunogen associated with a major histocompatibility complex molecule on the surface of an antigen-presenting cell, such as a macrophage or macrophage-like cell (Sharon, 1983). The polyclonal activating mitogen acted in part by binding the T-lymphocyte receptor for specific antigen (Kanellopoulos *et al.*, 1985). A mitogen activates many lymphocyte clones and is referred to as a polyclonal activator, while an immunogen activates only those clones that have specific receptors capable of binding antigenic determinants of the immunogen.

Lymphocytes obtained from guinea pigs on the fifth day of either a first or second tick infestation with two female and one male *D. andersoni* were assessed for their ability to proliferate *in vitro* when cultured with the B-lymphocyte polyclonal activator *Escherichia coli* lipopolysaccharide (LPS) or the T-lymphocyte mitogens PHA and Con A (Wikel, 1982b). *In vitro* responsiveness to PHA and Con A were reduced by the fifth day of both infestations, while proliferation in the presence of LPS was not altered. Suppression of T-lymphocyte responses to Con A was greater during the initial infestation (79%) than on the fifth day of a second infestation (49%). Proliferation induced by PHA was similarly reduced by 61% and 50% respectively. Tick infestation selectively reduced T-lymphocyte responses to mitogens. Why was the reduction less intense during the second infestation? One explanation could be that antibodies induced by the primary tick feeding neutralized immunosuppressive factors introduced during the second infestation; therefore, the greater intensity of T-lymphocyte responses to mitogens observed during the second exposure.

*In vitro* responses of rabbit lymphocytes to Con A were also depressed when cells were collected during infestation with *I. ricinus* adults (Schorderet and Brossard, 1994). Concanavalin A responses returned to normal levels after termination of each infestation. The lowest responses to Con A occurred during the third infestation. Lymphocytes collected from rabbits given high and low levels of tick infestations responded in an antigen-specific manner to

*I. ricinus* salivary gland and integumental extracts (Schorderet and Brossard, 1994). During each tick feeding, the high infestation group was exposed to 25 female and 25 male *I. ricinus* and the low infestation group was infested with five females and five males. Lymphocyte proliferative responses to salivary gland extract (SGE) remained significant from the first exposure through day 51 for the high infestation group and day 57 for the low infestation rabbits. No significant proliferation in the presence of SGE was observed during the third exposure to ticks for lymphocytes obtained from the high infestation group. Significant proliferative responses occurred during the third exposure to ticks for the low infestation group. *In vitro* lymphocyte proliferative responses to integumental extract were significantly lower than those to SGE for both treatment groups. The reduced reactivity to SGE, integumental extract and Con A during the third infestation was attributed to tick induced immunosuppression (Schorderet and Brossard, 1994).

Saliva of *I. dammini* significantly inhibited normal murine spleen cells from proliferating *in vitro* in the presence of Con A or PHA (Urioste et al., 1994). Pre-incubation of lymphocytes was not essential for the inhibition of mitogen responsiveness. Addition of mitogen and a 1:100 dilution of *I. dammini* saliva at the same time was as effective in reducing mitogen responsiveness as a two hour incubation with the same dilution of saliva prior to the addition of mitogen. Tick saliva was not cytotoxic to murine splenocytes. This suppression was attributed to either a direct effect of tick saliva upon T lymphocytes and/or cytokines produced by those cells (Urioste et al., 1994). Suppression of lymphocyte responsiveness to Con A and PHA was due to a protein with a molecular mass of 5 kDa or greater (Urioste et al., 1994).

What effect do salivary gland extracts prepared during the course of feeding have on B and T lymphocyte *in vitro* proliferative responses induced by mitogens? Salivary gland extracts were prepared from female *D. andersoni* daily during the course of engorgement and tested for their ability to modify normal mouse splenocyte *in vitro* responsiveness to Con A or LPS (Ramachandra and Wikel, 1992). Salivary gland extract from unfed ticks was not inhibitory for Con-A-induced proliferation, but SGEs prepared on days one to nine of engorgement suppressed Con A responses by 33.3 to 66.4% for the 0.5 µg group and by 43.2 to 92.6% for the 1.0 µg group. Pre-incubation of splenocytes with the same SGEs enhanced splenocyte responsiveness to LPS for the 0.5 µg treatment group from 6.3% for day one SGE to 59.0% for day nine SGE (Ramachandra and Wikel, 1992). The 1.0 µg treatment group responses to LPS were depressed up to and including day three and enhanced from days four to nine. Suppression of mitogen responses was not due to cytotoxicity of SGEs for splenocytes. Suppression of Con A and enhancement of LPS responses were not due to shifts in the dose–response curves for optimum concentrations of mitogens.

The effects of tick infestation on lymphocyte proliferation were assessed.

Purebred *Bos taurus* cows and calves were given up to four 10 day infestations with ten female and five male *D. andersoni* (Wikel and Osburn, 1982). Each subsequent infestation was separated by 10 days. Peripheral blood lymphocytes isolated at the termination of each infestation were tested for their ability to proliferate *in vitro* in the presence of an SGE prepared from female *D. andersoni* on the fourth day of engorgement or upon exposure to PHA. Antigen-specific lymphocyte responses to SGE by infested bovines were detected at the end of both third and fourth infestations. Peripheral blood lymphocytes of cows and calves receiving third or fourth infestations were significantly suppressed in their ability to proliferate *in vitro* in the presence of the T-lymphocyte mitogen PHA by a maximum of 47% (Wikel and Osburn, 1982). Low-level infestation of *B. taurus* cows and calves impaired T-lymphocyte responses to the polyclonal activator PHA, while specific reactivity with salivary-gland-derived molecules was evident. Therefore, the suppression induced by tick feeding appears to be a balance between reduced responsiveness and anti-parasite defences.

Purebred *B. taurus* infested with *Boophilus microplus* displayed a marginal decrease in the number of peripheral blood T lymphocytes, beginning with a second experimental infestation and lasting until the end of a fourth, final, exposure (Inokuma *et al.*, 1993). *In vitro* responses of peripheral blood lymphocytes to PHA were consistently less than those of similar cells from uninfested, control, cattle from the second infestation to the fourth infestation. Saliva induced by pilocarpine application to *B. microplus* inhibited the *in vitro* proliferation of normal bovine peripheral blood lymphocyte to PHA by 44 to 49%. The protein concentration of saliva was 73 µg ml$^{-1}$. Pilocarpine was not inhibitory to lymphocyte proliferation at the concentration found in the saliva tested. *Boophilus microplus* SGE at 50 µg ml$^{-1}$ inhibited PHA-induced blastogenesis of normal bovine lymphocytes by 32 to 46%. Lower concentrations of SGE were not inhibitory. The suppressive effects of *B. microplus* infestation were expressed during adult feeding and for a continued period after termination of tick exposure. The inhibitory factor was not identified, but it was speculated that a suppressive molecule other than prostaglandin $E_2$ (PGE$_2$) might be present.

Suppression of normal bovine lymphocyte *in vitro* responsiveness to PHA when cultured in the presence of *B. microplus* saliva was subsequently attributed to PGE$_2$ (Inokuma *et al.*, 1994). Earlier reports stated that PGE$_2$ in tick saliva might be an immunosuppressant (Ribeiro, 1987a; Ribeiro *et al.*, 1985). *Boophilus microplus* saliva was collected into capillary tubes placed over the mouthparts of partially engorged females at 35°C for 16 hours in the dark (Inokuma *et al.*, 1994). Capillary tubes contained heparinized bovine blood, plasma or washed red blood cells at 40% v/v in plasma. A second method of saliva collection involved membrane feeding of 35 newly moulted adult *B. microplus* feeding on bovine plasma for 24 hours at 35°C (Inokuma *et al.*, 1994). Fluids remaining in the capillary tubes or chambers were assayed for $PGE_2$ by radioimmunoassay at the end of feeding. Saliva used to suppress normal lymphocyte responses to PHA contained 33 ng $PGE_2$ and 1.30 mg protein $l^{-1}$. The lymphocyte culture medium contained 10% heat-inactivated fetal calf serum. The primary half-life of $PGE_2$ in blood was reported to be less than one minute (Trang, 1980). Contribution of saliva proteins to the observed immunosuppression cannot be ruled out, particularly in light of the identification of an immunosuppressive protein in the saliva of *I. dammini* (Urioste *et al.*, 1994). Bergman *et al.* (1995) described a protein in *D. andersoni* salivary glands that inhibited *in vitro* proliferation of T lymphocytes induced by Con A. In addition, the immunosuppressive effect of standard $PGE_2$ was less than that of *B. microplus* saliva (Inokuma *et al.*, 1994). Saliva prostaglandins and proteins could both contribute to tick modulation of host immunocompetence, and differences in immunosuppressants might occur among the ixodid species. The relative importance of $PGE_2$ in modulation of host immune function *in vitro* needs to be determined by comparison of the immunosuppressive properties of untreated tick saliva and saliva depleted of $PGE_2$ by immmunoprecipitation.

Peripheral blood lymphocytes from purebred *B. indicus* and purebred *B. taurus* were suppressed in their *in vitro* responsiveness to Con A and enhanced in their reactivity to LPS when cultured with *D. andersoni* SGEs prepared from females daily during the course of engorgement (Ramachandra and Wikel, 1995). The magnitude of SGE-induced suppression of Con A responsiveness was similar for lymphocytes from either *B. indicus* or *B. taurus*. Suppression of Con A responses at the 0.5 µg per well concentration ranged from 11.0 to 68.8% for *B. indicus* and 14.4 to 64.2% for *B. taurus*. An examination of counts per minute of radiolabel incorporated into cultured lymphocytes revealed a marked difference between cattle of these two genetic backgrounds. Lymphocytes derived from *B. indicus* were a mean 34.5% more reactive than similar cells of *B. taurus* origin. *Dermacentor andersoni* SGEs enhanced responsiveness of B lymphocytes to LPS by 8.6 to 249.5% for cells from *B. indicus* and by 20.0 to 490.8% for cells of *B. taurus* origin (Ramachandra and Wikel, 1995). *Bos indicus* lymphocyte responsiveness to LPS *in vitro* in the absence of SGEs was 42.9% greater than similar cells of *B. taurus* origin. *Bos indicus* appeared to be able to develop more vigorous immune responses in the presence of tick SGEs than did *B. taurus*, possibly explaining in part the reported 'innate' resistance of cattle of *B. indicus* genetic composition to ticks (Ramachandra and Wikel, 1995).

Immunoregulatory pathways involve a complex network of cytokine interactions (Kroemer *et al.*, 1993). Pathogen-mediated modulation of cytokine networks arose to facilitate survival in the presence of host immune defences (Marrack and Kappler, 1994). Macrophage and T-lymphocyte functions are linked through a variety of cytokines essential for macrophage activation and the regulation of growth and differentiation of B and T lymphocytes (Vilcek and Le, 1994). Macrophage-elaborated interleukin-1

(IL-1) is primarily an inflammatory cytokine of importance in pathogenesis of many infectious diseases (Dinarello, 1994). Interleukin-1 and tumour necrosis factor alpha (TNF) are closely related with TNF involved in mediating disease states, the activation of natural killer (NK) cells and activation of cellular cytotoxicity (Tracey, 1994).

Murine T lymphocytes are divided into $T_H1$ and $T_H2$ subpopulations based upon their cytokine elaboration profiles (Mossman and Coffman, 1989). The $T_H1$ lymphocytes secrete interleukin-2 (IL-2) and interferon-gamma (IFN-G) that mediate delayed type hypersensitivity and inflammatory responses. $T_H2$ cells elaborate interleukin-4 (IL-4), interleukin-5 (IL-5) and interleukin-10 (IL-10) involved in B-cell differentiation and antibody production. $T_H2$ production of IL-10 inhibits synthesis of $T_H1$ cytokines and IL-4 inhibits macrophage activation by IFN-$\gamma$, while IFN-$\gamma$ in turn prevents proliferation of $T_H2$ cells (Mossman and Moore, 1991; DeMaeyer and De Maeyer-Guignard, 1994). Pathogens rarely directly modulate IL-2, rather they cause changes in the ratio of helper T-lymphocyte subsets (Marrack and Kappler, 1994). Local responses are those most likely targeted by infectious agents, thus avoiding general immunosuppression of the host (Marrack and Kappler, 1994).

Salivary gland extracts were prepared daily from engorging female *D. andersoni* and assessed for their ability to affect cytokine production by normal murine macrophages and T lymphocytes (Ramachandra and Wikel, 1992). Macrophage-derived cytokines evaluated were IL-1 and TNF. T-lymphocyte cytokines studied were IL-2 and IFN-$\gamma$. $T_H2$ lymphocyte production of IL-4 was subsequently analysed (unpublished observation). Bioassays were used to determine cytokine levels, since increased transcription of cytokine mRNA can occur without the release of cytokine protein (Yao *et al.*, 1994). *Dermacentor andersoni* SGEs significantly suppressed IL-1 elaboration from 89.8% on day zero to 61.6% on day five, while day six to nine SGEs were not significantly suppressive (Ramachandra and Wikel, 1992). Macrophage production of TNF was significantly inhibited by SGEs prepared on days zero to nine. Suppression of TNF ranged from 40.7 to 94.7%. Elaboration of IL-1 and TNF by normal macrophages was affected in a differential manner by *D. andersoni* SGEs, indicating the presence of more than one immunosuppressive factor in SGE, multiple receptor–ligand interactions of SGEs on the macrophage surface or complex interactions with pathways of IL-1 and TNF production.

Lymphocyte elaboration of IL-2 was inhibited from 14.1 to 31.9% by days zero to nine SGEs, providing a relatively uniform level of suppression for all samples tested (Ramachandra and Wikel, 1992). Modulation of IFN-$\gamma$ was relatively consistent for day zero to five SGEs, ranging from 25.2% on day zero to 43.5% on day five (Ramachandra and Wikel, 1992). Day seven SGE inhibited IFN-$\gamma$ production by 57.0%, but production on days six, eight and nine was inhibited only 8.7 to 15.6%. Elaboration of IL-4 was not affected by the presence of SGEs (unpublished observation). All SGEs were not toxic to macrophages, lymphocytes or cytokine bioassay indicator cells.

Interleukin-1 production by *B. indicus*- and *B. taurus*-derived macrophages was suppressed in a similar manner upon exposure to SGEs prepared from female *D. andersoni* on days five to nine of engorgement (Ramachandra and Wikel, 1995). *Bos indicus* macrophage elaboration of IL-1 was suppressed 92.5% by day five SGE and suppression did not drop below 82.8% through day nine. *Bos taurus* macrophage production of IL-1 was reduced by 91.6% by day five SGE and 82.7% by day nine SGE, while day four SGE only reduced IL-1 production by 5.9%. Although macrophages of both genetic backgrounds were suppressed in their ability to produce IL-1 by similar percentages, *B. taurus* macrophages produced 45.6% less IL-1 in the presence of LPS and 43.0% less IL-1 in the absence of LPS than did similar cells collected from *B. indicus*. Macrophages of *B. indicus* and *B. taurus* were inhibited by days zero to seven SGEs in their production of TNF upon stimulation with LPS (Ramachandra and Wikel, 1995). The most intense suppression of TNF elaboration occurred during the first days of engorgement.

*Dermacentor andersoni* SGEs suppressed murine macrophage and $T_H 1$ lymphocyte function, while $T_H 2$ lymphocyte function appeared to remain intact. Bovine macrophage production was reduced by tick SGEs. Modulation of macrophage function could impair antibody and cell-mediated immune function, including the expansion of T-lymphocyte clones essential for regulation of the immune response and delayed type hypersensitivity reactions. Suppressed levels of IL-1 would impair the activation of T lymphocytes through the loss of signals required for neutrophil mobilization from the bone marrow, chemotaxis, and degranulation; production of IL-2 and expression of IL-2 receptors by T lymphocytes; and, B-lymphocyte maturation, proliferation and immunoglobulin synthesis. Reduced TNF production would impair expression of classes I and II MHC molecules; antiviral responses; and, activation of polymorphonuclear leukocytes. Impaired IL-2 production would reduce the autocrine expansion of antigen-specific T-lymphocyte clones and B-lymphocyte differentiation. Suppression of IFN-γ elaboration would reduce induction and expression of classes I and II MHC molecules; macrophage and NK cell activation; antiviral activity; and, B-lymphocyte differentiation and production of immunoglobulins. The reduced *in vitro* proliferative responses to mitogens by T lymphocytes derived from infested laboratory animals or cattle or suppression of normal cell responsiveness *in vitro* to SGEs could be explained by the observed changes in cytokine elaboration.

Tick suppression of host macrophage and $T_H 1$ lymphocyte function impairs a number of pathways involved in development and expression of acquired resistance to tick feeding. Antibodies and delayed type hypersensitivity reactions were involved in the acquisition and expression of resistance to ticks (Wikel, 1982a, 1996; Brossard *et al.*, 1991). Macrophages or 'macrophage-like' cells have a central role in immunoregulatory and effector functions. Tick suppression of host immune defences would enhance tick feeding and coincide with the time when pathogens would be transmitted to the host. The balance between tick modulation of host immunity and remaining host defences allows the host to function in an environment filled with insults that require some measure of immune competence. The tick would not want the host to be overwhelmed by other infectious agents, thus losing its source of food. Tick immunosuppression has been studied *in vitro* and in the context of the systemic immune response of the host. The next stage of analysis needs to be the characterization of tick-induced modulation of local immune function at the bite-site. Magnitude of tick-induced local immunosuppression is likely greater than that observed systemically.

Effector functions of peripheral blood NK cells collected from healthy humans were reduced upon exposure to SGE prepared from *Dermacentor reticulatus* allowed to engorge on mice for six days (Kubes et al., 1994). Tenfold dilution of day six SGE reduced the inhibitory activity, and 100-fold dilution resulted in negligible inhibition of NK cell function. Inhibition of NK cell function was not due to SGE cytotoxicity. Salivary gland extract prepared from unfed ticks was not inhibitory to NK cell function. Natural killer cells represented a non-specific defence that can lyse virus-infected target cells in a MHC unrestricted manner (Trinchieri, 1989). Reduced NK cell function could facilitate infection of the host with tick-borne disease causing agents, particularly viruses.

Hosts repeatedly infested with pathogen-free *D. andersoni* were resistant to infection with *Francisella tularensis* transmitted by *D. andersoni*, indicating that resistance to tick feeding altered transmission of a tick-borne pathogen (Wikel, 1980). *Clethrionomys glareolus*, bank vole, acquired resistance to *I. ricinus* infestation interfered with subsequent tick transmission of *B. burgdorferi* (Dizij et al., 1994). The mechanism(s) responsible for altered pathogen transmission to hosts resistant to tick infestation were not identified. Does repeated infestation with pathogen-free ticks induce antibodies or effector cells capable of neutralizing host immunosuppressants introduced by the feeding tick?

A research objective in the laboratory of the authors is the development of a vaccine capable of inhibiting the action of tick-derived immunosuppressants introduced into the host during feeding. An 'anti-immunosuppressant' vaccine has the potential of enhancing host defences against any tick-borne pathogen and increasing resistance to tick feeding.

## ACKNOWLEDGEMENTS

Research of S.K.W. is supported by the US Department of Agriculture, Oklahoma Center for Advancement of Science and Technology, Pfizer Animal Health, and Oklahoma Agricultural Experiment Station Project Number OKL02174.

# REFERENCES

Allen, J.R. (1973) Tick resistance: basophils in skin reactions of resistant guinea pigs. *International Journal for Parasitology* 3, 195–200.

Allen, J.R., Doube, B.M. and Kemp, D.H. (1977) Histology of bovine skin reactions to *Ixodes holocyclus*, Neuman. *Canadian Journal of Comparative Medicine* 41, 26–35.

Allen, J.R., Khalil, H.M. and Graham, J.E. (1979) The location of tick salivary gland antigens, complement and immunoglobulin in the skin of guinea pigs infested with *Dermacentor andersoni* larvae. *Immunology* 38, 467–472.

Arlian, L.G. (1989) Biology, host relations, and epidemiology of *Sarcoptes scabiei*. *Annual Review of Entomology* 34, 139–161.

Bergman, D.K., Ramachandra, R.N. and Wikel, S.K. (1995) *Dermacentor andersoni*: Salivary gland proteins suppressing T-lymphocyte responses to Concanavalin A *in vitro*. *Experimental Parasitology* 81, 262–271.

Bissonnette, E.Y., Rossignol, P.A. and Befus, A.D. (1993) Extracts of mosquito salivary gland inhibit tumor necrosis factor alpha release from mast cells. *Parasite Immunology* 15, 27–33.

Brossard, M. and Fivaz, V. (1982) *Ixodes ricinus* L.: mast cells, basophils and eosinophils in the sequence of cellular events in the skin of infested or reinfested rabbits. *Parasitology* 85, 583–592.

Brossard, M. and Girardin, P. (1979) Passive transfer of resistance in rabbits infested with adult *Ixodes ricinus* L.: humoral factors influence feeding and egg laying. *Experientia* 35, 1395–1396.

Brossard, M., Monneron, J.P. and Papatheodorou, V. (1982) Progressive sensitization of circulating basophils against *Ixodes ricinus* L. antigens during repeated infestations of rabbits. *Parasite Immunology* 4, 355–361.

Brossard, M., Rutti, B. and Haug, T. (1991) Immunological relationships between host and ixodid ticks. In: Toft, C.A., Aeschliman, A. and Bolic, L. (eds), *Parasite–Host Associations: Coexistence or Conflict*. Oxford University Press, Oxford, pp. 177–200.

Brown, S.J. (1985) Immunology of acquired resistance to ticks. *Parasitology Today* 1, 165–171.

Champagne, D.E. (1994) The role of salivary vasodilators in bloodfeeding and parasite transmission. *Parasitology Today* 10, 430–433.

Corbett, R., Banks, K., Hinrichs, D. and Bell, T. (1975) Cellular immune responsiveness in dogs with demodectic mange. *Transplantation Proceedings* 7, 557–559.

Cross, M.L., Cupp, M.S., Cupp, E.W., Galloway, A.L. and Enriquez, F.J. (1993a) Modulation of murine immunological responses by salivary gland extract of *Simulium vittatum* (Diptera: Simuliidae). *Journal of Medical Entomology* 30, 928–935.

Cross, M.L., Cupp, M.S., Cupp, E.W., Ramberg, F.B. and Enriquez, F.J. (1993b) Antibody responses of BALB/c mice to salivary antigens of hematophagous black flies (Diptera: Simuliidae). *Journal of Medical Entomology* 30, 725–734.

DeMaeyer, E. and DeMaeyer-Guignard, J. (1994) Interferons. In: Thomson, A. (ed.), *The Cytokine Handbook*, 2nd edn. Academic Press, San Diego, California, pp. 265–288.

Dessaint, J.-P.L. and Capron, A.R. (1993) Survival strategies of parasites in their immunocompetent hosts. In: Warren, K.S. (ed.), *Immunology and Molecular Biology of Parasitic Infections*, 3rd edn. Blackwell Scientific Publications, Oxford, pp. 87–99.

Dinarello, C.A. (1994) Interleukin-1. In: Thomson, A. (ed.), *The Cytokine Handbook*, 2nd edn. Academic Press, San Diego, California, pp. 31–56.

Dizij, A., Arndt, S., Seitz, H.M. and Kurtenbach, K. (1994) *Clethrionomys glareolus* acquired resistance to *Ixodes ricinus*: a mechanism to prevent spirochete inoculation? In: Cevenini, R., Sambri, V. and LaPlaca, M. (eds), *Advances in Lyme Borreliosis Research*. Bologna, pp. 228–231.

Dvorak, H.F., Dvorak, A.M., Simpson, B.A., Richardson, H.B., Leskowitz, S. and Karnovsky, M.J. (1970) Cutaneous basophil hypersensitivity. II. A light and electron microscopic description. *Journal of Experimental Medicine* 132, 558–582.

Falk, E.S. (1980) Serum immunoglobulin values in patients with scabies. *British Journal of Dermatology* 102, 57–61.

Falk, E.S. (1981) Serum IgE before and after treatment for scabies. *Allergy* 36, 167–174.

Fivaz, B.H. (1989) Immune suppression induced by the brown ear tick *Rhipicephalus appendiculatus* Neumann, 1901. *Journal of Parasitology* 75, 946–952.

Friend, W.G. and Smith, J.J.B. (1971) Feeding in *Rhodnius prolixus*: mouthpart activity and salivation, and their correlation with changes in electrical resistance. *Journal of Insect Physiology* 17, 233–243.

George, J.E., Osburn, R.L. and Wikel, S.K. (1985) Acquisition and expression of resistance by *Bos indicus* and *Bos indicus* × *Bos taurus* calves to *Amblyomma americanum* infestation. *Journal of Parasitology* 71, 174–182.

Gern, L., Schaible, U.E. and Simon, M.M. (1993) Mode of inoculation of Lyme disease agent *Borrelia burgdorferi* influences infection and immune response in inbred strains of mice. *Journal of Infectious Diseases* 167, 971–975.

Goldstein, I.M. (1988) Complement: biologically active products. In: Gallin, J.I., Goldstein, I.M. and Snyderman, R. (eds), *Inflammation: Basic Principles and Clinical Correlates*. Raven Press, New York, pp. 55–74.

Goodman, J.W. (1991) The immune response. In: Stites, D.P. and Torr, A.I. (eds), *Basic and Clinical Immunology*, 7th edn. Appleton and Lange, Norwalk, Connecticut, pp. 34–44.

Hancock, B.W., Ward, A.M. and Path, M.R.C. (1974) Serum immunoglobulin in scabies. *Journal of Investigative Dermatology* 63, 482–484.

Inokuma, H., Kerlin, R.L., Kemp, D.H. and Willadsen, P. (1993) Effects of cattle tick (*Boophilus microplus*) infestation on the bovine immune system. *Veterinary Parasitology* 47, 107–118.

Inokuma, H., Kemp, D.H. and Willadsen, P. (1994) Prostaglandin E2 production by the cattle tick (*Boophilus microplus*) into feeding sites and its effect on the response of bovine mononuclear cells to mitogen. *Veterinary Parasitology* 53, 293–299.

Kanellopoulos, J.M., DePetris, S., Leca, G. and Crumpton, M.J. (1985) The mitogenic lectin from *Phaseolus vulgaris* does not recognize the T3 antigen of human T-lymphocytes. *European Journal of Immunology* 15, 478–486.

Konishi, E. (1990) Distribution of immunoglobulin G and E levels to salivary gland extracts of *Aedes albopictus* (Diptera: Culicidae) in several age groups of a Japanese population. *Journal of Medical Entomology* 27, 519–522.

Kozin, F. and Cochrane, C.G. (1992) The contact system of plasma: biochemistry and pathophysiology. In: Gallin, J.I., Goldstein, I.M. and Snyderman, R. (eds), *Inflammation: Basic Principles and Clinical Correlates*, 2nd edn. Raven Press, New York, pp. 103–122.

Kroemer, G., deAlboran, I.M., Gonzalo, J.A. and Martinez-A, C. (1993) Immunoregulation by cytokines. *Critical Reviews in Immunology* 13, 163–191.

Kubes, M., Fuchsberger, N., Labuda, M., Zuffova, E. and Nuttall, P.A. (1994) Salivary gland extracts of partially fed *Dermacentor reticulatus* ticks decrease natural killer cell activity *in vitro. Immunology* 82, 113–116.

Lavoipierre, M.M.J., Dickerson, G. and Gordon, R.M. (1959) Studies on the methods of feeding of blood sucking arthropods. I. The manner in which triatomine bugs obtain their blood meal, as observed in the tissues of the living rodent, with some remarks on the effects of the bite on human volunteers. *Annals of Tropical Medicine and Parasitology* 53, 235–250.

Marrack, P. and Kappler, J. (1994) Subversion of the immune system by pathogens. *Cell* 76, 323–332.

Mbow, M.L., Rutti, B. and Brossard, M. (1994) IFN-G, IL-2 and IL-4 mRNA expression in the skin and draining lymph nodes of BALB/c mice repeatedly infested with nymphal *Ixodes ricinus* ticks. *Cellular Immunology* 156, 254–261.

McSwain, J.L., Essenberg, R.C. and Sauer, J.R. (1982) Protein changes in the salivary glands of the female lone star tick, *Amblyomma americanum*, during feeding. *Journal of Parasitology* 68, 100–106.

Mims, C.A. (1987) Microbial strategies in relation to the immune response. In: *The Pathogenesis of Infectious Disease*. Academic Press, San Diego, California, pp. 152–178.

Mossman, T.R. and Coffman, R.L. (1989) $T_H1$ and $T_H2$ cells: differential patterns of lymphokine secretion lead to different functional properties. *Annual Review of Immunology* 7, 145–173.

Mossman, T.R. and Moore, K.W. (1991) The role of IL-10 in crossregulation of $T_H1$ and $T_H2$ responses. *Immunology Today* 12, A49–A53.

Muller-Eberhard, H.J. (1988) Complement: chemistry and pathways. In: Gallin, J.I., Goldstein, I.M. and Snyderman, R. (eds), *Inflammation: Basic Principles and Clinical Correlates*. Raven Press, New York, pp. 21–53.

Nithiuthai, S. and Allen, J.R. (1984a) Significant changes in epidermal Langerhans cells of guinea-pigs infested with ticks (*Dermacentor andersoni*). *Immunology* 51, 133–141.

Nithiuthai, S. and Allen, J.R. (1984b) Effects of ultraviolet irradiation on the acquisition and expression of tick resistance in guinea-pigs. *Immunology* 51, 153–159.

Nithiuthai, S. and Allen, J.R. (1985) Langerhans cells present tick antigens to lymph node cells from tick sensitized guinea-pigs. *Immunology* 55, 157–163.

Oliver, J. H. Jr, Owsley, M.R., Hutcheson, H.J., James, A.M., Chen, C., Irby, W.S., Dotson, E. M. and McLain, D.K. (1993) Conspecificity of the ticks *Ixodes scapularis* and *Ixodes dammini* (Acari: Ixodidae). *Journal of Medical Entomology* 30, 54–63.

Paine, S.H., Kemp, D.H. and Allen, J.R. (1983) In vitro feeding of *Dermacentor andersoni* (Stiles): effects of histamine and other mediators. *Parasitology* 86, 419–428.

Pinnas, J.B., Lindberg, R.E., Chen, T.M.W. and Meinke, G.C. (1986) Studies of kissing bug-sensitive patients: evidence for the lack of cross-reactivity between *Triatoma protracta* and *Triatoma rubida* salivary gland extracts. *Journal of Allergy and Clinical Immunology* 77, 364–370.

Playfair, J.H.L. (1993) Overview: parasitism and immunity. In: Lachmann, P.J., Peters, K., Rosen, F.S. and Walport, M.J. (eds), *Clinical Aspects of Immunology*. Blackwell Scientific Publications, Oxford, pp. 1439–1454.

Porteu, F. and Nathan, C. (1990) Shedding of tumor necrosis factor receptors by activated human neutrophils. *Journal of Experimental Medicine* 172, 559–607.

Ramachandra, R.N. and Wikel, S.K. (1992) Modulation of host immune responses by ticks (Acari: Ixodidae): effect of salivary gland extracts on host macrophages and lymphocyte cytokine production. *Journal of Medical Entomology* 29, 818–826.

Ramachandra, R.N. and Wikel, S.K. (1995) Effects of *Dermacentor andersoni* (Acari: Ixodidae) salivary gland extracts on *Bos indicus* and *Bos taurus* lymphocytes and macrophages: *in vitro* cytokine elaboration and lymphocyte blastogenesis. *Journal of Medical Entomology* 32, 338–345.

Ratzlaff, R.E. and Wikel, S.K. (1990) Murine immune responses and immunization against *Polyplax serrata* (Anoplura: Polyplacidae). *Journal of Medical Entomology* 27, 1002–1007.

Ribeiro, J.M.C. (1987a) Role of saliva in blood-feeding by arthropods. *Annual Review of Entomology* 32, 463–478.

Ribeiro, J.M.C. (1987b) *Ixodes dammini*: salivary anti-complement activity. *Experimental Parasitology* 64, 347–353.

Ribeiro, J.M.C. (1989) Vector saliva and its role in parasite transmission. *Experimental Parasitology* 69, 104–106.

Ribeiro, J.M.C. and Spielman, A. (1986) *Ixodes dammini*: salivary anaphylatoxin inactivating activity. *Experimental Parasitology* 62, 292–297.

Ribeiro, J.M.C., Makoul, G.T., Levine, J., Robinson, D.R. and Spielman, A. (1985) Antihemostatic, antiinflammatory and immunosuppressive properties of the saliva of a tick, *Ixodes dammini*. *Journal of Experimental Medicine* 161, 332–344.

Roehrig, J.T., Piesman, J., Hunt, A.R., Keen, M.G., Happ, C.M. and Johnson, B.J.B. (1992) The hamster immune response to tick transmitted *Borrelia burgdorferi* differs from the response to needle-inoculated, cultured organisms. *Journal of Immunology* 149, 3648–3653.

Salmon, J.K., Armstrong, C.A. and Ansel, J.C. (1994) The skin as an immune organ. *Western Journal of Medicine* 160, 146–152.

Schorderet, S. and Brossard, M. (1993) Changes in immunity to *Ixodes ricinus* by rabbits infested at different levels. *Medical and Veterinary Entomology* 7, 186–192.

Schorderet, S. and Brossard, M. (1994) Effects of human recombinant interleukin-2 on resistance and on the humoral and cellular response of rabbits infested with adult *Ixodes ricinus* ticks. *Veterinary Parasitology* 54, 375–387.

Scott, D.W., Farrow, B.R.H. and Schultz, R.D. (1974) Studies on the therapeutic and immunologic aspects of generalized demodectic mange in the dog. *American Animal Hospital Association Journal* 10, 233–244.

Sharon, N. (1983) Lectin receptors as lymphocyte surface markers. *Advances in Immunology* 34, 213–298.

Shen, H.D., Chen, C.C., Chang, H.N., Chanf, L.Y., Tu, W.C. and Han, S.H. (1989) Human IgE and IgG antibodies to mosquito proteins detected by the immunoblot technique. *Annals of Allergy* 63, 143–146.

Sher, A. (1988) Vaccination against parasites. Special problems imposed by the adaptation of parasitic organisms to the host response. In: Englund, P.T. and Sher,

A. (eds), *The Biology of Parasitism: A Molecular and Immunologic Approach*. Alan R. Liss, New York, pp. 169–182.

Sher, A. and Scott, P.A. (1993) Mechanisms of acquired immunity against parasites. In: Warren, K.S. (ed.), *Immunology and Molecular Biology of Parasitic Infections*, 3rd edn. Blackwell Scientific Publications, Oxford, pp. 35–50.

Sonenshine, D.E. (1991) *Biology of Ticks*, Volume 1. Oxford University Press, Oxford, pp. 165–175.

Theodos, C. and Titus, R.G. (1993) Salivary gland material from the sand fly *Lutzomyia longipalpis* has an inhibitory effect on macrophage function *in vitro*. *Parasite Immunology* 15, 481–487.

Titus, R.G. and Ribeiro, J.M.C. (1990) The role of vector saliva in transmission of arthropod-borne disease. *Parasitology Today* 6, 157–160.

Tracey, K. (1994) Tumor necrosis factor-alpha. In: Thomson, A. (ed.), *The Cytokine Handbook*, 2nd edn. Academic Press, San Diego, California, pp. 289–304.

Trang, L.E. (1980) Prostaglandins and inflammation. *Seminars in Arthritis and Rheumatism* 9, 153–190.

Trinchieri, G. (1989) Biology of natural killer cells. *Advances in Immunology* 47, 187–376.

Urioste, S., Hall, L.R., Telford, S.R. III and Titus, R.G. (1994) Saliva of the Lyme disease vector, *Ixodes dammini*, blocks cell activation by a non-prostaglandin $E_2$-dependent mechanism. *Journal of Experimental Medicine* 180, 1077–1085.

Vilcek, J. and Le, J. (1994) Immunology of cytokines: an introduction. In: Thomson, A. (ed.), *The Cytokine Handbook*, 2nd edn. Academic Press, San Diego, California, pp. 1–19.

Warburg, A., Saraiva, E., Lanzaro, G.C., Titus, R.G. and Neva, F. (1994) Saliva of *Lutzomyia longipalpis* sibling species differs in its composition and capacity to enhance leishmaniasis. *Philosophical Transactions of the Royal Society of London, Series B* 345, 223–230.

Ward, P.A., Dvorak, H.F., Cohen, S., Yoshida, T., Data, R. and Selvaggio, S.S. (1975) Chemotaxis of basophils by lymphocyte-dependent and lymphocyte-independent mechanisms. *Journal of Immunology* 114, 1523–1527.

Whelen, A.C. and Wikel, S.K. (1993) Acquired resistance of guinea pigs to *Dermacentor andersoni* mediated by humoral factors. *Journal of Parasitology* 79, 908–912.

Wiggins, R.C., Giclas, P.C. and Henson, P.M. (1981) Chemotactic activity generated from the fifth component of complement by plasma killikrein of the rabbit. *Journal of Experimental Medicine* 153, 1391–1404.

Wikel, S.K. (1979) Acquired resistance to ticks. Expression of resistance by C4 deficient guinea pigs. *American Journal of Tropical Medicine and Hygiene* 28, 586–590.

Wikel, S.K. (1980) Host resistance to tick-borne pathogen by virtue of resistance to tick infestation. *Annals of Tropical Medicine and Parasitology* 74, 103–104.

Wikel, S.K. (1982a) Immune responses to arthropods and their products. *Annual Review of Entomology* 27, 21–48.

Wikel, S.K. (1982b) Influence of *Dermacentor andersoni* infestation on lymphocyte responsiveness to mitogens. *Annals of Tropical Medicine and Parasitology* 76, 627–632.

Wikel, S.K. (1985) Effects of tick infestation on the plaque-forming cell response to a thymic dependent antigen. *Annals of Tropical Medicine and Parasitology* 79, 195–198.

Wikel, S.K. (1989) Mite antigens recognized by antibodies from rabbits infested with *Psoroptes cuniculi*. *Medical Science Research* 17, 455–456.

Wikel, S.K. (1996) Host immunity to ticks. *Annual Review of Entomology* 41, 1–22.

Wikel, S.K. and Osburn, R.L. (1982) Immune responsiveness of the bovine host to repeated low-level infestations with *Dermacentor andersoni*. *Annals of Tropical Medicine and Parasitology* 76, 405–414.

Wikel, S.K., Graham, J.E. and Allen, J.R. (1978) Acquired resistance to ticks. IV. Skin reactivity and *in vitro* lymphocyte responsiveness to salivary gland antigen. *Immunology* 32, 257–263.

Wikel, S.K., Ramachandra, R.N. and Bergman, D.K. (1994) Tick-induced modulation of the host immune response. *International Journal for Parasitology* 24, 59–66.

Willadsen, P. (1980) Immunity to ticks. *Advances in Parasitology* 18, 293–313.

Wilson, A.B. and Clements, A.N. (1965) The nature of the skin reaction to mosquito bites in laboratory animals. *International Archives of Allergy* 26, 294–314.

Yao, Q., Cua, D., Seinsintaffer, J., Dimacalli, M. and Stohlman, S. (1994) Increased TNF transcription in the absence of protein release by mouse hepatitis virus infected macrophages. *FASEB Journal* 8, A252.

# 6

# Digestion and Fate of the Vertebrate Bloodmeal in Insects

## Michael J. Lehane

*School of Biological Sciences, University of Wales, Bangor LL57 2UW, UK*

Of the estimated 1 to 10 million species of insect only 300 to 400 are regular blood feeders. They can be divided into two categories depending on the design of the midgut and associated structures (Fig. 6.1). In the first group, typified by the lice, fleas and blood-sucking bugs, the midgut is a simple tube in which both storage and digestion of the meal occur. In the second group, exemplified by the blood-sucking flies, the midgut proper is still a simple tube but it is associated with one or more diverticula of foregut origin in which food, including the bloodmeal, may be stored prior to digestion (Gooding, 1972). In some insects, such as the tsetse fly, there is a single diverticulum which is commonly termed the crop. In others, such as the mosquitoes, there may be up to three diverticula. The distribution of the meal between diverticula and midgut depends on the species in question and the composition of the meal (Trembley, 1952; Megahed, 1958) but the most common, but certainly not the exclusive pattern, is that sugar-based meals pass to the diverticula and blood to the midgut. The most convincing proposal for the biological significance of diverticula is that they protect midgut proteases from inhibitors in nectar. If these enter the gut in number they may interfere with digestion of a bloodmeal which may become available immediately after a nectar meal is taken. So the diverticula give a means of segregating the majority of the inhibitors and allowing their gradual transfer so that of any one time quantities too small to seriously interfere with digestive proteases will be present in the midgut (Bailey, 1952; Gooding *et al.*, 1973). In addition, the crop appears to function as a 'spare tank' – when no bloodmeals are available

---

© CAB INTERNATIONAL 1996. From Wikel, S.K. (ed.) *The Immunology of Host–Ectoparasitic Arthropod Relationships*. CAB INTERNATIONAL, Wallingford.

the insect survives and avoids dehydration on small quantities of sugar meal transferred from crop to midgut.

On the basis of their method of digestion blood-sucking insects can once again be divided into two categories (Fig. 6.1). The first group, typified by the bugs and higher Diptera, use a continuous system of digestion. Here the midgut is divided into functionally distinct zones (Fig. 6.2) through which the bloodmeal is gradually passed. In the continuous system much of the meal will have been completely processed and defecated before some has even entered the digestive portion of the midgut. In the second group, exemplified by the mosquitoes, sandflies and fleas, the bloodmeal in a batch manner. In these insects there is comparatively little differentiation of the midgut into separate functional regions and digestion occurs more or less simultaneously over the entire surface of the food bolus.

Insects do not produce the mucus covering so typical of the vertebrate intestine. Instead the food is normally separated from the simple epithelium

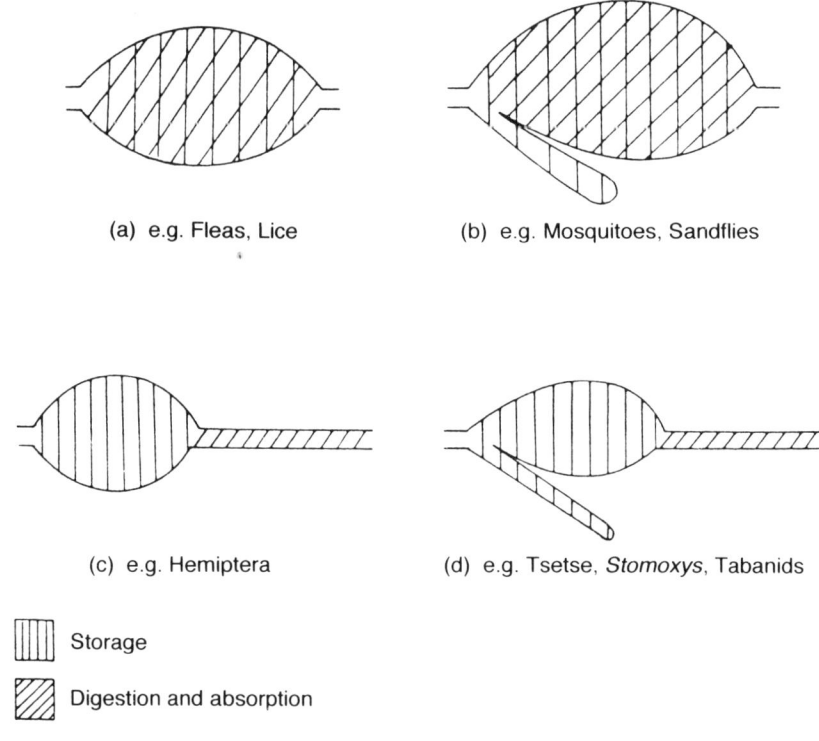

(a) e.g. Fleas, Lice

(b) e.g. Mosquitoes, Sandflies

(c) e.g. Hemiptera

(d) e.g. Tsetse, *Stomoxys*, Tabanids

Storage

Digestion and absorption

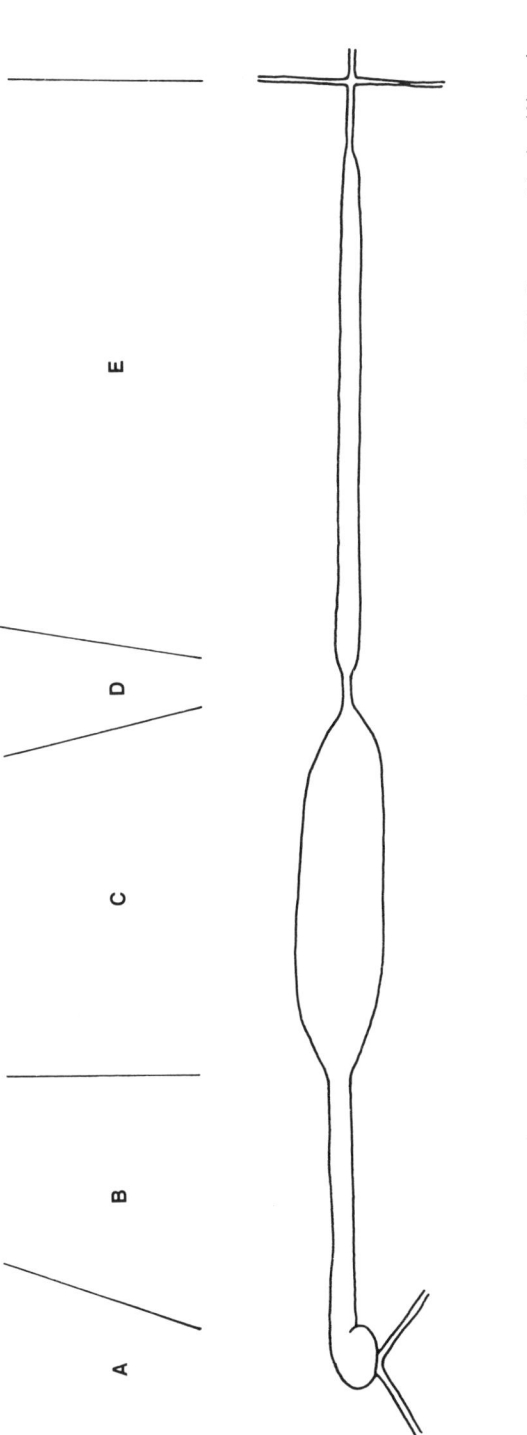

**Fig. 6.2.** The midgut of *S. calcitrans* is ultrastructurally and functionally divisible into seven zones. The first three lie within the proventriculus (A) and are responsible for producing the multi-layered peritrophic matrix. Then comes the thoracic midgut (B) whose function has not been investigated. Next comes the reservoir zone (C) which stores and dehydrates the bloodmeal. The opaque zone (D) is responsible for the secretion of the bulk of the soluble digestive enzymes. Digestion and absorption occur in the lipoid zone (E). In addition there is a foregut diverticulum called the crop, in which mainly sugar meals, but on occasion bloodmeals, are stored.

from which the midgut is formed, by the peritrophic matrix (= peritrophic membrane). The peritrophic matrix (PM) is produced in one of two ways. The adult mosquitoes, blackflies, sandflies and tabanids produce a type I PM. This is absent in the unfed fly and is only produced in response to blood-feeding although it can be artificially induced by midgut distension (Berner et al., 1983). Type I PM is secreted by cells along the full length of the midgut. This is the most common method of production and is widespread among many phyla besides insects (Peters, 1992). It often takes up to 48 hours after the bloodmeal is taken for type I membranes to be fully formed although there is considerable interspecific variation. For example, in anopheline mosquitoes the PM is partly built from preformed material stored in secretory granules in the unfed insect (Billingsley and Rudin, 1992) while in culicine mosquitoes the initial appearance of the PM takes longer because it is synthesized *de novo* (Hecker and Rudin, 1979). Such interspecific differences may be important in determining vectorial capacity. The fully constructed type I PM is often formed of several ultrastructurally discernible laminae. When produced artificially in the absence of food these PMs are less highly ordered suggesting the food has a template function. Nevertheless, the degree of organization shown by type I PMs produced in the absence of food show that other factors must also be important in determining PM organization.

The adult muscids, tsetse flies and hippoboscids produce a type II PM. Type II PMs are produced by a special press-like organ normally called the proventriculus (more correctly termed the cardia) which is formed by the extreme anterior end of the midgut and extreme posterior end of the foregut. Ultrastructurally, type II PMs are more highly ordered than type I PMs and, from an ultrastructural viewpoint, the food-facing surface appears to be very similar in all type II PMs and to bear a resemblance to epicuticle (Binnington et al., 1995). The type II PM is continuously formed by the proventriculus although production rate is species dependent and is affected by temperature and nutrition. Apart from a brief period in the early teneral stage type II PM forms a continuous sleeve running the full length of the midgut in both unfed and fed flies. In most non-blood-feeding higher Diptera the type II PM is formed of two or three concentric sleeves. Some blood feeders have two sleeves (e.g. *Haematobia exigua*) but the tsetse flies and *S. calcitrans* are unusual in having only a single sleeve. This reduction may be related to the blood-sucking habit. Further evidence for a reduced requirement of PM in blood-sucking insects is seen in rates of production; thus larvae of *Eristalis tenax* may produce PM at a rate of 18 mm per hour (Waterhouse, 1954), adult *Calliphora erythrocephala* at over 7 mm per hour (Becker, 1978), while the blood-feeding *G. m. morsitans* produces PM at only 0.95 mm per hour (Lehane and Msangi, 1991). All this suggests PMs may play a reduced role in blood-sucking insects and it is interesting to note that digestion of the bloodmeal will occur normally in the mosquito *An. stephensi* in the absence of a PM (Berner et al., 1983) and, in another mosquito *Ae. aegypti*, in the presence of an unusually thick PM

(Billingsley and Rudin, 1992). These are very interesting results and on this evidence the authors argue that the PM is a vestigeal structure conferring no advantage or disadvantage. However, Terra and Ferreira (1994) rightly point out that this can only be concluded if the efficiency of digestion, as measured by efficiency of food conversion into body mass and growth rate, is shown not to be affected by lack of PM. I would add that before such a conclusion evidence is needed that lack of PM has no selective advantage when a pathogen-contaminated meal is encountered.

Although much detailed work is still required on PM composition it is clear it is non-cellular and formed of a complex mixture of proteins, glycoproteins, glycosaminoglycans and the structural polysaccharide chitin. The PM of *G. m. morsitans* is better studied than most and will be cited as an example although it should be remembered that there are clearly interspecific differences in PM composition (e.g. Berner *et al.*, 1983). *G. m. morsitans* PM is a single concentric tube formed of three concentric layers, each layer formed of GlcNAc-hexuronic and Gal-GlcNAc-containing glycosaminoglycans, together with chitin (Lehane *et al.*, 1995). The epithelium-facing layer of the PM is sulphated and the charged sites these form have a mean interspace distance of 53 nm. While the function significance of this has not been experimentally investigated, the interspace distance of *c.* 53 nm is remarkably similar to that seen in vertebrate glomerular basement membrane where such charged sites are known to play a key role in determining the filtration characteristics of the kidney (Rennke *et al.*, 1975; Kanwar and Farquhar, 1979). The density of charged sites varies with pH (Lehane *et al.*, 1995) and so such a system might be used by insects to vary the filtration characteristics of the PM either along the length of the gut in continuous digesters or in time in batch digesters. Glycoproteins possessing O-linked oligosaccharides are present in all three layers of *G. m. morsitans* PM. The food-facing layer of the PM contains the widest variety of glycoprotein oligosaccharide constructs – the histochemical evidence suggests the presence of GlcNac and α-linked GalNAc possibly as GalNAcα1,3GalNAc, the latter apparently distal to Galβ1-4GlcNAc. Lectin binding suggests that the central layer of the PM contains GalNAcα-1,3Galβ1,3GlcNAc and that the epithelium-facing layer contains GalNAc and Galβ1,4GlcNAc. The evidence for N-linked oligosaccharides is more equivocal. Two-dimensional electrophoresis showed the PM also contained a range of proteins most of which require relatively harsh treatment for their solubilization (Lehane *et al.*, 1995).

Little experimental work has been performed on the function of the PM although much speculation exists as to its function. Type II PMs form a permanent barrier separating bacteria and virus from the epithelium, some protozoa (see below) and helminths are capable of actively penetrating the PM. Peters *et al.* (1983) showed the PM of *C. erythrocephala* contains lectins capable of binding bacteria; lectins are also associated with tsetse PM (Lehane and Msangi, 1991). As the PM is continuously produced and shed in the faeces lectin-binding of bacteria and other material to the PM may be an important element of intestinal defence. Arguing against this, most blood-sucking insects will encounter pathogens most regularly while feeding on non-blood foods such as sugars; under these feeding circumstances those insects producing type I PM are assumed not to produce PM (Peters, 1992)! With a putative defence function for PM in mind it should also be noted that type I PM is only produced after the bloodmeal is taken and so there is a brief period when the meal and any pathogen contained will be in direct contact with the epithelium and even then type I PM may not form a completely encircling barrier (Billingsley, 1990).

It has been suggested that the importance of the PM lies in its spatial division of the gut lumen into two separate compartments (Terra, 1988). This may be important in increasing the efficiency of digestion by spatially separating the various digestive processes and there is evidence that enzymes are found in different concentrations on the two sides of the PM (Eguchi et al., 1982; Graf and Briegel, 1982). The spatial division may enable the conservation of digestive enzymes by their recirculation in the ectoperitrophic space. It has also been suggested that the division may increase the efficiency of absorption by producing an unstirred layer next to the epithelium.

From its position between the food and the midgut epithelium it is clear that PM must function as a filter, although type I PM may not entirely surround the food bolus suggesting the filter function is not absolute (Billingsley, 1990). The charge density on the PM may play a key role in determining its filtration characteristics (see above). Under experimental conditions the effective pore size in *S. calcitrans* is around 9 nm and so, judged on a size basis, this PM will permit the passage of globular proteins up to c. 140 kDa (Miller and Lehane, 1990). Similar pore sizes have been determined for several non-blood-sucking insects (Binnington et al., 1995).

The blood-sucking bugs do not have a PM but instead produce a layer surrounding the microvilli termed the lumenal apical membrane (LAM) and a system extending into the lumen called the extracellular membrane layers (ECMLs). LAM and ECMLs are together termed the perimicrovillar membrane and probably consist of a single contiguous membrane per cell. They may fulfil the PM functions in bugs (Billingsley, 1990).

Blood-sucking insects show considerable variation in the size of the bloodmeal they ingest – the size being roughly correlated with the nature of the insect/host relationship. Lehane (1991) identified three loose groups of blood-sucking insect based on their spatial relations with the host: *temporary ectoparasites* which are largely free-living insects which only visit the host for long enough to take a bloodmeal, e.g. tabanids, mosquitoes, blood-feeding bugs and blackflies; *periodic ectoparasites* which spend considerably longer on the hosts than is required merely to obtain a bloodmeal but which nevertheless spend a significant amount of time away from the host, e.g.

many fleas and Pupipara; *permanent ectoparasites* which live out their lives virtually entirely on the surface of the host animal, e.g. lice, sheep keds and tungid fleas. While many factors can influence meal size including temperature, insect age, mating status, stage of the gonotrophic cycle, previous feeding history, nature of the host, etc., there is a clear difference in the size of bloodmeal taken by insects in each of these groups with the largest meals taken by temporary ectoparasites and the smallest by permanent ectoparasites.

Temporary ectoparasites routinely take huge bloodmeals, commonly one-and-a-half times their unfed body weight in adult insects (Table 6.1). Some take considerably larger meals; the first instar nymph of *Triatoma infestans* takes meals 10 times its unfed body weight! These huge meals allow the insect to minimize the number of visits to the host. This is advantageous in three ways: first, if the time saved can better be spent in reproductive activity; second, to take maximum advantage of each encounter if hosts are difficult to find; and third, and most importantly, because visiting the host is a dangerous activity. The latter two advantages are clearly not available to permanent ectoparasites for whom such large meals would be dangerous as they would impede the movement of the insect through the host pelage. So permanent ectoparasites take more frequent, smaller bloodmeals which, nevertheless, are still 20–30% of the unfed body weight for anopluran lice for example (Murray and Nichols, 1965) (Table 6.1).

The exceptional size of these bloodmeals, particularly in temporary ectoparasites, not only presents mechanical problems for the insect but also impairs its manoeuvrability. The morphology and physiology of the insects is designed to deal with these problems. The midgut and crop storage regions are capable of considerable stretching and the intercellular junctions and underlying muscular coats are suitably modified to allow for this (Billingsley, 1990). Thus in *Ctenophthalmus* species, *Echidnophaga gallinacea* and *Xenopsylla*

**Table 6.1.** Meal size in blood-sucking insects.

| Species | Average meal size for an adult female insect (% unfed body weight) |
|---|---|
| Aedes aegypti | 109 |
| Anopheles quadrimaculatus | 122 |
| Culex quinquefasciatus | 140 |
| Stomoxys calcitrans | 110 |
| Glossina m. morsitans | 170 |
| Pediculus humanus | 30 |
| Cimex lectularius | 130 |
| Triatoma infestans | 210 |

(*Source*: Lehane, 1991.)

*cheopis* collagen-like fibres are arranged in a hexagonal lattice in the basal lamina (Richards and Richards, 1968; Reinhardt *et al.*, 1972) which may be a response to the particularly extensive stretching which occurs in the flea midgut. In addition, the abdominal wall of blood-sucking insects is elastic or has folds allowing the considerable distension required to accommodate these large bloodmeals. Overdistension is prevented by stretch receptors in the abdominal wall (Maddrell, 1963; Gwadz, 1969). The stretching involved in taking these huge meals may introduce an element of 'leakiness' to the midgut epithelium (Houk and Hardy, 1979; Modespacher *et al.*, 1986) which may help explain the presence of intact host proteins in insect haemolymph (see below).

In dealing with these huge meals insects also make use of the fact that blood is *c.* 80% water (Table 6.2). Most of this water is not required by the insects and they have developed extraordinarily efficient means of rapidly excreting it, thus restoring their manoeuvrability. Specific regions of the midgut epithelium are often adapted for this purpose. In tsetse flies and triatomine bugs water movement across the anterior midgut epithelium is linked to the ouabain-sensitive $Na^+$–$K^+$-ATPase which is so typical of secretory epithelia; the pump molecules involved appear highly conserved across phyla (Reeves and Yamanaka, 1993). The pump is located in the basolateral plasma membranes of the epithelium which allows directed fluid transport to occur (Farmer *et al.*, 1981; Peacock, 1982; MacVicker *et al.* 1993, 1994). Such pumps work by generating an osmotic gradient across the epithelium which drags water passively after it. Such systems are so efficient in blood-sucking insects that tsetse flies, for example, can shed 40% of the weight of the meal in the first 30 min following feeding (Gee, 1975).

**Table 6.2.** Major constituents of host blood are unevenly distributed in the different compartments of the bloodmeal. The major constituents, except nucleic acids, remain largely the same across host species.

|  | Water (g 100 ml$^{-1}$) | Lipid (g 100 ml$^{-1}$) | Carbohydrate (g 100 ml$^{-1}$) | Protein (g 100 ml$^{-1}$) | Nucleic acid (g 100 ml$^{-1}$) |
|---|---|---|---|---|---|
| *Man* | | | | | |
| Whole blood | 80 | 0.7 | 0.01 | 20.5 | |
| Erythrocytes | 72 | 0.6 | 0.01 | 36.8 | 0.1 |
| Plasma | 94 | 0.6 | 0.01 | 7.41 | 0.1 |
| *Chicken* | | | | | |
| Whole blood | 87 | | 0.17 | | |
| Erythrocytes | 72 | | 29 | 29 | 4.2 |
| Plasma | 94 | 0.5 | | 3.6 | |

(*Source*: Altman and Dittmer, 1971.)

There is a range of antihaemostatic molecules produced in the salivary glands of blood-sucking insects (reviewed in Law *et al.*, 1992). In addition there are poorly researched antihaemostatic factors released by the insect midgut (reviewed by Gooding, 1972). These, in combination with the dehydration of the meal, have a strong bearing on the consistency and characteristics of the bloodmeal at the time digestion commences.

The major constituents of the bloodmeal are quite uniform across the range of hosts (Table 6.2). The exception is in the high levels of nucleic acids only found in birds and reptiles because of their nucleated blood cells. Proteins are the most abundant constituent of blood but the majority are not immediately accessible to the insect being locked away in erythrocytes (Table 6.2). Consequently one of the most important events in digestion in blood-sucking insects is haemolysis. Despite the central importance of haemolysis to blood-sucking insects we know very little about it. Haemolysis is partly achieved by physical means in some fleas and mosquitoes which have cuticular spines and armatures in the foregut which puncture red cells releasing their contents. Chemical means of haemolysis are used in other insects but the haemolysins involved are poorly understood. A salivarian haemolysin of an unknown chemical nature is reported present in the bedbug *Cimex* (Sangiorgi and Frosini, 1940). The haemolysin in the triatomine bug *R. prolixus* is a basic peptide rather than a phospholipase-like enzyme (de Azambuja *et al.*, 1983). In the mosquito *Ae. aegypti* proteases and possibly phospholipases are reportedly involved in haemolysis (Geering, 1975). In the tsetse fly the haemolysin is reportedly proteinaceous but is not one of the insect's proteases (Gooding, 1977). A different means of haemolysis is reported in *S. calcitrans* where disruption of the erythrocyte membrane is thought to be brought about by the detergent-like activities of free fatty acids (Spates *et al.*, 1982) in conjunction with trypsin activity (Kirch *et al.*, 1991).

To deal with their largely proteinaceous diet the majority of blood-sucking insects produce a range of alkaline digestive proteases. *G. m. morsitans* is a typical example possessing six alkaline proteases – aminopeptidase, carboxypeptidase A, carboxypeptidase B, trypsin, protease IV (trypsin-like) and protease VII (chymotrypsin-like) (Gooding and Rolseth, 1976). The activities of these six proteases is sufficient to account for the complete hydrolysis of the bloodmeal proteins. The hydrolytic activity of protease IV and trypsin produce peptides with basic amino acids at their carboxy terminus and these are the preferred substrate of carboxypeptidase B which cleaves the terminal amino acid. It is likely that protease VII produces the substrates acted on by carboxypeptidase B which cleaves terminal aromatic amino acids. The aminopeptidase is membrane bound and cleaves the amino acids from the peptide products of trypsin and chymotrypsin digestion. It is likely that this array of enzymes produced by the tsetse will prove to be typical of blood-sucking insects. For example, specific substrate and inhibitor studies show that *S. calcitrans* produces a similar battery of enzymes including molecules with trypsin-like, chymotrypsin-like, aminopeptidase, and carboxypeptidase A and B activities (Houseman et al., 1987). Hemiptera are exceptional in using digestive enzymes active at acid pH (Terra and Ferreira, 1994) which may be evidence of a sap-sucking or, more probably, a seed-feeding ancestry.

The trypsins of mosquitoes have been looked at in some detail. Using SDS–PAGE three to six major trypsin bands have been found in *Aedes* spp. and one to three major bands in *Anopheles* spp. Amino acid sequencing work revealed a surprising degree of variability in the structure of the trypsins found in the six mosquito species investigated (Graf and Briegel, 1985; Graf et al., 1991). Cross-reactivity tests with immune sera confirmed the presence of two and possibly three immunologically distinct families of trypsins. None of these showed cross-reactivity with bovine trypsin. Using isoelectric focusing 20 or so trypsin isoenzymes were found in *Ae. aegypti*. It was suggested that this microheterogeneity shown by *Aedes* spp. might be due to autolysis of digestive trypsins. Contrastingly, the immunologically distinct trypsins of *Anopheles* spp. show far fewer bands suggesting they are more stable.

Precise mechanisms controlling digestive enzyme secretion are poorly understood. Nevertheless it is clear that paracrine and prandial (Lehane et al, 1995) mechanisms are probably of most importance and that soluble proteins act as potent stimulators of digestive enzyme secretion and synthesis. New synthesis can be regulated at either the transcriptional level, as in mosquitoes (Muller et al. 1993), or at the translational level, as in *S. calcitrans* (Moffatt et al., 1995). Considerable interest is now being shown in the genetic mechanisms controlling digestive enzyme secretion in blood-sucking insects. One reason is that these studies may provide suitable promoters for the induction of 'refractory' genes in transgenic mosquitoes. Muller et al. (1993) identified a trypsin gene family in *Anopheles gambiae* and showed that the expression of two distinct members of this family (Antryp 1 and Antryp 2) is induced in response to blood-feeding. The transcriptional control of the two trypsin genes appears to be different. Thus, whereas Antryp 1 mRNA could be detected in unfed mosquitoes, Antryp 2 was under tighter control and could only be found in fed females. The two trypsins appear to fulfil different digestive roles with recombinant Antryp 1 being extremely active against both serum albumin and haemoglobin and Antryp 2 being preferentially active against haemoglobin. Barillas-Mury and Wells (1993) characterized the upstream regulatory region of the bloodmeal-induced late trypsin gene from *Aedes aegypti* and identified a cluster of five repeat sequences, within 200 nucleotides upstream of the cap site, homologous to the yeast GCN4 binding site. GCN4 is responsible for derepression-mediated control of general amino acid biosynthesis in yeast suggesting a protein similar to GCN4 may regulate expression of the late trypsin gene in mosquitoes.

The absorption of digested materials in the midgut of blood-sucking insects is a subject which has hardly been touched upon. One of the few studies undertaken looked at the absorption of lipids from the posterior midgut of *Stomoxys calcitrans*, the results suggesting that many of the amino acids absorbed may be rapidly converted to fats in the midgut epithelium enabling the facilitated diffusion of the amino acids liberated from the bloodmeal (Lehane, 1977). The evidence for absorption of some intact proteins from the bloodmeal is discussed below.

Haematophagous insects vary in the degree to which they depend on blood for their full nutrition. Many insects, including mosquitoes, blackflies, ceratopoginids, sandflies, biting muscids, horseflies and fleas, either utilize foods in addition to blood in the adult stage or have larval stages which utilize foods other than blood. For other insects such as the streblids, hippoboscids, nycteribiids, triatomine and cimicid bugs, lice and tsetse flies, blood is the sole nutrient resource used in the insect's entire life. Because blood is deficient in certain nutrients these latter insects harbour symbiotic microorganisms which supplement the insects' nutrition. In tsetse flies, for example, these symbionts are probably the fimbriated, gram negative bacteroids described by Ma and Denlinger (1972). These symbionts are often housed in specialized organs termed mycetomes. Physical, pharmacological, enzymatic and immunological methods have been used experimentally to produce symbiont-free insects (e.g. Pell and Southern, 1976; Nogge, 1978). In general symbiont-free juveniles develop less readily to adults while symbiont-free adult females are sterile (Nogge, 1978). Sterility can be partially reversed by supplementing the diet with various vitamins the most essential being the B group vitamins, thiamine and pyridoxine along with vitamin H (biotin), folic and pantothenic acids (Nogge, 1981).

Particular blood-sucking insects thrive better on some hosts rather than others. For example, the mosquito *Culex pipiens* will produce 82 eggs per milligram of ingested canary blood compared to 40 eggs per milligram of human blood (Woke, 1937). Similarly, *G. austeni* shows considerably higher fecundity when fed on rabbit blood rather than goat blood, although longevity is similar on both (Jordan and Curtis, 1968). There can even be variations between adults and juvenile hosts. Thus *Xenopsylla cheopis* fed on adult mice produce far more eggs than those fed on baby mice (Buxton, 1948). Many of us who keep blood-sucking insects in culture know they thrive better when blood from two or more different hosts is used presumably because different bloods are providing different quantities of materials essential to an overall balanced diet. It is well known that host hormone levels can be essential in regulating the reproductive biology of some blood-sucking insects (Rothschild and Ford, 1973).

Blood-sucking insects transmit some of the most debilitating diseases of humans and their domesticated animals. They are also a drain on the agricultural industry by directly causing meat and milk losses, damage to hides and wool, etc. For these reasons humans use various methods to control insects and prevent transmission of disease. Recently interest has been directed to the potential use of the components of the host immune system carried into the insect in the bloodmeal, as a controlling mechanism. Two strategies are being discussed – attack on the parasite while it is in the insect by use of transmission blocking vaccines or a direct attack on the insect itself.

A question worth asking before trying to design a vaccine directed against the blood-sucking insect is whether target tissues are necessarily restricted to those available in the insect gut? The evidence suggests not. Several reports show that immunoglobulin G, specific antibody or $F_{ab}$ fragments are present in the haemolymph of blood-sucking insects including some mosquitoes (Hatfield, 1988; Ramasamy et al, 1988; Lackie and Gavin, 1989), tsetse flies (Nogge and Giannetti, 1979), *Bovicola ovis* (Eisemann et al., 1994) and *Haematobia irritans exigua* (Allingham et al., 1992). The results of Schlein and Lewis (1976) suggest the same is true for *S. calcitrans* although Pruett and Thomas (1986) were unable to confirm this. The most thorough study is that of Allingham et al. (1992) who showed that intact, functional immunoglobulin G could be detected in the haemolymph of 94% of the buffalo flies fed on heparinized ovine or bovine blood. The authors suggest the concentration detected ($10^{-8}$M) is sufficient to contemplate the targeting of low abundance molecules in the haemolymph which are essential to the flies well-being, e.g. hormones or cell surface receptor molecules.

Several target molecules are being considered as candidates for transmission blocking vaccines. Much interest has centred on Pfs25 – a 25 kDa, epidermal-growth-factor-like, cysteine-rich, lipid-anchored antigen present in zygote/early ookinete stages of *Plasmodium falciparum* (Kaslow et al., 1988). Monoclonal antibodies directed against the antigen can prevent development of the parasite in the mosquito (Kaslow et al., 1991) building confidence that it may be a suitable target for a transmission blocking vaccine. Although the antigen may be expressed in small amounts by gametocytes, the antigen is not detectable in sera from exposed individuals living in endemic areas (Kaslow, 1993). Consequently, in practical use it may be expected that repeated vaccinations would be required because no natural boosting effect will occur. But, as pointed out by Kaslow (1993), this may cease to be a concern with the advent of slow release particle technology which may make the concept of natural boosting or multiple vaccination obsolete. An advantage of the antigen not being expressed in the vertebrate is that the antigens have not, presumably, been exposed to immunoselective pressure and in consequence antigenic diversity and/or poor immunogenicity should not be a problem. Field trials of the vaccine candidate are now in planning stages (Collins and Paskewitz, 1995) and it may well form a valuable component of a multivalent vaccine construct designed to control malaria.

Chitinases are produced by several parasitic organisms including *Plasmodium*, *Leishmania*, *Trypanosoma*, *Brugia* and *Wuchereria*. The helminths use the chitinase to escape from the larvel integument while in the vertebrate host. Whether helminths also use their chitinase in the vector remains to be seen. In contrast, the protozoa use the chitinase to compromise chitinous materials in the vector. The best studied example is *Plasmodium* in the mosquito. Sieber *et al* (1991) showed ultrastructurally that ookinetes crossing the mosquito PM disrupted chitin fibrils suggesting the presence of a chitinase. Huber *et al.* (1991) went on to show chitinase in *P. gallinaceum* ookinetes held under *in vitro* conditions for 15 hours. This chitinase became detectable in the culture supernatant from 20 hours. This is the time at which *P. gallinaceum* ookinetes would be penetrating the PM. This strongly suggests a role for the chitinase in PM transmigration and this was firmly established by Shahabuddin *et al.* (1993). These authors used the chitinase inhibitor allosamidin to specifically block chitinase activity in ookinetes held either *in vitro* or *in vivo* in mosquitoes. *In vitro* the ookinetes progressed successfully to the oocyst stage while *in vivo* no oocysts formed. Addition to *in vivo* experiments of fungal chitinase (which is not susceptible to allosamidin) or polyoxin D (which inhibits chitin synthase) permitted oocysts to develop. Based on this evidence it seems clear that chitinases are used by *Plasmodium* ookinetes to enable them to penetrate mosquito PM. For this reason chitinases are potential candidates as target molecules for transmission blocking vaccines. Cloning and sequencing of the gene is the next step. Several plant and fungal chitinases have been cloned and sequenced (Fuhrman *et al.*, 1992) revealing, unfortunately, that chitinases are a disparate group of molecules showing no clear consensus sequences. This has impeded the cloning of *Plasmodium* chitinase but chitinase purification should reveal sufficient sequence information for cloning to be achieved in the near future (Shahabuddin, 1995).

The chitinase produced by the ookinete is a proenzyme which relies upon the mosquito proteases for its full activation – a process which may ensure the coordination of PM penetration with the mosquito's digestive cycle. Blocking of midgut trypsin using leupeptin or antitrypsin antibodies blocks malaria development while addition of fungal chitinase reverses the outcome suggesting the blocking step is in chitinase-dependent crossing of the PM (Shahabuddin and Kaslow, 1994). Consequently it has been proposed that the midgut proteases may be another target for a transmission blocking vaccine. However, development of vaccines against trypsins is likely to be difficult for the reasons discussed by Lehane (1994). He pointed out that anopheline mosquitoes produce more than one immunologically distinct family of trypsins (Graf *et al.*, 1991) complicating the task at hand. Proteolytic enzymes are also poor immunogens, probably because they attack the antibody-making machinery approaching them and because the complexes formed between the abundant vertebrate antiproteolytic molecules and proteases provide a different antigenic target to the native enzymes (McKerrow *et al.*, 1991). In addition, blood-sucking insect digestive proteases by their nature will effectively digest antibodies – the question is would this be achieved before they are neutralized by the antigen. It seems questionable if antibodies would win this armed struggle when the fight is on the home territory of the digestive enzymes. Tellam *et al.* (1994) developed an antiprotease vaccine for larval *Lucilia cuprina*, the causative agent of blowfly strike, but after *in vitro* and *in vivo* trials the authors concluded that serine proteases were unlikely to be effective antigens in that instance. Nevertheless the assertion of Shahabuddin and Kaslow (1994), that antibody directed at trypsin prevents *Plasmodium* development, is evidence that antiprotease vaccines may abolish *Plasmodium* development in mosquitoes. However, such a vaccine may fail for other reasons. Ideally any vaccine directed at the mosquito will not reduce fecundity but vaccines successfully targeted at proteases almost certainly will. This will place the insect under a strong selection pressure – a pressure not imposed by a vaccine directed at the chitinase which might conceivably increase mosquito egg output.

While some progress has been made in development of transmission-blocking vaccines, very little progress has been made towards the development of a vaccine targeted directly against any of the blood-sucking insects. Perhaps the most promising data are those of Alger and Cabrera (1972) who have shown an increase in the mortality rate of *Anopheles stephensi* fed on rabbits immunized with mosquito antigen. Some interesting work has also been performed on the ectoparasitic larvae of *Lucilia cuprina* which cause sheep strike. East *et al.* (1993) have shown that vaccines directed against the protein components of the PM of larval *L. cuprina* lower their average weight by 50% compared to those grown on control sheep. But the greatest success has been achieved not against insects but against ticks (see Chapter 12). The most advanced work is against *Boophilus microplus* where a vaccine raised against a gut-based, membrane-bound 83 kDa glycoprotein has led to a commercial vaccine (Willadsen and McKenna, 1991). The vaccine causes severe damage to the tick gut leading to decreased survival, decreased feeding success and fecundity.

For a variety of reasons, largely centred on the possibility that insects or the diseases they transmit can be ameliorated by manipulation of the insect alimentary canal, the midgut of blood-sucking insects is attracting increasing attention. I hope this review makes it clear that while we may be fortuitous in achieving that end with the little information we have to hand a tremendous amount of basic research is still required before the digestive processes of blood-sucking insects are clearly understood.

# REFERENCES

Alger, N.E. and Cabrera, E.J. (1972) An increase in the death rate of *Anopheles stephensi* fed on rabbits immunized with mosquito antigen. *Journal of Economic Entomology* 65, 165–168.

Allingham, P.G., Kerlin, R.L., Tellam, R.L., Briscoe, S.J. and Standfast, H.A. (1992) Passage of host immunoglobulin across the midgut epithelium into the hemolymph of blood-fed buffalo flies *Haematobia irritans exigua*. *Journal of Insect Physiology* 38, 9–17.

Altman, P.L. and Dittmer, D.S. (1971) Blood and other body fluids. In *Respiration and Circulation*. Federation of American Societies of Experimental Biology, Bethesda, Maryland.

Bailey, L. (1952) The action of the proventriculus of the worker honeybee. *Journal of Experimental Biology* 29, 310–327.

Barillas-Mury, C. and Wells, M.A. (1993) Cloning and sequencing of the bloodmeal-induced late trypsin gene from the mosquito *Aedes aegypti* and characterization of the upstream regulatory region. *Insect Molecular Biology* 2, 7–12.

Becker, B. (1978) Determination of the formation rate of peritrophic membranes in some Diptera. *Journal of Insect Physiology* 24, 529–533.

Berner, R., Rudin, W. and Hecker, H. (1983) Peritrophic membranes and protease activity in the midgut of the malaria mosquito, *Anopheles stephensi* (Liston) (Insecta, Diptera) under normal and experimental conditions. *Journal of Ultrastructural Research* 83, 195–204.

Billingsley, P.B. (1990) The midgut ultrastructure of haematophagous arthropods. *Annual Review of Entomology* 35, 219–248.

Billingsley, P.F. and Rudin, W. (1992) The role of the mosquito peritrophic membrane in bloodmeal digestion and infectivity of *Plasmodium* species. *Journal of Parasitology* 78, 430–440.

Binnington, K., Lehane, M.J. and Beaton, C.D. (1996) The peritrophic membrane. In: Locke, M. (ed.), *The Microscopic Anatomy of Invertebrates–Insect Structure*. John Wiley & Sons, New York.

Buxton, P.A. (1948) Experiments with lice and fleas. I. The baby mouse. *Parasitology* 39, 119–124.

Collins, F.C. and Paskewitz, S.M. (1995) Malaria: Current and future prospects for control. *Annual Review of Entomology* 40, 195–219.

de Azambuga, P., Guimaraes, J.A. and Garcia, E.S. (1983) Haemolytic factor from the crop of *Rhodnius prolixus*, evidence and partial characterization. *Journal of Insect Physiology* 29, 833–837.

East, I.J., Fitzgerald, C.J., Pearson, R.D., Donaldson, R.A., Vuocolo, T., Cadogan, L.C., Tellam, R.L. and Eisemann, C.H. (1993) *Lucilia cuprina* – inhibition of larval growth induced by immunization of host sheep with extracts of larval peritrophic membrane. *International Journal of Parasitology* 23, 221–229.

Eguchi, M., Iwamoto, A. and Jamauchi, K. (1982) Interrelation of proteases from the midgut lumen, epithelia and peritrophic membrane of the silkworm, *Bombyx mori* L. *Comparative Biochemistry and Physiology* 72A, 359–363.

Eisemann, C.H., Pearson, R.D., Donaldson, R.A. and Cadogan, L.C. (1994) Ingestion of host antibodies by *Bovicola ovis* on sheep. *International Journal of Parasitology* 24, 143–145.

Farmer, J., Maddrell, S.H.P. and Spring, J.H. (1981) Absorption of fluid by the midgut of *Rhodnius*. *Journal of Experimental Biology* 94, 301–316.

Fuhrman, J.A., Lane, W.S., Smith, R.F., Piessens, W.F. and Perler, F.B. (1992) Transmission-blocking antibodies recognize microfilarial chitinase in brugian lymphatic filariasis. *Proceedings of the National Academy of Sciences* 89, 1548–1552.

Gee, J.C. (1975) Diuresis in the tsetse fly *Glossina austeni*. *Journal of Experimental Biology* 63, 381–390.

Geering, K. (1975). Haemolytic activity in the blood clot of *Aedes aegypti*. *Acta tropica* 32, 145–151.
Gooding, R.H. (1972) Digestive processes of haematophagous insects. I. A literature review. *Quaestiones Entomologicae* 8, 5–60.
Gooding, R.H. (1977) Digestive processes of haematophagous insects. XIV Haemolytic activity in the midgut of *Glossina morsitans morsitans* Westwood (Diptera: Glossinadae). *Canadian Journal of Zoology* 55, 1899–1905.
Gooding, R.H. and Rolseth B.M. 1976. Digestive processes of haematophagous insects. XI Partial purification and some properties of six proteolytic enzymes from the tsetse fly *Glossina morsitans morsitans* Westwood (Diptera: Glossinidae). *Canadian Journal of Zoology* 54, 1950–1959.
Gooding, R.H., Cheung, A.C. and Rolseth, B.M. (1973) The digestive processes of haematophagous insects. III Inhibition of trypsin by honey and the possible functions of the oesophageal diverticula of mosquitoes (Diptera). *Canadian Entomologist* 105, 433–436.
Graf, R. and Briegel, H. (1982) Comparison between aminopeptidase and trypsin activity in blood fed females of *Aedes aegypti*. *Revue Suisse Zoologie* 89, 845–850.
Graf, R. and Briegel, H. (1985). Isolation of trypsin isozymes from the mosquito *Aedes aegypti* (L.). *Insect Biochemistry* 15, 611–618.
Graf, R., Boehlen, P. and Briegel, H. (1991). Structural diversity of trypsin from different mosquito species feeding on vertebrate blood. *Experientia* 47, 603–609.
Gwadz, R.W. (1969) Regulation of bloodmeal size in the mosquito. *Journal of Insect Physiology* 15, 2039–2044.
Hatfield, P.R. (1988) Detection and localization of antibody ingested with a mosquito bloodmeal. *Medical and Veterinary Entomology* 2, 339–345.
Hecker, H. and Rudin, W. (1979) Normal versus $\alpha$ amantin induced cellular dynamics of the midgut epithelium in female *Aedes aegypti* L. (Insecta: Diptera) in response to blood feeding. *Journal of Cell Biology* 19, 160–167.
Houk, E.J. and Hardy, J.L. (1979) *In vivo* negative staining of the midgut continuous junction in the mosquito *Culex tarsalis* (Diptera: Culicidae). *Acta Tropica* 36, 267–275.
Houseman, J.G., Campbell, F.C. and Morrison, P.E. (1987) A preliminary characterization of digestive proteases in the posterior midgut of the stablefly *Stomoxys calcitrans* (L.) (Diptera: Muscidae). *Insect Biochemistry* 17, 213–218.
Huber, H., Cabib, E. and Miller, L.H. (1991) Malaria chitinase and penetration of the mosquito peritrophic membrane. *Proceedings of the National Academy of Science USA*, 88, 2807–2810.
Jordan, A.M. and Curtis, C.F. (1968) The performance of *Glossina austeni* when fed on lop-eared rabbits and goats. *Transactions of the Royal Society of Tropical Medicine and Hygiene* 62, 123–124.
Kanwar, Y.S. and Farquhar, M.G. (1979) Anionic sites in the glomerular basement membrane: *in vivo* and *in vitro* localization in lamina rarae by cationic probes. *Journal of Cell Biology* 81, 137–153.
Kaslow, D.C. (1993) Transmission blocking immunity against malaria and other vector borne diseases. *Current Opinions in Immunology* 5, 557–565.
Kaslow, D.C., Quakyi, I.A., Syin, C., Raum. M.G., Keister, D.B., Coligan, J.E., McCutchan, T.F. and Miller, L.H. (1988) A vaccine candidate from the sexual stage of human malaria that contains EGF-like domains. *Nature* 333, 74–76.

Kaslow, D.C., Isaacs, S.N., Quakyi, I.A., Gwadz, R.W., Moss, B. and Keister, D.B. (1991) Induction of *Plasmodium falciparum* transmission-blocking antibodies by recombinant vaccinia virus. *Science* 252, 1310–1313.

Kirch, H.J., Spates, G., Drokeskey, R., Kloft, W.J. and De Loach, J.R. (1991) Mechanism of haemolysis of erythrocytes by haemolytic factors from *Stomoxys calcitrans* (L.) Diptera: Muscidae). *Journal of Insect Physiology* 37, 851–861.

Lackie, A.M. and Gavin, S. (1989) Uptake and persistence of ingested antibody in the mosquito *Anopheles stephensi*. *Medical and Veterinary Entomology* 3, 225–230.

Law, J.H., Ribeiro, J.M.C. and Wells, M.A. (1992) Biochemical insights derived from insect diversity. *Annual Reviews of Biochemistry* 61, 87–111.

Lehane, M.J. (1977). Transcellular absorption of lipids in the midgut of the stablefly, *Stomoxys calcitrans*. *Journal of Insect Physiology* 23, 945–954.

Lehane, M.J. (1991) *Biology of Blood-sucking Insects*, 1st ed. Chapman and Hall, 288pp.

Lehane, M.J. (1994) Digestive enzymes, hemolysins and symbionts in the search for vaccines against bloodsucking insects. *International Journal of Parasitology* 24, 27–32.

Lehane, M.J. and Msangi, A.R. (1991) Lectin and peritrophic membrane-development in the gut of *Glossina m. morsitans* and a discussion of their role in protecting the fly against trypanosome infection. *Medical and Veterinary Entomology* 5, 495–501.

Lehane, M.J., Blakemore, D., Williams, S. and Moffatt, M.R. (1995) Mini review: Regulation of digestive enzyme levels in insects. *Comparative Biochemistry and Physiology. B* 110, 285–289.

Ma, W.C. and Denlinger, D.L. (1972) Secretory discharge and microflora of milk gland in tsetse flies. *Nature London* 247, 301–303.

MacVicker, J.A.K., Billingsley, P.F. and Djamgoz, M.B.A. (1993) ATPase activity in the midgut of the mosquito, *Anopheles stephensi* – biochemical characterization of ouabain-sensitive and ouabain-insensitive activities. *Journal of Experimental Biology* 174, 167–183.

MacVicker, J.A.K., Billingsley, P.F., Djamgoz, M.B.A. and Harrow, I.D. (1994) Ouabain-sensitive Na+/K+-ATPase activity in the reservoir zone of the midgut of *Stomoxys calcitrans* (Diptera, Muscidae). *Insect Biochemistry and Molecular Biology* 24, 151–159.

Madrell, S.H.P. (1963) Control of ingestion in *Rhodnius prolixus* Stal. *Nature* 198, 210.

Megahed, M.M. (1958) The distribution of blood, water and sugar solutions in the mid gut and oesophagela diverticulum of *Culicoides nubeculosus* Meigen (Diptera: Ceratopoginidae). *Bulletin of the Society of Entomology. Egypt* 42, 339–355.

McKerrow, J.H., Newport, G. and Fishelson, Z. (1991) Recent insights into the structure and function of a larval protease involved in host infection by a multicellular parasite. *Proceedings of the Society of Experimental Biology and Medicine* 197, 119–124.

Miller, N. and Lehane, M.J. (1990) In-vitro perfusion studies on the peritrophic membrane of the tsetse-fly *Glossina morsitans morsitans* (Diptera: Glossinidae). *Journal of Insect Physiology* 36, 813–818.

Modespacher, U.P., Rudin, W., Jenni, L. and Hecker, H. (1986) Transport of peroxidase through the midgut epithelium of *Glossina m. morsitans* (Diptera: Glossinidae). *Tissue Cell* 18, 429–436.

Moffatt, M.A., Blakemore, D. and Lehane, M.J. (1995) Studies on the synthesis and secretion of trypsin in the midgut of *Stomoxys calcitrans*. *Comparative Biochemistry and Physiology B* 110, 291–300.

Muller, H.M., Crampton, J.M. Dellatorre, A., Sinden, R. and Crisanti, A. (1993) Members of a trypsin gene family in *Anopheles gambiae* are induced in the gut by bloodmeal. *EMBO Journal* 12, 2891–2900.

Murray, M.D. and Nichols, D.G. (1965) Studies on the ectoparasites of seals and penguins. I. The ecology of the louse *Lepidophthirus macrorhini* Enderlein on the southern elephant seal, *Mirounga leonina* (L.). *Australian Journal of Zoology* 13, 437–454.

Nogge, G. (1978) Aposymbiotic tsetse flies, *Glossina morsitans morsitans*, obtained by feeding on rabbits immunized specifically with symbionts. *Journal of Insect Physiology* 24, 299–304.

Nogge, G. (1981) Significance of symbionts for the maintenance of an optimal nutritional state for successful reproduction in haematophagous arthropods. *Parasitology* 82, 101–104.

Nogge, G. and Gianetti, M. (1979) Midgut absorption of undigested albumin and other proteins by the tsetse *Glossina morsitans morsitans*. *Journal of Medical Entomology* 16, 263.

Peacock, A.J. (1982) Effects of sodium transport inhibitors on diuresis and mid gut (Na+ and K+) ATPase in the tsetse fly *Glossina morsitans*. *Journal of Insect Physiology* 28, 553–558.

Pell, P.E. and Southern, D.I. (1976) Effect of the coccidiostat, sulphaquinoxline, on symbiosis in the tsetse fly, *Glossina* species. *Microbios Letters* 2, 203–211.

Peters, W. (1992) *Peritrophic Membrane*. Springer Verlag.

Peters, W., Kolb, H. and Kolb-Bachofen, V. (1983) Evidence for a sugar receptor (lectin) in the peritrophic membrane of the blowfly larva, *Calliphora erythrocephala* Mg. (Diptera). *Journal of Insect Physiology* 29, 275–280.

Pruett, J.H. and Thomas, D.B. (1986) Post-bloodmeal analysis of *Stomoxys calcitrans* (L.) haemolymph for the presence of bovine immunoglobulin G. *Journal of Insect Physiology* 32, 9–16.

Ramasamy, M.S., Ramasamy, R., Kay, B.H. and Kidson, C. (1988) Anti-mosquito antibodies decrease the reproductive capacity of *Aedes aegypti*. *Medical and Veterinary Entomology* 2, 87–93.

Reeves, S.A. and Yamanaka, M.K. (1993) Cloning and sequence-analysis of the alpha-subunit of the cat flea sodium-pump. *Insect Biochemistry and Molecular Biology* 23, 809–814.

Reinhardt, C., Schulz, U., Hecker, H. and Freyvogel, T.A. (1972) Xur Ultrastruktur des Mitteldarmepithels bei flohen (Insecta, Siphonaptera). *Revue Suisse Zoologie* 79, 1130–1137.

Rennke, H.G., Cotran, R.S. and Venkatachalam, M.A. (1975) Role of molecular charge in glomerular permeability: tracer studies with cationized ferritins. *Journal of Cell Biology* 67, 638–646.

Richards, A.G. and Richards, P.A. (1968) Flea *Ctenophthalmus*: heterogeneous hexagonally arranged layer in the midgut. *Science* 160, 423–425.

Rothschild, M. and Ford, B. (1973) Factors influencing the breeding of the rabbit flea (*Spilopsyllus cuniculi*) a spring-time accelerator and a kairomone in nesting rabbit urine (with notes on *Cediopsyla simplex*, another 'hormone bound' species). *Journal of Zoology* 170, 87–137.

Sangiorgi, G. and Frosini, D. (1940) Di un principio emolitico ('Cimicina') nella saliva del *Cimex lectularius. Pathologica* 32, 189–191.
Schlein, Y. and Lewis, C.T. (1976) Lesions in haematophagous flies feeding on rabbits immunized with fly tissues. *Physiological Entomology* 1, 55–59.
Shahabuddin, M. (1995) Chitinase as a vaccine. *Parasitology Today* 11, 46–47.
Shahabuddin, M. and Kaslow, D.C. (1944) *Plasmodium* parasite chitinase and its role in malaria transmission. *Experimental Parasitology* 79, 85–88.
Shahabuddin, M., Toyoshima, T., Aikawa, M. and Kaslow, D.C. (1993) Transmission-blocking activity of a chitinase inhibitor and activation of malarial parasite chitinase by mosquito protease. *Proceedings of the National Academy of Sciences USA*, 90, 4266–4270.
Sieber, K.P., Huber, M., Kaslow, D.C., Banks, S.M., Torii, M., Aikawa, M. and Miller, L.H. (1991). The peritrophic membrane as a barrier: its penetration by *Plasmodium gallinaceum* and the effect of a monoclonal antibody to ookinetes. *Experimental Parasitology* 72, 145–156.
Spates, G.E., Stipanovic, R.D., Williams, H. and Holman, G.M. (1982) Mechanisms of haemolysis in a blood-sucking dipteran, *Stomoxys calcitrans. Insect Biochemistry* 12, 707–712.
Tellam, R.L., Eisemann, C.H. and Pearson, R.D. (1994) Vaccination of sheep with purified serine proteases from the secretory and excretory material of *Lucilia cuprina* larvae. *International Journal of Parasitology* 24, 757–764.
Terra, W.R. (1988) Physiology and Biochemistry of insect digestion an evolutionary perspective. *Brazilian Journal of Medical and Biological Research* 21, 675–734.
Terra, W. and Ferreira, C. (1994) Insect digestive enzymes: properties compartmentalisation and function. *Comparative Biochemistry and Physiology* 109, 1–62.
Trembley, H.L. (1952) The distribution of certain liquids in the oesophagela diverticula and stomach of mosquitoes. *American Journal of Tropical Medicine and Hygiene* 1, 693–710.
Waterhouse, D.F. (1954) The rate of production of the peritrophic membrane in some insects. *Australian Journal of Biological Science* 7, 59–72.
Willadsen, P. and McKenna, R.V. (1991) Vaccination with 'concealed' antigens: myth or reality? *Parasite Immunology* 13, 605–616.
Woke, P.A. (1937) Comparative effects of the blood of man and of canary on egg production of *Culex pipiens* Linn. *Journal of Parasitology* 23, 311–313.

# 7

# Immune Responses to Fleas, Bugs and Sucking Lice

## Carl J. Jones

*Department of Veterinary Pathobiology, College of Veterinary Medicine, University of Illinois, Urbana, Illinois 61801, USA*

The insects that are going to be discussed in this chapter are members of three separate orders: Siphonaptera, the fleas; Hemiptera, particularly the bedbugs and assassin bugs; and Anoplura, the sucking lice. All of these are primarily solenophagous, feeding from small blood vessels. Of the three orders, only the Siphonaptera have complete metamorphosis, and spend their immature stages developing off the host. However, many species of fleas are entirely dependent on host-derived substances, with adults feeding primarily on blood while larvae feed on blood from adult faeces as well as detritus from the nest of hosts. The true bugs under discussion will be members of the suborder Hemiptera, which undergo gradual metamorphosis. The families under discussion, Cimicidae and Reduviidae, contain most blood-sucking species in this order, the rest are predominantly plant feeders. Anoplura also undergo gradual metamorphosis, with all mobile stages parasitic. In some texts Anoplura and Mallophaga, the chewing lice, are considered suborders of Pthiraptera.

Because blood-feeding evolved independently in these groups, we should assume that there are differing chemical mediators which can be used to achieve the common goals of increased blood-feeding efficiency and decreased danger to the blood seeker from host behavioural or physiological response. Pruritic associations between hosts and permanent ectoparasite will result in different degrees of co-adaptation than may evolve between hosts and those micropredators whose ancestors were true predators. In an optimal situation, host stimulation by parasites results in minimal damage to the host at

© CAB INTERNATIONAL 1996. From Wikel, S.K. (ed.) *The Immunology of Host–Ectoparasitic Arthropod Relationships.* CAB INTERNATIONAL, Wallingford.

relatively low levels of infestation; while when parasite populations surpass a certain level, the physiological expense to the host necessitates a reaction to prevent a decline in host fitness. Selection should benefit the parasite that stimulates its host to minimize damage itself at low parasite population levels, yet fosters a habitat more suitable to the parasite in question than to its competitors (Jones, 1989). As population levels of a parasite grow progressively higher, the probability of multiple and deleterious effects on the hosts increases as antigens from parasites, or lost energy, affect the host's physiological capacity for normal functions. There are, of course, cases in which an undirected response results in severe damage to the host. Much of what we know about the three groups in this chapter is the result of studies which were initiated because of hypersensitivity responses which cause damage or discomfort to humans or their pets and livestock. In addition, increased food intake and need for protein to replace damaged or lost tissue, decreased hair size and quality, and immune responses to secondary infections (Nelson, 1984, 1989) can all result from excessive parasite burdens.

## SIPHONAPTERA – FLEAS

Fleas are monophyletic and must have originated during the Jurassic, having probably been dispersed throughout Pangea (Traub, 1980). The family Pulicidae are of African origin, but Traub (1980) has postulated that one branch drifted to South America presumably with hosts and *Pulex* species currently existing in South America originated from this stock.

Flea genera and species within genera do not all exhibit the same degree of dependence on individual hosts or on a single species of host. Some adult fleas remain on the host only for as long as is required to take a bloodmeal and spend most of their post-prandial time off the host. Animals which spend most of their time in nests or burrows are likely hosts for these flea species.

The cat flea, *Ctenocephalides felis felis*, exemplifies a lifestyle in which the parasites spend most of their time on the host. They rarely leave the host even though relatively little time is spent feeding. Dryden (1989) demonstrated that 58% of female *C. felis felis* reside on the same animal for up to 113 days. Fleas like *C. felis felis* are able to feed and move freely on a single host signifying a long-term relationship with a single host specimen, although they are able to reproduce successfully after feeding on more than one species of host. Whether this is the result of modifications of host immune response or the parasite's behavioural and physical abilities to evade grooming behaviour, is unknown. The third feeding strategy, which is used by some fleas in the genus *Tunga*, is one in which females attach permanently to the host while feeding. Some become embedded in the host either as the result of host tissues responding to the physical irritation they cause or, less often, by their own activity. These feed primarily by penetrating the blood vessels as do the other two groups, but are not limited to this food source, and do not feed as rapidly as the other two groups.

## Specificity

Many species of fleas are opportunistic feeders, but some appear to have evolved such close ties to a host species that it is doubtful that they could successfully survive without that host species. In a series of classic studies, Rothschild and Ford (1964, 1966) showed that the reproductive cycle of the rabbit flea, *Spilopsyllus cuniculi*, is under control of the reproductive cycle of the host rabbit. The female flea matures during the last 10 days of the doe's pregnancy and transfers to the newborn rabbits within seven days of parturition. When they have mated successfully on the new hosts, oviposition begins. Fleas staying on the doe undergo regression of the eggs and oocyte resorption. During the next host reproductive cycle the fleas go through the same process. Somatotropin has been postulated to be a copulation factor and luteinizing hormone and progestins secreted by the rabbit are egg regression factors.

That this is a highly specific relationship may be further inferred by cutaneous papules, modules and excoriations found on the ear pinnae of cats which had been hunting rabbits infested with *S. cuniculi* (Studdert and Arundel, 1988). Physical transfer of this species to a new host is relatively easy to accomplish, but the severity of the symptoms on the cats indicated that they had much higher response levels to salivary components of rabbit fleas than to the cat fleas with which they were infested.

Benjamini *et al.* (1961) gave a preliminary description of the progress of sensitivity to flea feeding by guinea pigs, the complete description developed to describe the interactions between arthropods and vertebrate hosts was published by Larrivee *et al.* (1964) from data developed through observation of guinea pigs fed upon by *C. felis felis*.

I. Formally, stage one is the induction of response with no abnormal changes observable in the skin or microscopically.
II. Delayed hypersensitivity showing an intense infiltration by mononuclear cells approximately 24 hours after the bite.
III. Both immediate and delayed hypersensitivity exists simultaneously. Eosinophilia begins within 20 minutes after the initial bite.
IV. Immediate hypersensitivity only.
V. A complete lack of observed skin reactivity. No abnormalities observed either cellular or microscopic changes in the skin.

## Response

In guinea pigs, delayed hypersensitivity to feeding by *Pulex irritans*, *P. simulans*, and *C. felis felis* develops within 12–24 hours after feeding (Hudson *et al.*, 1960). Five to seven days of feeding were required for both immediate hypersensitivity and delayed hypersensitivity reactions to be demonstrable. Host sensitivity, as measured by severity of response, appeared, generally, to be independent of the species of flea. Strongest response period was 7 to 14 days after feeding began.

Benjamini *et al.* (1963) advanced the theory that the substances from *C. felis felis* causing hypersensitivity responses by guinea pigs were haptenic in nature. They found that a dialysable portion of the saliva collected through membranes from fleas seeking bloodmeals caused the response. This material was very heat tolerant, requiring more than four hours at 95°C for breakdown. They did, however, admit that *in vivo* feeding may differ substantially from feeding in distilled water through a membrane. As will be discussed later, there are apparent differences in the substances released during probing and those released during feeding by solenophagous arthropods. It is highly probable that the substance they described is used to aid the arthropod in finding blood vessels, because when Michaeli *et al.* (1966) incubated collagen with *C. felis felis* saliva, they found that the active product was not dialysable. They got an assortment of dialysable haptens at high and low pHs, i.e. at 10.6 and 3.5. The immunogenicity of the reaction products resulting from the reaction of saliva and collagen was highest, surprisingly, with saline rather than an adjuvant. At a neutral pH, collagen and saliva did not react, but still caused immunogenicity when injected into a naive host. Previously they had shown (Michaeli *et al.*, 1965) that both neutral and acid-soluble collagens of flea-bite-exposed guinea pigs could induce hypersensitivity when injected into naive guinea pigs. At that time, they thought that flea haptens stayed with the collagen fraction of the skin of bitten hosts.

Halliwell (1984) proved that canine flea sensitivity is not dependent upon the reaction of collagen with a hapten from flea saliva. He showed that the antigen causing hypersensitivity has to be present as a complete compound in fleas because the release of histamine to create the immune hypersensitivity wheal has to result from an antigen bridging two adjacent molecules of IgE on a mast cell. A hapten cannot do this, and collagen comes to the bite-site too slowly to serve to create the immediate hypersensitivity. Halliwell's research demonstrated that flea allergy involves both type I, IgE, and a type IV, cell-mediated hypersensitivity response, and that there are at least two proteins involved, at least one of whose antigenic factors had a molecular mass of greater than 75 kilodaltons (kDa).

Cutaneous basophil hypersensitivity and late onset IgE response may coexist simultaneously in dogs. Halliwell and Schemmer (1987) demonstrated cutaneous basophil hypersensitivity in dogs which were allergic to cat fleas. Basophil infiltration into the dermis was highest 4–18 hours after hypersensitive dogs were intradermally injected with a mixture of cat flea antigens. Granules of both eosinophils and basophils were found accumulating in intercellular spaces within one hour after injection. At four hours, 8–62% of the cells were eosinophils and 2–12% of the cells were basophils. By eight hours after injection the cellular composition had changed slightly and 12–48% were eosinophils while the basophils decreased to 0–12.3% of the total cellular infiltrate. By 18 hours basophils had increased to about 10% of the cells and between 15 and 56% of the cellular infiltrate consisted of eosinophils. Basophils were almost completely absent by 48 hours after injection. In these studies there was an increase in IgG and IgE concentrations in animals treated with flea extract for hypersensitization.

Using immunoblotting techniques to quantify IgE and other antibodies from canine sera, Greene *et al.* (1993) found *C. felis felis* allergens with molecular masses of 25, 40 and 58 kDa in about 40% of the dogs they examined, leading them to postulate that immediate hypersensitivity may be the most important mechanism in the pathogenesis of canine flea allergy. They found 15 different allergens from whole body extracts of *C. felis felis*. McKeon and Opdebeeck (1994) found no pattern to sensitivity and allergic reactivity of dogs to fleas with IgG and IgE antibody responses. They believe intradermal skin tests, at least when they are responses to crude extracts, are inappropriate for the evaluation of an animal's response to flea allergens even when the reactions to this material are positive. They found strong responses to molecules with molecular masses of 224 and 297 kDa. Although there was strong response to bands at lower molecular masses, below 66 kDa there was far more variation in the response, some bands showing weak response in some dogs and strong response in other dogs, suggestive of the variation found by Greene *et al.* (1993).

When *Xenopsylla cheopis*, the oriental rat flea, was allowed to feed on guinea pigs, the hosts developed cutaneous basophilic sensitivity (Johnston and Brown, 1985). Feeding resulted in increased blood basophil levels. Although blood eosinophil levels remained approximately normal, dermal eosinophilia and a maculopapular rash developed. By 24 hours after the initial feeding, the cellular infiltrates were 52% neutrophils, 32% eosinophils and 11% basophils. After a challenge feeding 14 days later, the infiltrate into the dermis consisted of 59% eosinophils and only 7% neutrophils, while basophils had risen to 30% of the cellular infiltrate. Mast cells were showing up in measurable amounts of 4–7%. There were diffused mononuclear fibroblast cell infiltrates at the sites of both the primary and secondary feeding lesions, later superseded by an eosinophil response. There were no effects on the feeding success of the flea as a result of these cellular changes. Brown (1989), discussing the same system, noted that granulocytes concentrate in the upper dermis near mast cell congregations, postulating that chemotaxis probably is the result of non-specific degranulation of mast cells caused by contact with unspecified salivary secretions in the naive hosts. Eosinophil accumulations in the tissues were often surrounded by neutrophils. Generally, skin lesions were weak and dermal disruption and oedema were minor components of the response. Within the first seven days after the first challenge feeding, the size of the lesions increased as did the the number of fibroblasts and macrophages along with the eosinophil concentrations. There were still some basophils remaining along the epidermal/dermal border. In rats, basophils are the only leukocytes which respond to *X. cheopis* feeding (Vaughan *et al.*, 1989). They found no visible histopathology at the bite-sites, and the basophilic response had no apparent effect on either feeding behaviour or survival of the fleas.

As mentioned previously, female fleas in the genus *Tunga* remain attached to a single site 'on' one host throughout their reproductive life. Infestation of mice by *Tunga monositus* resulted in no immunopathologic processes which damaged the hosts (Lavoipierre *et al.*, 1979). In a fascinating report, the authors showed that fleas fed on fluid exudate for the first 24 hours after attachment, then changed their feeding to take advantage of the neutrophilic cellular infiltrate, ingesting primarily neutrophils up to and including the tenth day of infestation. From days 10 to 14, they took advantage of the accumulation of collagen, fibroblasts, and macrophage which accumulated at the site of attachment. This generally was the end of the neosomatic stage of development. After the 14th day, the fleas' mouthparts penetrated a blood vessel and began solenophagous blood-feeding. Only then did oogenesis begin. Throughout this phase, internal haemorrhage in the host appeared to cause damage to the fleas. Neutrophilia surrounding the mouthparts at times impaired the feeding process. Additionally, after day 25 the vascular bed of the hosts began to diminish. This continued until about day 42 when haemorrhages started to form in the feeding site and pool feeding started. By this time, the fleas' mouthparts were frequently encased in a mass of fibrin and collagen containing fibroblasts and disintegrating neutrophils. Normally, fleas were sloughed by the host by day 56, but the process took as long as 90 days. Between days 10 and 56 or 90, fleas generally deposited about 500 eggs. If, however, there was a delay in insemination, they were able to delay reproduction and delay death. A similar course of response was followed after *Tunga penetrans* females were placed on mice. However, *T. penetrans* started blood-feeding within hours of attachment, not within days.

## Clinical Response

Scheidt (1988) wrote that there is more money and energy spent on fleas or flea bite dermatitis and problems surrounding fleas than on any other problem in veterinary medicine.

Nesbitt and Schmitz (1978) described flea bite allergy dermatitis as 'the primary lesion is a papule accompanied by erythema. Secondary lesions result from self-excoriation with breaking of hair, local alopecia, and occasional areas of acute serous dermatitis.' As flea allergy dermatitis goes on over longer periods, alopecia may become diffuse, there may also be benign overgrowth of the prickle cell layer of skin (also known as acanthosis), and there is likely to be hyperkeratinization of the dog's skin. Relatively few dogs respond to flea salivary secretions (Baker and O'Flanagan, 1975). However, of those that do respond, all seem to develop flea bite hypersensitivity. Atopic dogs may be genetically predisposed to flea bite allergy. In clinical studies in Florida more than 70% of atopic dogs gave positive responses to flea allergens.

Challenge exposure to virgin *C. felis felis* feeding by dogs (Gross and Halliwell, 1985) induced flea bite hypersensitivity. They observed an initial wheal with a perivascular eosinophilic infiltrate, at least superficially, plus oedema. Twenty-four hours later, there was a mononuclear infiltrate, continuing eosinophilia, and some infiltration of histiocytes. Mast cells were found around blood vessels as early as 15 minutes after fleas began feeding and were present until the end of the trials.

Cats which are hypersensitive to *C. felis felis* bites develop feline Miliary dermatitis (McDougal, 1986). Common responses include perivascular dermatitis, hyperplasia, spondylosis and dermal infiltrates with neutrophils, mononuclear cells, eosinophils and mast cells. Cats frequently get secondary responses as well.

Kristensen and Kieffer (1978) tested 143 animals which demonstrated signs of different sorts of pruritic skin disease. In a clinical setting, 32% of the dogs and 65% of the cats showed flea allergy hypersensitivity during skin tests. This may have underestimated the percentage of sensitive cats since Baker (1974) determined that intradermal tests for cat dermatitis were usually negative at least in part because the dermis is too thin to produce measurable swellings in cats, and Moriello and McMurdy (1989) demonstrated that a positive skin test to whole flea extracts is not diagnostic for flea allergy dermatitis in felines.

## Desensitization

Flea bite allergies and hypersensitivity have led to much of the research on the cat flea and its relationship to pets. Many workers have attempted desensitization procedures, and there are dozens of 'successful' procedures described in the veterinary literature. In humans, spontaneous desensitization of previously allergic patients is reported to *Pulex* spp. (Barriga, 1981). The use of prednisone in canines with delayed and immediate hypersensitivity response to flea bites results in significant decreases in response.

Using *C. felis felis* salivary secretions at a rate of 100,000 fleas per 100 µg of saliva, Benjamini and Feingold (1960) injected 3 µg of material into guinea pigs which had previously been sensitized by intradermal injection of whole flea extract or by IP injection with oral secretion and adjuvant. Because they elicited no dermal response to the extract alone, they postulated that the oral secretion itself contained an incomplete antigen that could only cause a response when combined with a carrier molecule. Later work would show that this hapten theory was incorrect.

Some systemic allergic phenomena (Benjamini et al., 1960) can occur in guinea pigs after *C. felis felis* feeding. Both flea bites and whole flea extracts sensitize guinea pigs to the other, i.e. using real flea bites created sensitivity to whole flea extracts and using whole flea extracts created sensitivity to flea feeding. This simple experiment laid the groundwork for desensitization studies in which whole flea extracts could be used, the basis of most veterinary clinical desensitization programmes.

Michaeli and Goldfarb (1968) used a hyposensitizing preparation with 10 mg salivary gland equivalents in a 1% sodium alginate solution to desensitize dogs. They felt that this solution contained the haptenic fraction discussed by Benjamini et al. (1960). In their studies, they used weekly subcutaneous or intradermal injections to cure 77% of hypersensitive animals. If a later recurrence of sensitivity occurred, they were able to cure that with a 'booster shot'. Whole flea extract hyposensitization of dogs was assessed as 80.6% successful in studies by Nesbitt and Schmitz (1978), but Halliwell (1981) used a double-blind clinical study to test the effect of ground flea injections against flea allergic dermatitis and found that the success rates he was able to show were extremely marginal. Often clinicians depend on the judgement of the patients' owners in cases of hypersensitivity. When the relative objectivity of owners who want their dog to improve is combined with a conscientious effort to eliminate a pest flea population, the results may be less than scientific.

There is excellent evidence for natural hyposensitization in dogs. Dogs exposed to fleas at rates of one to three times per week become allergic to flea bites in about eight weeks (Halliwell, 1984). More frequent or continual exposure resulted in short-lived responses that were less intense and had delayed onset. Halliwell and Longino (1985) determined that there were high levels of IgE and IgG in dogs with clinical signs of hypersensitivity to *C. felis felis*. Flea exposed dogs that do not have apparent allergic response have IgE levels near background and very low IgG levels. They believe that chronic flea exposure may result in partial or complete tolerance in a certain portion of the canine population. Treatment of dogs with whole flea extracts for hyposensitization (Schemmer and Halliwell, 1987) results in increases in both IgG and IgE concentrations.

Although Opdebeeck and Slacek (1993) found significant elevation of anti-flea antibodies in sera, their injection of gut membrane antigens of *C. felis felis* failed to protect against infestation with no effect on flea fecundity.

Hayasaki *et al.* (1993) showed that six of nine dogs demonstrating flea allergic dermatitis responded favourably to injections of whole flea antigens. There were slight increases in indirect haemagglutination titres in three of the six dogs which responded positively, and Prausnitz–Kustner (P–K) responses decreased in two.

Obviously the status of immune responses to fleas is at the stage where advances can be made fairly quickly in the laboratory, if research funds can be found. Moving toward utilization of the information which will soon be available will require collaborative efforts on the part of molecular biologists, government regulators, veterinary clinicians and industry. Unlike many other areas of basic research, there is tremendous opportunity for studies that can result in products with immediate application. These studies may provide sufficient financial, as well as humanitarian, benefit that it seems likely that many commercial products will be available to control flea bite hypersensitivity very early in the 21st century.

## HEMIPTERA — THE TRUE BUGS

The evolution of Cimicidae probably started on bats and spread to other vertebrates, movement to the other vertebrates probably occurred as a consequence of early cave habitation by both hominids and birds (Usinger, 1966). Reduviidae probably were ancestrally predatory. Some species are still cannibalistic (Wood, 1976). In fact, adaptations of the salivary venoms of some of the original ancestral forms probably resulted in the anaesthetic components of saliva of modern reduviids (Waage, 1979).

The bugs have piercing–sucking mouthparts. There is a supporting sheath, usually jointed, which is folded ventrally when kissing bugs or bed bugs are not feeding, preparing to feed, or using their mouthparts in self-defence. Feeding is solenophagous but when used in defence, the mouthparts can deliver a painful bite. When feeding, they penetrate the vessel, feed, and remove their mouthparts without excessive damage to the vessel. The process has been described as being very similar to cannulating a blood vessel.

Because bugs feed nocturnally they feed for a few minutes at a time, rarely longer. They are able to withdraw their mouthparts at a very rapid rate if the host moves during the feeding process. Prostaglandins, peptides, and nitrovasodilators may be among the solutions present in arthropods' salivary glands which are used to increase their blood-feeding efficiency (Ribeiro, 1987), and much of the research demonstrating the presence of these chemical mediators in arthropods had its origins in studies carried out on Reduviidae.

## Host Response

Ryckman (1979) gave an excellent historical overview of the initial studies of the effects of Cimicidae and Reduviidae on humans. Only symptomatic treatments are available for relief of human hypersensitivity responses to Cimicidae and Reduviidae, these are used primarily for relief from the responses caused by proteolytic substances injected by the micropredators. All five stages of reactivity are shown to bedbug bites by humans. After long exposure, there is loss of sensitivity by a portion of the population although some remain hypersensitive for years. Both immediate and delayed hypersensitivity have been demonstrated. Humans develop papular lesions with a central vesicle 2–3 cm$^2$ surrounding the puncture mark. Urticaria with oedema is possible, and unilateral nodular to bulbous haemorrhagic lesions may form on the hands and feet of some individuals. There is frequently itching at the site of the bite and a generalized response characterized by hives and itching may spread over the whole body. Severe cases may result in fainting, nausea, vomiting, or other extreme responses including shock. The physical diagnostic characteristics may include some of those mentioned above, as well as angioneurotic oedema, hypotension and oedema of the tongue and larynx. IgE responses to salivary antigens from reduviid bugs have been specifically implicated in incidences of anaphylaxis (Pinnas *et al.*, 1978).

## Cimicidae, the Bedbugs

Asthma attacks from July to September were attributed to feeding by *Cimex lectularis* (Sternberg, 1929) on a patient with a positive P–K test for bedbugs. Once the insects were eliminated from the individual's bed, there were no more allergy attacks. Although the seasonality of the preceding may seem slightly to imply the involvement of other seasonally prevalent antigenic substances, other early research indicated not only strong responses to bedbug bites, but a high degree of antigenic specificity in response to bites (Hase, 1916 as reported in Benjamini and Feingold, 1970). Patients with strong responses to *C. lectularis* demonstrated no response to *C. rotundatus*, indicating that the salivary component causing the allergy was extremely species specific. However, Usinger (1966) found some cross-reactivity to bites by *Hesperocimex sonorensis*, *Leptocimex duplicatus*, and *Cimex pilosellus* in an individual previously sensitized to *Cimex lectularis*. The pattern of cross-reactivity, as measured by the strength of response, corresponded generally with the taxonomic relationships among bedbugs.

## Reduviidae, Assassin Bugs and Ambush Bugs

Delayed hypersensitivity responses to chemical mediators from *Rhodnius prolixus* developed over a period of time in volunteers serving as bloodmeal sources (Lavoipierre et al., 1959), but the authors were unable to carry out observations for long enough to determine how and whether immediate hypersensitivity responses might have developed. Fox and Bayona (1968) found circulating precipitating antibodies to three salivary components of *R. prolixus* in the rabbit, but little interest in research on antibody response has been shown, probably for one of two reasons: (i) immediate hypersensitivity is rarely seen to triatomine bugs; (ii) reduviid bugs have proven a rich source for studies on a variety of chemical mediators, and research in that area has been so productive as to preclude other interests. Anticoagulants, haemolytic factors, and histamine and other putative mediators have all been found in the saliva of these bugs. Anti-inflammatory agents such as antihistamine, antiserotonin, pyrase, and anticomplement may all be found in the salivary glands of *R. prolixus* alone (Ribeiro, 1987), and it seems likely that parallel evolution would create opportunities for a variety of different molecules to serve similar purposes within the family. Salivary glands of arthropods may be possible sources for chemicals with interesting and useful pharmacological properties according to Ribeiro et al. (1990). The reduviid bugs may be a particularly rich source.

Although it is well known that there is substantial immune response to reduviid bites, the most intriguing research on this group has been focused on the ways in which this group evades the immune response while feeding by both slowing down mammalian response time and optimizing their ability to withdraw blood from the host rapidly.

## Reduviid Salivary Components

As early as 1964, Hellman and Hawkins (1964, 1965) described a blood-clotting factor in *Rhodnius prolixus* which, by acting as an antagonist of clotting factor VIII extended the possible feeding time of *R. prolixus* by delaying thrombin formation. They studied the anticoagulant and fibrinolytic properties of *R. prolixus* of salivary secretions in guinea pigs, as well as the calcium clotting time of horse serum to determine the anticoagulant activity. In their studies anticoagulant activity could be dialysed out and survive in 80°C for 30 minutes with very little damage effected by mild acids or bases. Fibrinolytic activity was hurt by acid but not by base and could not be dialysed out.

Many of the effects of compounds found in the saliva of *R. prolixus* have the effect of deterring platelet aggregation, thereby delaying serotonin, histamine or heparin release, depending on the host. An apyrase discovered in the saliva of *R. prolixus* splits phosphate groups from ATP and has a similar effect on ADP (Ribeiro and Garcia, 1980). Since ADP is a primary inducer of platelet aggregation, by exhaustion of the ADP the apyrase plays a major role in the prevention of platelet aggregation. About 0.04 units of apyrase activity are released by a single bug during probing or normal feeding. Additional effects (Ribeiro and Garcia, 1981a) of *R. prolixus* salivary secretions that act against ADP-induced aggregation include disruption of collagen-induced platelet aggregation. A 19 kDa collagen-induced platelet aggregation inhibitor characterized from *Triatoma pallidipennis* by Noeske-Jungblut et al. (1994) inhibits ATP release from human platelets, but had no apparent effects on other platelet aggregation mediators.

Ribeiro and Sarkis (1982) outlined further activities of salivary secretion of *R. prolixus*, including inhibition of arachidonic acid-induced-platelet aggregation and induction of aortic contraction in rabbit preparations. Further, the salivary secretions contain a heat, acid, and protease insensitive, non-dialysable, substance with antagonistic effects on thromboxane $A_2$. The latter has an apparent molecular mass of 39 kDa, and loss of activity after mild oxidative procedures led them to suggest that it may be a carbohydrate. The antithromboxane $A_2$ activity probably helps to prevent the host from mounting an effective haemostatic defence as well as affecting the ability to call basophils to the feeding site. This secretion may be biologically important because it allows large amounts of blood to be obtained in a very short period of time with low cost to the parasite (Ribeiro and Garcia, 1981b). Both antiserotonin and antihistaminic activity from *R. prolixus* salivary components antagonize serotonin effects in the rat uterus preparations and histamine induced contractions in guinea pig ileum preparation (Ribeiro, 1982).

A salivary vasodilator protein weighing about 16.5 kDa was isolated from *R. prolixus* (Ribeiro et al., 1990). The protein has reactive nitrogen groups which induce dose-dependent transient relaxation of rabbit aorta preparations. Nitrosyl haemoproteins, haemoproteins with a Fe(III) haem which reversibly binds nitric oxide, are the primary vasodilator compounds present in *R. prolixus* salivary secretions (Ribeiro et al., 1993). Nitric oxide also exhibits antiplatelet activity.

## Cellular Responses to Reduviidae

Among the responses to feeding by Reduviidae which are of interest from both a host and a parasite's perspective are cellular responses. Brown and Rosalsky (1984) studied the effects of *T. protracta* on the kinetics of basophils and eosinophils in the peripheral blood of guinea pigs. A slowly developing basophilia, peaking at day five, followed the first feeding on naive hosts. A challenge feeding 14 days later elicited an immediate anamnestic response (within 24 hours) in which basophil populations were more than 120%

higher than naive controls. The basophil population levels in the blood declined rapidly after the peak, and were normal by the time of the third feeding, after which they remained approximately steady for the first 24 hours, but rose to the same levels as after the second feeding within five days. This peak was followed by a rapid decline. There was no response by eosinophils to the first feeding, either. Eosinophil populations were 66% higher in exposed than naive guinea pigs after the second feeding. Eosinophil blood levels had only begun to decline by the time the final challenge feeding was made, 28 days after the first. The short-lived response incurred by this feeding began to decline almost immediately and eosinophil populations in the blood were not significantly different from controls within 10 days. The authors hypothesize that the rapid drop after the third feeding is the result of emigration of basophils from blood into other tissues, where they would play a role in cutaneous responses to feeding. The anamnestic nature of the responses to the second and third feedings demonstrate immunological mediation.

## Immune-mediated Behavioural Responses to Reduviidae

Host behaviour can modify the feeding success and reproduction of some bugs. Irritation caused by salivary secretions during *Triatoma infestans* feeding causes changes in chicken and mouse behaviour that results in host mediation of bloodmeal size, and is based on the size of the micropredator population (Schofield, 1982). Decreased numbers of eggs were deposited by bugs which had fed on unrestrained animals. There appears to be a threshold for bloodmeal size below the (hypothetical) repletion threshold above which fed females were unwilling to return to the host, but were unable to oviposit a full complement of viable eggs. The response to all stages of *T. infestans* was very similar in the living mouse *in vivo*; however, higher antibody concentrations resulted from mature bug feeding than from the feeding of any of the immature stages upon the hosts (Volf *et al.*, 1993). Since *T. infestans* serves as a vector of *Trypanosoma cruzi*, it is interesting to note that infection of vertebrates by *T. cruzi* results in severely diminished humoral and cellular responses to mitogens which are unrelated to the trypanosome (Bitkowska *et al.*, 1982). This may very well be an instance in which the disease and the vector are in a mutualistic relationship. If responses to salivary secretions were lessened in an infected host, then the reduviid would be more likely to achieve repletion, the trypanosome would be more likely to be ingested by the arthropod, and an undamaged insect would be more likely to transmit the pathogen to a new host after optimal reproduction had occurred.

## Clinical Responses

*Triatoma-infestans*-sensitized humans displayed skin reactions to both *T. infestans* and *Dipetalogaster maxima* even with no previous known exposure to *D. maxima* (Costa *et al.*, 1981). Response to *D. maxima* is weaker than to *T. infestans* at first, but continued exposure results in a stronger, more comparable, reaction. Human anaphylaxis to *T. protracta* is IgE mediated (Marshall and Street, 1982). Immunotherapy can be used to raise specific IgG levels to compete with IgE over a period of one year, with strong improvement in patient response to antigens. The antigen(s) which result in the immune response were produced by salivary glands maintained in cell culture (Marshall, 1982). If this were possible with more arthropods, it would be highly beneficial, but little cell culture has been reported with arthropods of medical or veterinary medical importance. Binding of IgE antibodies to *T. protracta* is not inhibited by salivary gland extracts from other *Triatoma* spp. nor by *R. prolixus* extracts (Marshall *et al.*, 1986).

The major allergens from *Triatoma* species are being studied by several scientists, but at this time it seems possible that there will be a wide array of antigens with differences among species in the same genus. Chapman *et al.* (1986) found two major isoallergens in *T. protracta*, both at 18–20 kDa. They were believed to be different isoelectric forms of the same allergen because cross-reaction inhibition was complete for them. The antigens which Marshall *et al.* (1986) found were responsible for human anaphylactic response to *Triatoma* species, as determined by gel electrophoresis, are all of low molecular masses. There are six proteins with molecular masses between 17 and 25 kDa that may be responsible (Chapman *et al.*, 1986), while a glycoprotein of approximately 70 kDa appears to elicit no response from hypersensitive humans.

There are bands of mouse serum antibodies to *T. infestans* salivary components at 100 kDa, 120 kDa, where there were two double bands, and 80 kDa (Volf *et al.*, 1993). Glycan structures in the main components of *T. infestans* saliva at 18–35 kDa did not react with antibodies of mice, even though a bi- or tri-antennary complex glycan showed affinity to pea lectins.

## ANOPLURA, THE SUCKING LICE

Early forms of sucking lice were probably found on rodents as early as the Palaeocene (Traub, 1980). By the Cretaceous, lice resembling the current Anoplura had probably arisen in North America (Kim and Ludwig, 1978). Parallel evolution may account for the presence of *Pediculus* species in both Africa and South American monkeys. It is postulated by some authors, however, that secondary infestations from New World humans may have been the ancestors of the species of anoplurans found on South American monkeys (Traub, 1980). *Pthirus* and *Pediculus* species probably originated much earlier than the divergence of hominid and monkey stocks. Both chimpanzees and gorillas have their own branches of both genera.

Lice use their mouthparts not only to feed but also at least partially to attach. Rather than being able to withdraw quickly when the host becomes distressed or moves, the lice are able to evade host grooming response because they are dorsoventrally flattened and can hide among the hairs, in skin folds or flat on the skin of mammalian hosts. The mouthparts of lice, except when in use, are invaginated (folded into the head). The act of feeding is the result of the evagination of a circle of oral teeth which point forward. These teeth protrude from the cavity as stylets and are pressed against the host's skin. These hold it in place as the proboscis is evaginated and placed into the skin and as the search for a vessel containing blood begins. Feeding speed and frequency vary among species, with most lice feeding to repletion in minutes. Lice generally need to feed frequently, young nymphs can starve in as little as 5 hours, while dehydration is of concern for all stages.

## Host Responses

Wikel (1982) discusses the early observations of Nuttall, Moore *et al.* concerning the presence of systemic responses to *Pediculus humanus*, mentioning that some individuals tolerated extremely high burdens for years while others showed both immediate and delayed hypersensitivity responses. Spontaneous desensitization of previously allergic patients may occur (Barriga, 1981), although the mechanism is not described. There is little evidence that, on humans at least, the most effective resistance is behavioural resistance driven by pruritus. Although it is possible that there are self-limiting factors in human lice populations. There is evidence (Lang, 1976) that female head lice populations are decreased under conditions of high density because of damage to young females acquired via excessive matings before proper sclerotization can take place. Although Nelson *et al.* (1977) were discussing lice affecting mice and cattle, their observation that 'acquired resistance to lice might prove to be a chronic inflammatory response combined with immune effector elements' is probably accurate across all species of hosts and parasites. Of the arthropods in this chapter, the sucking lice are the parasites in which population dynamics matter most for a single host, since all life stages of all the species are found on the host. The study of interactions between the parasite population size and host immune response has proven to be especially beneficial for scientists working with rodents and cattle.

Probably because of the expense relative to rodents, research on cattle lice has been driven by economic need for control of lice populations and frequently limited to economic damage thresholds. There are five species of lice associated with domestic cattle (Matthysse, 1946; Meleney and Kim, 1974). Cattle lice can be a severe problem for beef or dairy cattle throughout much of the world due to decreases in cattle weight gains and milk production coupled with the need for additional feed to maintain lice populations and increased time on feed for feedlot animals. Primarily a cool season problem, populations result in economic losses which were estimated as high as 126.3 million dollars in 1981 (Drummond et al., 1981). Lice populations on range cattle under severe Midwestern winter field conditions may be high enough to severely weaken animals and predispose them to morbidity and mortality from other causes (Campbell, 1988) such as pneumonia. Weight gain differences among cattle with high or low populations of more than one simultaneous infestation of cattle lice have been difficult to study, at least partially because of the problem of obtaining sufficient replicated data which could be correlated with high lice populations. No differences among treatments were seen in the studies of Utech et al. (1969), Ely and Harvey (1969) or Kettle (1974). However, in a replicated, multiple year, trial in cattle feedlots, Gibney et al. (1985) examined weight gains according to lice population levels finding significant differences between weight gains of cattle heavily infested with multiple species of lice and weight gains of uninfested cattle with no differences in feed consumption. This would suggest that feed and nutrients consumed by the cattle were actually utilized by the lice populations, rather than converted to cattle weight gain. This situation should result in longer feeding periods for infested cattle to achieve the same weight gain as uninfested cattle.

Most cattle affected by the presence of lice are not in danger of dying (Nelson et al., 1970). However, in North America, *Haematopinus eurysternus* can be a cause of severe, and terminal, anaemia (Petersen et al., 1953; Shemanchuk et al., 1960; Collins and Dewhirst, 1965) in range cattle, with anoxia a possible result of erythroblast mitosis disruption caused by louse toxin(s) in some cattle. Similar toxic response appears to appear in mice hosts of *Polyplax serrata* (Bell et al., 1966), and has been shown to be dependent upon the ability of a mouse strain to mount an effective resistance response (Clifford et al., 1967). *Haematopinus eurysternus* populations also damage capillary and arteriolar flow via vasoconstriction and vascular cuffing, often resulting in lice population movement to unaffected skin (Nelson et al., 1977). Economic loss to other species of lice on cattle has been frequently reported, but precise data are difficult to obtain. Nutrition has an important effect, the most reliable figures for loss to *Linognathus vituli* (Cummins and Graham, 1982) show that infestations can be damaging to undernourished calves, although compensatory weight gain may take place after lice population pressure is released. Other studies with *L. vituli* (Kettle, 1974; Cummins and Tweedle, 1977; Callinan, 1980; Chalmers and Charleston, 1980) have found essentially no negative response to infestation.

Capable of building up to tremendous populations on restrained cattle,

*Bovicola bovis* (biting lice) population dynamics are apparently regulated by external factors such as temperature and grooming response, as well as by build-up of their own excretory products or other self-induced depopulation phenomena (Mock, 1974). Antigens of this species show cross-reactivity with gross antigenic preparations from *L. vituli* specimens (Mock, 1974) leaving doubt as to whether the immune response was directed toward inhalants or arthropod enzymes resulting from unknown exposures.

Cattle lice cause obvious signs of distress. Behavioural changes such as increases in grooming and aggressive rubbing are common, making it more difficult to work infested cattle (Smith and Roberts, 1956). In lice-infested swine, similar behaviour (specifically increases in rubbing and scratching behaviour) results in decreased feeding activity (Davis and Williams, 1986). The effectiveness of these behavioural techniques in modulating populations of *B. bovis*, is evident from the data of Geden *et al.* (1990) in which cattle held in free stalls had less than half the populations of stanchioned cattle.

## Lice Population Dynamics

Lloyd and Kumar (personal communication) have artificially infested calves with known numbers of *B. bovis* from other cattle and followed the population growth from January to March, in Laramie, Wyoming. Starting populations of less than 500 per calf resulted in indexed populations comparable to long-term natural populations within four weeks of application, even during periods when natural infestations are in decline. Coexisting populations of three species of lice may reach 17,000–30,000 of each species on a single animal with total lice burden of nearly 75,000 (Watson, unpublished data).

There have been few papers specifically correlating the degree of infestation and host nutritional status. Ely and Harvey (1969) found that the higher the plane of nutrition, the lower the *H. eurysternus* populations on yearlings in feedlots. Low plane nutrition in cattle may result in increases in spring populations of *B. bovis* (Utech *et al.*, 1969). Nelson (1984) postulated general host susceptibility changes to ectoparasites as common results of some dietary deficiencies.

How, or whether, host response to cattle lice modulates the population of the lice is unknown. Animal lice are generally host specific and in some cases may even specialize as to host body regions (Jensen and Roberts, 1966; Busvine, 1978; Chalmers and Charleston, 1980; Stock and Hunt, 1989), and colonization of a specific region may be influenced by the presence of other species of lice (Lewis *et al.*, 1967). Host region specificity may have some correlation to a general inability to develop artificial feeding techniques or alternative hosts for sucking lice. Laboratory models of cattle lice infestation are not available at this time because microclimatologic and skin or hair conditions needed for feeding stimuli or for survival have proven to be extremely difficult to duplicate. Mock (1974) posited that *in vitro* populations did not have the opportunity to move to optimal environments as happens on the host. During research at the University of Wyoming (Lloyd and Watson, unpublished data), sucking lice of cattle were unsuccessful in feeding through a variety of artificial membranes.

De Vaney *et al.* (1988) found population declines of three species of lice on cattle in Texas, starting approximately 8 weeks after infestation. Declines were presumed to be related to host response, but weather may have been a confounding factor. Studies on mouse resistance to lice (Bell *et al.*, 1962, 1966; Clifford *et al.*, 1967; Nelson *et al.*, 1972, 1979; Stewart *et al.*, 1976; Ratzlaff and Wikel, 1990) identified anamnestic, systemic, cell-mediated responses with genetic components.

In mice (Nelson *et al.*, 1972, 1979) resistance was correlated with increases in several cell types at the site of louse attachment. Nelson *et al.* (1972) noted an increase in the epidermal thickness during an initial infestation of lice on mice during weeks 1 to 4. Neutrophils, eosinophils, and lymphocytes were the primary cellular infiltrates. There was a hyperaemia during the first week and then a decrease in the number of vessels by week 5, a finding which did not change for some time. There was a decrease at that time of epidermal thickness as well. Arteriolar constriction occurred from the second week onwards. By 8 weeks, another increase in epidermal thickness had begun. The primary cellular invasion was by lymphocytes and monocytes. There was a sustained increase, however, in mast cells with fibroblast proliferation increasing and granulation of mast cells in tissues.

## Clinical Response

Ratzlaff and Wikel (1990) successfully used bacteria free whole lice soluble components for histological, lymphoid cell proliferation and immunization studies with the *Polyplax serrata*-mouse system. They demonstrated the presence of systemic anamnestic resistance inducible with subcutaneous injections of soluble components of lice. *In vivo*, the responses are independent of continuous louse exposure. Lymphoid cells from infested mice proliferated significantly more rapidly when incubated with soluble lice components than did cells from naive mice.

Volf *et al.* (1990) reported that titres of circulating antibodies to *Polyplax spinulosa* (in rats) correlated with the 'degree and duration' of infestation, although they were not studying antibody response, but rather attempting to characterize antigenic components of lice. Volf and Grubhoffer (1991) isolated one immunogenic glycoprotein (31 kDa) from the gamma globulin fraction of an antigen-specific antiserum using immunoaffinity chromatography.

# REFERENCES

Baker, K.P. (1974) Observations on allergic reactions to arthropod parasites. *Irish Veterinary Journal* April, 65–70.

Baker, K.P. and O'Flanagan, J. (1975) Hypersensitivity of dog skin to fleas – a clinical report. *Journal of Small Animal Practice* 16, 317–327.

Barriga, O.O. (1981) Immune reactions to arthropods. In: Barriga, O.O. (ed.), *The Immunology of Parasitic Infections*. University Park Press, Baltimore, pp. 283–317.

Bell, J.F., Jellison, W.L. and Owen, C.R. (1962) Effects of limb disability on lousiness in mice. I. Preliminary studies. *Experimental Parasitology* 12, 176–183.

Bell, J.F., Clifford, C.M., Moore, G.J. and Raymond, G. (1966) Effects of limb disability on lousiness in mice. III. Gross aspects of acquired resistance. *Experimental Parasitology* 18, 49–60.

Benjamini, E. and Feingold, B.F. (1960) Antigenic property of the oral secretion of fleas. *Nature* 188(4754), 959–960.

Benjamini, E. and Feingold, B.F. (1970) Immunity to arthropods. In: Jackson, G.J., Herman, R. and Singer, I. (eds), *Immunity to Parasitic Animals*, Volume 2: Appleton, Century Crofts, New York, pp. 1061–1134, 1217.

Benjamini, E., Feingold, B.F. and Kartman, L. (1960) Allergy to flea bites. III. The experimental induction of flea bite sensitivity in guinea pigs by exposure to flea bites and by antigen prepared from whole flea extracts of *Ctenocephalides felis felis*. *Experimental Parasitology* 10(2) 99–107.

Benjamini, E., Feingold, B.F. and Kartman, L. (1961) Skin reactivity in guinea pigs sensitized to flea bites: the sequence of reactions. *Proceedings of the Society for Experimental Biology and Medicine* 108(3), 700–702.

Benjamini, E., Feingold, B.F., Young, J.D., Kartman, L. and Shimizu, M. (1963) Allergy to flea bites. IV. In vitro collection and antigenic properties of the oral secretion of the cat flea, *Ctenocephalides felis felis* (Bouche). *Experimental Parasitology* 13, 143–154.

Bitkowska, E., Dzbenski, T.H., Szadziewska, M. and Wegner, Z. (1982) Inhibition of Xenograft rejection in the bug *Triatoma infestans* during infection with a protozoan, *Trypanosoma cruzi*. *Journal of Invertebrate Pathology* 40, 186–189.

Brown, S.J. (1989) Pathological consequences of feeding by hematophagous arthropods: comparison of feeding strategies. In: Jones, C.J. and Williams, R.E. (eds), *Proceedings of a Symposium: Physiological Interactions Between Hematophagous Arthropods and Their Vertebrate Hosts*. Entomological Society of America, Lanham, Maryland, pp. 4–14.

Brown, S.J. and Rosalsky, J.H. (1984) Blood leukocyte response in hosts parasitized by the hematophagous arthropods *Triatoma protracta* and *Lutzomyia longipalpis*. *American Journal of Tropical Medicine and Hygiene* 33(3), 499–505.

Busvine, J.R. (1978) Evidence from double infestations for the specific status of human head and body lice (Anoplura). *Systematic Entomology* 3, 1–8.

Callinan, A.P.L. (1980) Effects of artificially induced infestations of the cattle louse, *Linognathus vituli*. *Australian Veterinary Journal* 56, 484–486.

Campbell, J.B. (1988) Arthropod induced stress in livestock. *Veterinary Clinical Nutrition: Food Animal Practitioner* 4, 551–555.

Chalmers, K. and Charleston, W.A.G. (1980) Cattle lice in New Zealand: observations on the biology and ecology of *Damalinia bovis* and *Linognathus vituli*. *New Zealand Veterinary Journal* 28, 214–216.

Chapman, M.D., Marshall, N.A. and Saxon, A. (1986) Identification and partial purification of species specific allergens from *Triatoma protracta* (Heteroptera: Reduviidae). *Journal of Allergy Clinical Immunology* 78, 436–442.

Clifford, C.M., Bell, J.F, Moore, G.J. and Raymond, G. (1967) Effects of limb disability on lousiness in mice. IV. Evidence of genetic factors in susceptibility to *Polyplax serrata*. *Experimental Parasitology* 20, 56–67.

Collins, R.C. and Dewhirst, L.W. (1965) Some effects of the sucking louse, *Haematopinus eurysternus*, on cattle on unsupplemented range. *Journal of the American Veterinary Medical Association* 146, 129–132.

Costa, C.H.N., Costa, M.T., Weber, J.N., Gilks, G.F., Castro, C. and Mardsen, P.D. (1981) Skin reactions to bug bites as a result of Xenodiagnosis. *Transactions of the Royal Society of Tropical Medicine and Hygiene* 75(3), 405–408.

Cummins, L.J. and Graham, J.F. (1982) The effect of lice infestation on the growth of Hereford calves. *Australian Veterinary Journal* 58, 194–196.

Cummins, L.J. and Tweedle, N.E. (1977) The influence of light infestations of *Linognathus vituli* on the growth of young cattle. *Australian Veterinary Journal* 53, 591–592.

Davis, D.P. and Williams, R.E. (1986) Influence of hog lice, *Haematopinus suis*, on blood components, behavior, weight gain and feed efficiency of pigs. *Veterinary Parasitology* 22, 307–314.

De Vaney, J.A., Rowe, L.D. and Craig, T.M. (1988) Density and distribution of three species of lice on calves in central Texas. *Southwest Entomology* 13, 125–135.

Drummond, R.O., Lambert, D., Smalley, H.E., Jr and Terrill, C.E. (1981) Estimated losses of livestock to pests. In: Pimentel, D. (ed.), *CRC Handbook of Pest Management in Agriculture*, Volume I. CRC Press, Boca Raton, Florida, USA.

Dryden, M.W. (1989) Host association, on-host longevity and egg production of *Ctenocephalides felis felis*. *Veterinary Parasitology* 34, 117–122.

Ely, D.G. and Harvey, T.L. (1969) Relation of ration to short-nosed cattle louse infestations. *Journal of Economic Entomology* 62, 341–344.

Fox, I. and Bayona, I.G. (1968) Circulating precipitating antibodies in the rabbit from the bites of *Rhodnius prolixus* as shown by agar-gel tests. *Journal of Parasitology* 54, 1239–1240.

Geden, C.J., Rutz, D.A. and Bishop, D.R. (1990) Cattle lice (Anoplura, Mallophaga) in New York: seasonal population changes, effects of housing type on infestations of calves, and sampling efficiency. *Journal of Economical Entomology* 83, 1435–1438.

Gibney, V.J., Campbell, J.B, Boxler, D.J., Clanton, D.C. and Deutscher, G.H. (1985) Effects of various infestation levels of cattle lice (Mallophaga: Trichodectidae and Anoplura: Haematopinidae) on feed efficiency and weight gains of beef heifers. *Journal of Economical Entomology* 78, 1304–1307.

Greene, W.K., Carnegie, R.L., Shaw, S.E., Thompson, R.C.A. and Penhale, W.J. (1993) Characterization of allergens of the cat flea, *Ctenocephalides felis*: detection and frequency of IgE antibodies in canine sera. *Parasite Immunology* 15, 69–74.

Gross, T.L. and Halliwell, R.E.W. (1985) Lesions of experimental flea bite hypersensitivity in the dog. *Veterinary Pathology* 22, 78–81.

Halliwell, R.E.W. (1981) Hyposensitization in the treatment of flea bite hypersensitivity: results of a double blind study. *Journal of the American Animal Hospital Association* 17, 249–253.

Halliwell, R.E.W. (1984) Factors in the development of flea-bite allergy. *Veterinary Medicine* October, 1273–1278.

Halliwell, R.E.W. and Longino, S.J. (1985) IgE and IgG antibodies to flea antigen in differing dog populations. *Veterinary Immunology and Immunopathology* 8, 215–223.

Halliwell, R.E.W. and Schemmer, K.R. (1987) The role of basophils in the immunopathogenesis of hypersensitivity to fleas (*Ctenocephalides felis*) in dogs. *Veterinary Immunology and Immunopathology* 15, 203–213.

Hase, A. (1916) Weitere Beobachtungen uber die Lausplage. *Zentralblatt fur Bakteriologie* 77, 153–163.

Hayasaki, M., Akiyama, Y., Konno, K. and Ohishi, I. (1993) Immune treatment of flea allergic dermatitis of dogs with flea extract. *Journal of the Japanese Medical Association* 46, 866.

Hellmann, K. and Hawkins, R.I. (1964) Anticoagulant and fibrinolytic activities from *Rhodnius prolixus* stahl. *Nature* 201, 1008–1009.

Hellmann, K. and Hawkins, R.I. (1965) Prolixin-S and Prolixin-G: two anticoagulants for *Rhodnius prolixus* stahl. *Nature* 207, 265–267.

Hudson, B.W., Feingold, B.F. and Kartman, L. (1960) Allergy to flea bites. I. Experimental induction of flea-bite sensitivity in guinea pigs. *Experimental Parasitology* 9, 18–24.

Jensen, R.E. and Roberts, J.E. (1966) A model relating microhabitat temperatures to seasonal changes in the little blue louse (*Solenopotes capillatus*) populations. *Geography and Agriculture Experimental Station Bulletin* 55, 22.

Johnston, C.M. and Brown, S.J. (1985) *Xenopsylla cheopis*: Cellular expression of hypersensitivity in guinea pigs. *Experimental Parasitology* 59, 81–89.

Jones, C.J. (1989) Introduction: coevolution of hematophagous arthropods and their vertebrate hosts. In: Jones, C.J. and Williams, R.E. (eds), *Proceedings of a Symposium: Physiological Interactions Between Hematophagous Arthropods and Their Vertebrate Hosts*. Entomological Society of America, Lanham, Maryland, pp. 1–3.

Kettle, P.R. (1974) The influence of cattle lice (*Damalinia bovis* and *Linognathus vituli*) on weight gain in beef animals. *New Zealand Veterinary Journal* 22, 10–12.

Kim, K.C. and Ludwig, H.W. (1978) Phylogenetic relationships of Psocodea and taxonomic position of the parasitic Anoplura. *Annals of the Entomological Society of America* 71, 910–922.

Kristensen, S. and Kieffer, M. (1978) A study of skin diseases in dogs and cats. V. The intradermal test in the diagnosis of flea allergy in dogs and cats. *Nordisk Veterinaer Medicin* 30, 414–423.

Lang, J.D. (1976) Sex ratio of adult head lice under crowded conditions. *New York Entomological Society* 84, 243–245.

Larrivee, D.H., Benjamini, E., Feingold, B.F. and Shimizu, M. (1964) Histological studies of guinea pig skin: different stages of allergic reactivity to flea bites. *Experimental Parasitology* 15, 491–502.

Lavoipierre, M.M.J., Dickerson, G. and Gordon, R.M. (1959) Studies on the methods of feeding of blood-sucking arthropods. I. The manner in which triatomine bugs obtain their blood-meal, as observed in the tissue of the living rodent, with some remarks on the effects of the bite on human volunteers. *Annals of Tropical Medicine and Parasitology* 53, 235–250.

Lavoipierre, M.M.J., Radovsky, F.J. and Budwiser, P.D. (1979) The feeding process of a tungid flea, *Tunga monositus* (Siphonaptera: Tungidae), and its relationship to the host inflammatory and repair response. *Journal of Medical Entomology* 15(3), 187–217.

Lewis, L.F., Christenson, D.M. and Eddy, G.W. (1967) Rearing the long-nosed cattle louse and cattle biting louse on host animals in Oregon. *Journal of Economical Entomology* 60, 755–757.

Marshall, N.A. (1982) Allergy to *Triatoma protracta* (Heteroptera: Reduviidae). II. Antigen production *in vitro*. *Journal of Medical Entomology* 19(3), 253–254.

Marshall, N.A. and Street, D.H. (1982) Allergy to *Triatoma protracta* (Heteroptera: Reduviidae). I. Etiology, antigen preparation, diagnosis and immunotherapy. *Journal of Medical Entomology* 19(3), 248–252.

Marshall, N.A., Chapman, M.D. and Saxon, A. (1986) Species-specific allergens from the salivary glands of Triatominae (Heteroptera: Reduviidae). *Journal of Allergy and Clinical Immunology* 78(3), 430–435.

Matthysse, J.G. (1946) Cattle lice: their biology and control. *Cornell University Agricultural Experiment Station Bulletin 832*.

McDougal, B.J. (1986) Allergy testing and hyposensitization for three common feline dermatoses. *Modern Veterinary Practice* July/Aug, 629–633.

McKeon, S.E. and Opdebeeck, J.P. (1994) IgG and IgE antibodies against antigens of the cat flea, *Ctenocephalides felis*, in sera of allergic and non-allergic dogs. *International Journal for Parasitology* 24(2), 259–263.

Meleney, W.P. and Kim, K.C. (1974) A comparative study of cattle-infesting *Haematopinus*, with redescription of *H. quadripertusus* Fahrenholz, 1916 (Anoplura: Haematopinidae). *Journal of Parasitology* 60(3), 507–522.

Michaeli, D. and Goldfarb, S. (1968) Clinical studies on the hyposensitivity of dogs and cats to flea bites. *Australian Veterinary Journal* 44, 161–165.

Michaeli, D., Benjamini, E., De Buren, F.P., Larrivee, D.H. and Feingold, B.F. (1965) The role of collagen in the induction of flea bite hypersensitivity. *Journal of Immunology* 95(1), 162–170.

Michaeli, D., Benjamini, E., Miner, R.C. and Feingold, B.F. (1966) In vitro studies on the role of collagen in the induction of hypersensitivity to flea bites. *Journal of Immunology* 97(3), 402–406.

Mock, D.E. (1974) The cattle biting louse, *Bovicola bovis* (Linn.). I. In vitro culturing, seasonal population fluctuations, and role of the male. II. Immune response of cattle. PhD Dissertation, Cornell University, Ithaca, New York, 193pp.

Moriello, K.A. and McMurdy, M.A. (1989) The prevalence of positive intradermal skin test reactions to flea extract in clinically normal cats. *Companion Animal Practice* 19(3), 28–30.

Nelson, W.A. (1984) Review article: Effects of nutrition of animals on their ectoparasites. *Journal of Medical Entomology* 21(6), 621–635.

Nelson, W.A. (1989) Metabolic responses of livestock to hematophagous arthropod invasion. In: Jones, C.J. and Williams, R.E. (eds), *Physiological Interactions Between Hematophagous Arthropods and Their Vertebrate Hosts*. Misc. Publ. Entomological Society of America 71, pp. 15–21.

Nelson, W.A., Shemanchuk, J.A. and Haufe, W.O. (1970) *Haematopinus eurysternus*: Blood of cattle infested with the short nosed louse. *Experimental Parasitology* 28, 263–271.

Nelson, W.A., Clifford, C.M., Bell, J.F. and Hestekin, B. (1972) *Polyplax serrata*:

Histopathology of the skin of louse-infested mice. *Experimental Parasitology* 31, 194–202.

Nelson, W.A., Bell, J.F., Clifford, C.M. and Keirans, J.E. (1977) Review article: interaction of ectoparasites and their hosts. *Journal of Medical Entomology* 13(4–5), 389–428.

Nelson, W.A., Bell, J.F. and Stewart, S.J. (1979) *Polyplax serrata*: cutaneous cytologic reactions in mice that do (CFW strain) and do not (C57BL strain) develop resistance. *Experimental Parasitology* 48, 259–264.

Nesbitt, G.H. and Schmitz, J.A. (1978) Fleabite allergic dermatitis: a review and survey of 330 cases. *Journal of the American Veterinary Medical Association* 173, 282–288.

Noeske-Jungblut, C., Kratzschmar, J., Haendler, B., Alagon, A., Possani, L., Verhallen, P., Donners, P. and Scheuning, W. (1994) An inhibitor of collagen-induced platelet aggregation from the saliva of *Triatoma pallidipennis*. *Journal of Biological Chemistry* 269(7), 5050–5053.

Opdebeeck, J.P. and Slacek, B. (1993) An attempt to protect cats against infestation with *Ctenocephalides felis felis* using gut membrane antigens as a vaccine. *International Journal for Parasitology* 23(8), 1063–1067.

Petersen, H.W., Roberts, I.H., Becklund, W.W. and Kemper, H.E. (1953) Anemia in cattle caused by heavy infestations of the blood-sucking louse, *Haematopinus eurysternus*. *Journal of the American Veterinary Medical Association* 122, 373–376.

Pinnas, J.L., Chen, T.M.W. and Hoffman, D.R. (1978) Evidence for IgE-mediation of human sensitivity to reduviid bug bites. *Federation of American Societies for Experimental Biology, Federation Proceedings* 37, 1555.

Ratzlaff, R.E. and Wikel, S.K. (1990) Murine immune responses and immunization against *Polyplax serrata* (Anoplura: Polyplacidae). *Journal of Medical Entomology* 27(6), 1002–1007.

Ribeiro, J.M.C. (1982) The antiserotonin and antihistamine activities of salivary secretions of *Rhodnius prolixus*. *Journal of Insect Physiology* 28(1), 69–75.

Ribeiro, J.M.C. (1987) Role of saliva in blood-feeding by arthropods. *Annual Review of Entomology* 32, 463–478.

Ribeiro, J.M.C. and Garcia, E.S. (1980) The salivary and crop apyrase activity of *Rhodnius prolixus*. *Journal of Insect Physiology* 26, 303–307.

Ribeiro, J.M.C. and Garcia, E.S. (1981a) Platelet antiaggregating activity in the salivary secretion of the blood sucking bug *Rhodnius prolixus*. *Experientia* 37, 384–386.

Ribeiro, J.M.C. and Garcia, E.S. (1981b) The role of salivary glands in feeding in *Rhodnius prolixus*. *Journal of Experimental Biology* 94, 219–230.

Ribeiro, J.M.C. and Sarkis, J.J.F. (1982) Anti-thromboxane activity in *Rhodnius prolixus* salivary secretion. *Journal of Insect Physiology* 28(8), 655–660.

Ribeiro, J.M.C., Marinotti, O. and Gonzales, R. (1990) A salivary vasodilator in the blood sucking bug, *Rhodnius prolixus*. *British Journal of Pharmacology* 101, 932–936.

Ribeiro, J.M.C., Hazzard, J.M.H., Nussenzveig, R.H., Champagne, D.E. and Walker, F.A. (1993) Reversible binding of nitric oxide by a salivary heme protein from a bloodsucking insect. *Science* 260, 539–541.

Rothschild, M. and Ford, B. (1964) Breeding of the rabbit flea (*Spilopsyllus cuniculi* (Dale)) controlled by the reproductive hormones of the host. *Nature* 201, 103–104.

Rothschild, M. and Ford, B. (1966) Hormones of the vertebrate host controlling ovarian regression and copulation of the rabbit flea. *Nature* 211, 261–266.

Ryckman, R.E. (1979) Host reactions to bug bites (Hemiptera, Homoptera): a literature review and annotated bibliography. *California Vector Views* 26 (1/2), 1–24.

Scheidt, V.J. (1988) Flea allergy dermatitis. *Veterinary Clinics of North America: Small Animal Practice* 18(5), 1023–1042.

Schemmer, K.R. and Halliwell, R.E. (1987) Efficacy of alum-precipitated flea antigen for hyposensitization of flea-allergic dogs. *Seminars in Veterinary Medicine and Surgery (Small Animal)* 2(3), 195–198.

Schofield, C.J. (1982) The role of blood intake in density regulation of populations of *Triatoma infestans* (Klug) (Hemiptera: Reduviidae). *Bulletin of Entomological Research* 72, 617–629.

Shemanchuk, J.A., Haufe, W.O. and Thompson, C.M. (1960) Anemia in range cattle heavily infected with the short nosed sucking louse, *Haematopinus eurysternus* (Nitz.)(Anoplura: Haematopinidae). *Canadian Journal of Comparative Medicine* 24, 158–161.

Smith, C.L. and Roberts, I.H. (1956) Cattle lice. In: *Yearbook of Agriculture.* US Government Printing Office, Washington DC, pp. 307–310.

Sternberg, L. (1929) A case of asthma caused by the *Cimex lectularis* (bed bug). *Journal of Allergy and Clinical Immunology* 1, 83.

Stewart, S.J., Bell, J.F., Hestekin, B. and Moore, G.J. (1976) *Polyplax serrata*: effects of limb disability on lousiness in mice. VI. Lack of tolerance after neonatal exposure. *Experimental Parasitology* 40, 373–379.

Stock, T.M. and Hunt, L.E. (1989) Site specificity of three species of lice, Mallophaga, on the Willow Ptarmigan, *Lagopus lagopus*, from the Chilkat Pass, British Columbia. *Canadian Field-Naturalist* 103, 584–588.

Studdert, V.P. and Arundel, J.H. (1988) Dermatitis of the pinnae of cats in Australia associated with the European rabbit flea (*Spilopsyllus cuniculi*). *Veterinary Record* 123, 624–625.

Traub, R. (1980) The Zoogeography and evolution of some fleas, lice, and mammals. In: Traub, R. and Starcke, H. (eds), *Fleas*. A.A. Balkema, Rotterdam, pp. 93–173.

Usinger, R.L. (1966) *Monograph of Cimicidae,* Volume 7. Entomological Society of America, Thomas Say Foundation, 585pp.

Utech, K.B.W., Tierarzt H.B., Wharton, R.H. and Wooderson, L.A. (1969) Biting cattle-louse infestations related to cattle nutrition. *Australian Veterinary Journal* 45, 414–416.

Vaughan, J.A., Jerse, A.E. and Azad, A.F. (1989) Rat leucocyte response to the bites of rat fleas (Siphonaptera: Pulicidae). *Journal of Medical Entomology* 26(5), 449–453.

Volf, P. and Grubhoffer, L. (1991) Isolation and characterization of an immunogen from the louse *Polyplax spinulosa*. *Veterinary Parasitology* 38, 225–234.

Volf, P., Grubhoffer, L. and Matha, V. (1990) Antigenic characterization of rat louse *Polyplax spinulosa*. *Folia Parasitologica* 37, 275–278.

Volf, P., Grubhoffer, L. and Hosek, P. (1993) Characterization of salivary gland antigens of *Triatoma infestans* and antigen-specific serum antibody response in mice exposed to bites of *T. infestans*. *Veterinary Parasitology* 47, 327–337.

Waage, J.K. (1979) The evolution of insect/vertebrate associations. *Biological Journal of the Linnean Society* 12, 187–224.

Wikel, S.K. (1982) Immune responses to arthropods and their products. *Annual Review of Entomology* 27, 21–48.

Wood, S.F. (1976) Suspected cannibalism in survival of nymphs of *Triatoma protracta*. *Pan-Pacific Entomologist* 52, 264.

# 8

# Immune Responses to Mosquitoes and Flies

## R. Mark Sandeman

*School of Agriculture, Faculty of Science and Technology La Trobe University, Bundoora, Victoria, Australia 3083*

Dipterans which parasitize animals can be divided into two groups: those which are haematophagous as adults and require blood feeds usually to mature eggs; and those whose larvae infect animal tissues causing myiasis. The most important group in terms of both animal and human health are the haematophagous species because they are hosts and vectors of a wide range of viruses, bacteria, protozoan and metazoan parasites.

Unfortunately, haematophagous species are also more difficult to control via immune mechanisms since they usually feed for only a few minutes causing little obvious effect on their host except a local reaction. Feeding episodes may be repeated frequently by other individuals of the same species, but, even after developing an immune response, the host still has little opportunity to affect the insect while it is biting. In fact, it has been considered that the only form of immune-related defence was the development of hypersensitivity followed by the absolute protective effect of a tail or other extremity being applied forcefully soon after the insect lands. Exceptions exist to these general observations with species such as the sheep ked *Melophagus ovinus* being haematophagous but also an ectoparasite which lives permanently on its host causing relatively mild pathological symptoms (Nelson, 1988). Other haematophagous species can induce serious pathological effects though these often result from the induction of host hypersensitivity responses (Jones and Lloyd, 1987) or from the constant irritation of large numbers of feeding flies or mosquitoes (Dougherty *et al.*, 1994).

Myiasis flies are usually in contact with the host for a much longer period

---

© CAB INTERNATIONAL 1996. From Wikel, S.K. (ed.) *The Immunology of Host–Ectoparasitic Arthropod Relationships.* CAB INTERNATIONAL, Wallingford.

than haematophagous flies and as such should be more vulnerable to immune attack. However, *Lucilia cuprina*, the sheep blowfly, causes a relatively fast myiasis on sheep skin which is usually fatal if not treated within 3 to 5 days. Other myiasis flies such as the American screw worm, *Cochliomyia hominivorax*, the Eurasian fly, *Wohlfartia magnifica* and most other calliphorid or sarcophagid species have similar short life cycles with an acute pathology and apparently less opportunity for immune intervention.

The more common myiasis flies are the botflies which are obligate parasites with the larvae usually living systematically in the host. These fly larvae inhabit the respiratory system, the gut or live subcutaneously. Life cycles usually follow a similar pattern with the adult fly living for a few days or weeks, long enough to mate and find a new host for egg laying. The adult is free living and though it may require protein feeds especially to mature eggs these can be obtained from a variety of dead or living sources. In contrast, the maggots are parasitic undergoing all but the final phases of development in the host. Thus, the eggs hatch and the larvae undergo two moults before growing to a size where they leave the host to pupate. This parasitic phase usually occurs over several weeks to months in spring and summer depending on the species, and thus the parasite is in contact with the host's immune system for an extended period.

Pathological symptoms depend on the host as well as the fly species and its life cycle but as outlined above the calliphorids and sarcophagids usually only affect the skin and its underlying layers and are more directly pathogenic often causing life threatening wounding and direct toxicity (Broadmeadow *et al.*, 1984; Krafsur *et al.*, 1987; Lehrer *et al.*, 1988). Botflies tend to be less pathogenic though larger infections can cause significant morbidity and effects on growth and reproduction (Dart *et al.*, 1987; Cogley, 1989). In addition, specific production effects can occur depending on the species and include losses in value of leather due to warble breathing holes in the backs of cattle skins (Scholl, 1993), and loss of wool and wool quality due to sheep blowfly infections (Beck *et al.*, 1985).

The notion that immune responses might be stimulated to give direct protection against attack by haematophagous dipterans was given impetus in the 1970s by the work of Alger and Cabrera (1972) on mosquitoes and Schlein and Lewis (1976) on the stable fly. Despite the promise of these findings, further studies in this area have been limited mainly to immunization with crude extracts against a range of species. Few laboratories have extended the work and begun the isolation and analysis of individual antigens that is required to determine the potential of vaccines for controlling these biting dipterans.

In contrast, the development of vaccines against myiasis flies has proceeded quite rapidly especially against *Lucilia cuprina* which was not generally considered as a target of immunological responses until 1980 (O'Donnell *et al.*, 1980). Despite the slow start, the research that has been carried out is encouraging enough to allow the statement that vaccination against *Lucilia* is at least partially protective (70%) under experimental conditions and that field trials will proceed in the near future. The same situation has been reached with a vaccine against *Hypoderma lineatum* in cattle though the level of protection is greater than 90% and recombinant proteins are available (Moire *et al.*, 1994).

The parasitic dipteran species for which immune responses have been best studied are a diverse group with little in common except their importance in agriculture in developed countries and in human medicine in developed and developing countries. Haematophagous species of interest include the tabanids, *Simulium* and *Culicoides* and mosquitoes such as *Aedes*, *Anopheles* and *Culex*. Myiasis causing species include *Hypoderma*, *Dermatobia*, *Lucilia*, *Cochliomyia*, *Oestrus* and *Gasterophilus*. These species represent a diverse range of organisms with quite different life cycles, hosts and effects on their hosts. As a result, this review will divide the species into haematophagous and myiasis causing groups and concentrate on the immune responses they induce which often share common features rather than analyse the highly variable details of life cycle, pathological effects and agricultural and medical importance. In addition, because of the number of reviews encompassing this area, which have been published over the last few years (Ribeiro, 1987; Baron and Colwell 1991a; Sandeman, 1992; Barriga, 1993; Scholl, 1993; Kay and Kemp, 1994; Sandeman *et al.*, 1995b), I will concentrate on the more recent publications where possible.

# HAEMATOPHAGOUS DIPTERA

## Natural Immune Responses

### Hypersensitivity

Immune responses to haematophagous dipterans whose antigens are only briefly exposed to the host tend to be dominated by hypersensitivity responses including allergic (type I), Arthus (type III) and delayed reactions (type IV) (Frazier, 1987). The apparent propensity to induce hypersensitivity probably stems from the limited exposure of the host to the insect antigen, the site of exposure (the skin), and the nature of the antigens they secrete during the feeding process. The most common antigens are from the salivary gland and are injected to stimulate the flow of blood at the site (Ribeiro, 1995), through vasodilation (Ribeiro *et al.*, 1994) and antihaemostasis. The control of haemostasis is achieved through the inhibition of platelet aggregation (Champagne *et al.*, 1995) and inhibition of blood coagulation (Abebe *et al.*, 1994). The injected salivary antigens tend to stimulate local responses resulting in IgG and IgE production (Reunala *et al.*, 1994b) and induce $T_H 1$

cells, which mediated delayed type reactions (Ellis et al., 1986; Weir and Stewart, 1993). Further bites can result in large local reactions (Fadok and Greiner, 1990; Mason and Evans, 1991) with eosinophilia as a common consequence of chronic exposure to the bites of flies and mosquitoes (Ellis et al., 1986; Mason and Evans, 1991; Kerlin and Allingham, 1992) though the consequences of this reaction to the host or parasite have not been determined. Systemic complications are rare, possibly because the small amounts of antigens injected by the insect are usually confined to the local bite area by the host response and the feeding strategy of the dipteran.

The effect of these reactions on the future success of the same species of haematophagous diptera on the same host is difficult to measure. Certainly, through pain and irritation, the host is made aware of new bite sites though the insect has often completed its meal and left before effective responses are elicited. However, the irritation probably helps stimulate the recognition of specific types of biting insects and the resultant avoidance and control behaviour when these insects attack (Hart, 1994). The effectiveness of these types of physical responses has been illustrated by studies on the feeding success of other ectoparasitic species which are allowed to feed on restrained and unrestrained hosts (Murray, 1987). Restrained animals allow more parasites to feed and feeding is also more successful for individual insects. Other behaviour which apparently protects from biting flies is grouping into herds which reduces the individual's chances of being attacked (Mooring and Hart, 1993; Ralley et al., 1993). However, the same behaviour may make the animals more attractive to flies such as *Hypoderma* which lay eggs on as many animals as possible (Hart, 1994). This could partly explain the 'gadding' response or running away from warble flies which causes significant costs in cattle production (Georgi and Georgi, 1990). In fact, most forms of avoidance behaviour seem to place a cost on animal production (Wieman et al., 1992, Dougherty et al., 1994).

The development of hypersensitivity reactions also leads directly to reductions in growth rate and losses in production (Bean et al., 1987). In their most severe forms these reactions cause a pruritic dermatitis over large areas of the body (Connan and Lloyd, 1988; Fadok and Greiner, 1990; Mason and Evans, 1991). Hypersensitivity reactions also cause problems in humans, especially to mosquito bites and particularly in immunologically suppressed patients (Diven et al., 1988; Penneys et al. 1989). As a result, the control of hypersensitivity to dipteran bites is a major concern of the livestock industries and medicine in most parts of the globe.

One of the more frequently studied examples of dipteran hypersensitivity occurs in horses (Greiner et al., 1990; Ungar-Waron et al., 1990; Anderson et al., 1991) and other livestock (Yeruham et al., 1993) and is caused by midges of the genus, *Culicoides*. The major antigens of these midges are thought to be salivary proteins (Morrow et al., 1986). A range of proteins from whole midges have been shown to be recognized in skin tests after column chromatography and isoelectric focusing (Morrow *et al.*, 1986). The protective function of these reactions has not been investigated and in most cases the pathological consequences of the hypersensitivity responses would seem to easily outweigh any protective effect. Therapies which attempt to inhibit hypersensitivity by desensitizing or immunomodulating the responses have had limited effects against *Culicoides* sensitivity in horses (Barbet *et al.*, 1990). However, recent studies which suggest a role for platelet activating factor (PAF) rather than histamine in the reactions to *Culicoides* antigens (Foster *et al.*, 1995) may suggest alternative drug therapies. The best chance for controlling this allergy, short of insecticidal techniques, would appear to be selective breeding on the basis of MHC and other immune-response-associated genes (Halldorsdottir *et al.*, 1991; Marti *et al.*, 1992; Lazary *et al.*, 1994), though the practicality of such an approach in the horse industry is very doubtful.

Hypersensitivity to mosquito bites in humans is also apparently caused by a range of mainly salivary antigens (Wu and Lan, 1989; Brummer-Korvenkontio *et al.*, 1994) with antigens of molecular weights 21.5 and 36 kDa the most commonly recognized by both IgG and IgE antibodies after *Aedes communis* exposure (Reunala *et al.*, 1991; Brummer-Korvenkontio *et al.*, 1994). Neither the protective effect of these antibodies nor the hypersensitivity responses have been investigated and most patients would probably prefer the simple mosquito bites to any marginal effect against the mosquito from the hypersensitivity reactions. However, control of these reactions has proved difficult, desensitization has been reported as occurring naturally in human populations with high levels of exposure (Das *et al.*, 1991; Reunala *et al.*, 1994a) but desensitization as a clinical therapy is unreliable and can cause significant side effects without affecting the original reaction (Nakagawa and Gershwin, 1993). Antihistamines can reduce the symptoms of the type I responses and non-sedating drugs such as cetirizine are proving useful (Reunala *et al.*, 1991). Better desensitization procedures may have to await the isolation of the major sensitizing antigens or the development of specific cytokine or other therapies to block the stimulation of IgE responses (Tan *et al.*, 1992; Nakagawa and Gershwin, 1993).

As stated above, hypersensitivity reactions are apparently a natural sequel to haematophagous dipteran bites and in fact to all biting arthropods. Hypersensitivity reactions are commonly reported and investigated in humans and animals against the blackflies, *Simulium* spp. (Kitani *et al.*, 1991; Cross *et al.*, 1993a), tsetse flies, *Glossina* spp. (Ellis *et al.*, 1986; Matha and Weiser, 1988), *Haematobia irritans* (Kerlin and Allingham, 1992) and a range of tabanid species (Hoffman, 1986; Pythal and Rajan, 1993). The reactions have been described for many years though the actual molecular mechanisms involved are only now being elucidated. Why dipterans should induce such strong hypersensitivity reactions is not as yet clear but it is known that these reactions do not usually provide significant protection to the host. As a result interest has turned to new methods of controlling these pests and this has led to the analysis of other responses that might be directly active against the insect.

## Immune responses in resistant hosts

The large economic effects that result from the depredation of horn and buffalo flies, *Haematobia* spp., have resulted in a range of studies on resistant and susceptible cattle. In particular, the genetic basis of resistance to horn flies has been established with estimates of heritability of $0.59 \pm 0.1$ and $0.78 \pm 0.16$ in work reported by Brown *et al.* (1992), showing that resistance is a heritable and thus selectable trait. The role of the immune response in this resistance is not certain though in most resistances to parasites so far analysed in detail the immune response is at least partly responsible (Owen and Axford, 1991). However, in cattle there was no obvious increase in resistance with age from 2 to 12 year old (Steelman *et al.*, 1993), possibly reflecting a lack of an effective acquired immunological response. Although it should be noted that 2-year-old cattle would be expected to have already acquired a significant immune response given unrestricted environmental exposure. Recently, Tarn *et al.* (1994), suggested that the abundance of serum proteins of molecular weights of 54,000 and 76,000 varied between susceptible and resistant animals. The identity of the proteins has not been published but their proximity to IgG heavy chains and IgE or M heavy chains is interesting. However, a direct role for hypersensitivity or other immune responses in resistance to horn flies has not been investigated in resistant or susceptible strains.

Resistance was also examined in *Bos indicus* and *Bos taurus* breeds to the closely related buffalo fly. However, heritability estimates were quite low (0.06) for fly resistance (Mackinnon *et al.*, 1991) suggesting that the response to selection would be rather slow. Studies on hypersensitivity reactions to buffalo fly in cattle have also suggested that naturally acquired responses do not induce an effective anti-fly response (Kerlin and Allingham, 1992) despite high levels of antibodies against a range of antigens.

Resistance responses to other biting flies are at a similar level of understanding with a few isolated studies. The one general observation applicable to all these species is that hypersensitivity responses are ubiquitous suggesting that strongly allergenic and $T_H2$-cell-stimulating antigens are common in most fly species. In fact, respiratory allergies are also quite common to flies whether they are parasitic species or not (Kagen, 1986; Baldo and Panzani, 1988).

Despite the general nature of hypersensitivity responses the documented occurrence of natural acquired resistance to fly species is rare. However, it should be noted that the presence of natural resistance is not always easy to determine under field conditions and even in controlled experiments, variations in the response of outbred animals can be such that resistance can only be proved to exist by undertaking studies of large groups of animals from known genetic backgrounds and/or managed breeding programmes (Owen and Axford, 1991).

## Modulation of host responses

Salivary proteins and other compounds have been shown to aid haematophagous feeding by modulating the inflammatory and coagulation cascades and by inhibiting immune reactions. Ribeiro (1987) reported on the role these responses play in promoting blood finding and increasing the speed of feeding. Blood finding is suggested to be assisted by apyrase activity which catalyses the conversion of ADP and ATP to AMP and, as a result, blocks the ADP-dependent degranulation and aggregation of platelets (Mant and Parker, 1981; Ribeiro, 1987; Champagne et al., 1995). In the absence of this enzyme, haematophagous species take longer probing the skin before they start feeding (Ribeiro, 1987).

Other activities found in haematophagous insect saliva include anticoagulants such as an inhibitor of thrombin found in tsetse flies (Parker and Mant, 1979), and an inhibitor of factor Xa in *Simulium* spp. (Abebe et al., 1994). A multitude of salivary vasodilators have been characterized (Lerner et al, 1991; Ribeiro et al., 1994; Champagne, 1994) and factors with effects on cytokines and other direct immunomodulating activities. Salivary extracts from *Aedes aegypti* have been shown to inhibit the release of TNFα from rat mast cells (Bissonnette et al., 1993) and the production of IL-2 from T cells, these extracts also inhibited cell proliferation in the presence of IL-2 (Cross et al., 1994a). In similar experiments, salivary extracts of *Simulium vittatum* inhibited T- and B-cell responses to mitogens *in vitro* and to IL-2 and IL-4 (Cross et al., 1993b, 1994b) and affected the production of interleukins-5 and -10 but not IFN-γ or IL-2 and IL-4, though an effect on IL-4 processing was described (Cross et al., 1994b). Saliva from the sandfly *Lutzomyia longipalpis* has direct effects on antigen presentation by macrophages to T cells *in vitro* (Theodus and Titus, 1993).

Overall, these results suggest that the saliva of the parasitic flies has been selected to inhibit a specific set of responses in the local dermis. These include effects on inflammatory pathways through inhibition of platelet function, the coagulation cascade and the cytokine network resulting in the inhibition of haemostasis and coagulation, stimulation of vasodilation, inhibition of the kinin pathways and suppression of the inflammatory cell response. Responses against immune functions such as antigen presentation and T- or B-cell stimulation, though described, are harder to understand given the rapidity of feeding in most species and the apparently reduced opportunity for immune responses to affect the dipteran. Thus, the immunomodulatory effects may be by-products of inhibition of local inflammatory cytokine reactions rather than specific effects aimed at immunosuppression. The one immune response that might be able to affect the rapidly feeding haematophagous species is IgE-mediated mast cell degranulation and thus the effects on mast cells and IgE induction described above might be expected. However, the ubiquitous nature of allergic hypersensitivity to these flies and mosquitoes and the apparent lack of effect from these immediate reactions suggests that these responses are not particularly deleterious to feeding or survival.

## Vaccination

Vaccines against haematophagous dipterans were first seriously considered after Alger and Cabrera (1972) and Schlein and Lewis (1976) showed that rabbits which were immunized with extracts of mosquitoes and flies and then used in feeding trials could inhibit blood-feeding, fecundity or induce mortality in the feeding insects. This observation led to a range of similar studies on a number of fly and mosquito species.

### *Mosquitoes*

The control of malaria-carrying mosquitoes and other dipterans which carry viral, bacterial and parasitic diseases is a major problem in most parts of the world. Chemical insecticides are still the most efficacious and economical method of control despite the development of drug resistance and the adverse environmental impact of many insecticides. Direct vaccination against the mosquito vectors has shown some promise as a form of malaria control. Studies in rabbits and mice have demonstrated that immunization with mosquito (*Aedes aegypti*) homogenates resulted in effects on fecundity, mortality and feeding success in mosquitoes fed either directly on the hosts or in membrane feeding trials (Hatfield, 1988a; Ramasamy *et al.*, 1988). Further studies suggested that these anti-mosquito effects may be species specific with anti-fecundity effects evident against *Anopheles tesselatus* while mortality but not fecundity was affected in *Culex quinquefasciatus* fed on anti-*A. tesselatus* antibodies (Ramasamy *et al.*, 1992). In addition, extracts from either the head/thorax region or the abdomen or the midgut all produced the same effect in *A. tesselatus* (Sri-Krishnaraj *et al.*, 1993) suggesting that either the anti-fecundity effect was relatively non-specific with respect to the antigen or that an antigen important in reproductive physiology was present throughout the whole body of the insect.

The mechanism of these effective anti-mosquito responses is not known though the fate of antibody ingested with the bloodmeal has been investigated. IgG could be detected in the midgut for up to 48 hours after ingestion and as such was more persistent than other serum proteins (Irby and Apperson, 1989). Specific anti-midgut antibody has also been demonstrated to bind midgut antigens after a bloodmeal and to persist for up to 3 days (Hatfield, 1988b) suggesting that antibodies can reach and bind target gut antigens. Finally, the passage of IgG into the haemolymph has been analysed in some species (Vaughan and Azad, 1988). Three of four *Anopheles* spp. but not the *Aedes* or *Culex* spp. tested were found to pass IgG into the haemolymph within 3 hours of feeding. The ability to transfer IgG was found to correlate with the length of feeding time and the ability to concentrate the bloodmeal by pre-diuresis, the excretion of excess fluid through the anus, while feeding.

The importance of the passage of active antibody into the gut and haemolymph has not been established but by analogy with the vaccine developed against the tick *Boophilus microplus* (Willadsen *et al.*, 1989), the potential to attack antigens in the midgut may be crucial. Studies should now concentrate on isolating specific antigens (Lehane, 1994; Sauer *et al.*, 1994) in order to boost the anti-mosquito effects and definitively establish the efficacy or otherwise of this novel vaccine strategy.

An interesting observation, which grew out of this work and may be developed into an effective control strategy, is that host antibody can inhibit the development of malaria and other parasites in the mosquito. This transmission blocking response occurs if anti-mosquito antibodies are present in the blood feed. The antibodies apparently non-specifically block oocysts from invading midgut cells (Ramasamy and Ramasamy, 1900; Lal *et al.*, 1994) and can act against other mosquito infecting agents such as Ross River virus and Murray Valley encephalitis virus (Ramasamy *et al.*, 1990). Though this is an interesting side effect of immunization against mosquito midgut antigens, it does not appear to have the efficacy of direct anti-*Plasmodium* antibodies which can totally block parasite development in the mosquito after ingestion with a bloodmeal (Billingsley, 1994; Snewin *et al.*, 1995). Thus, unless the direct effects of mortality and anti-fecundity on the mosquito can be increased in vaccine regimes to significantly reduce vector populations, then it is likely that, for at least malaria control, vaccines directed against the parasite will be more effective.

## *Vaccines against other dipterans*

Vaccination studies against other haematophagous species have shown similar effects to those against mosquitoes. Thus, immunization against the stable fly, *Stomoxys calcitrans*, with gut antigens increased mortality and had an anti-fecundity effect (Webster *et al*, 1992) while immunization against the tsetse fly, *Glossina fuscipes*, showed a small but significant increase in mortality (Desquesnes, 1990). In an interesting attempt to prove the role of immunization against tsetse, Matha *et al.* (1989) passively transferred sera from rabbits immunized with a salivary gland extract, to naive rabbits and successfully transferred a 'killing effect'. However, the effect only lasted 72 hours despite the presence of antibody titres for 7 days after the injection. This, together with a lack of correlation between antibody titre and the 'killing effect', suggests that a reaction other than antibody activity was responsible, possibly some inflammatory reaction initiated by the transfer of the sera between the rabbits.

These results again show the difficulties that are inherent in controlling fast blood feeders which are in contact with the host for only a few minutes. However, it is clear that immune responses can affect feeding success even in mosquitoes – most of which feed in under 10 minutes. The development of these small effects or other novel responses into a viable vaccine response is the current challenge but one that will be assisted by the availability of molecular genetic and protein techniques and also by the rapidly increasing knowledge of immunological modulation by cytokines and related compounds.

## MYIASIS FLIES

## Hypoderma Species

The most obvious examples of natural resistance in which immune responses are involved, are those where resistance is acquired after a period of susceptibility. For example, a lasting immunity develops to the warble flies *Hypoderma bovis* and *H. lineatum* larvae after a single season of infection in cattle. This resistance has been analysed by a number of groups and a fairly extensive literature is available including recent reviews by Baron and Colwell (1991a) and Scholl (1994).

*Hypoderma* spp. larvae penetrate the skin of their hosts and undergo a migration in the subcutaneous tissues to the neck in the case of *H. bovis* and the back for *H. lineatum*. The larvae eventually migrate to the back near the backbone in both species and form a 'warble' or a large swelling with a central hole through which the larvae respire. Resistance develops after about three primary infections and results in a drop in the number of larvae forming warbles (Gingrich, 1980, 1982).

The larvae produce a range of proteases with the major activities being a collagenolytic chymotrypsin (hypodermin C) (Lecroisey *et al.*, 1978, 1987) and two trypsins, hypodermin A (Tong *et al.*, 1981) and hypodermin B (Lecroisey *et al.*, 1983). Other serine proteases are also produced (Schwinghammer *et al*, 1988) though the number of major antigenic components secreted and excreted by the larvae appears relatively small (Pruett *et al.*, 1990 and cf. Skelly and Howells, 1987) perhaps because of the large amounts of hypodermins relative to other antigens. Antibody responses in infected cattle develop mainly to the hypodermins (Pruett *et al.*, 1988; Baron and Colwell, 1991b), though all the protein bands in E/S run on SDS–PAGE and immunoblotted are recognized by antibodies after infection experience (Pruett *et al.*, 1988). However, antibody levels are generally not a reliable guide to the level of natural resistance and cellular reactions are thought to mediate the major killing mechanisms (Gingrich, 1982). This has been demonstrated recently by the stimulation of calf responses with tubercle bacillus cell walls (BCG) or bacterial endotoxin or endotoxin plus a crude larval antigen, one week prior to infection with *H. lineatum*. Lymphocyte responses were stimulated to the grub antigen and to mitogens and a significant drop in larval numbers was observed in all treated calves (Baron and Colwell, 1987). This work has since been extended by treating calves with monophosphoryl lipid (MPL) alone or in combination with the enzymes hypodermin A, B and C. Both the MPL alone and antigen-vaccinated groups produced fewer larvae than controls at a challenge infection though the lowest levels were achieved in calves which received MPL plus the enzymes (Baron and Colwell, 1991b). These demonstrations plus the findings of delayed hypersensitivity responses to specific antigens (Pruett and Barrett, 1984), correlations between antigen-specific lymphocyte responsiveness and larval mortality (Baron and Weintraub, 1987), and an association between migration inhibition factor (MIF) production, intradermal skin test sensitivity and resistance after *Hypoderma* spp. infections (Gingrich, 1982) all strongly suggest the involvement of cellular mechanisms in larval killing.

However, antibody involvement has not been excluded and vaccine studies aimed at inducing high levels of antibody have been successful (Pruett *et al*, 1987) though not universally (Chabaudie *et al*, 1991) and correlations between antibody titres and protection have not been observed (Pruett *et al.*, 1989). In natural infections, antibody responses are slow to develop with IgG antibodies appearing 4–6 weeks after infection and not reaching maximum levels until a week or so before the larvae leave the host (25–28 weeks post-infection) (Pruett and Barrett, 1985; Pruett *et al.*, 1987). Lymphocyte responses are also slow to develop with specific immune responses to hypodermin A not recorded until 6 months after recovery from a primary infection (Fisher *et al.*, 1991).

This slow immune development suggests the presence of immune evasion mechanisms. The strong proteases secreted by the larvae would be expected to be able to degrade a range of host substrates including immunological proteins and this has been demonstrated. The complement protein C3 is degraded by the trypsin enzymes hypodermin A and B (Boulard, 1989; Baron, 1990) with cleavage of both the $\alpha$ and $\beta$ chains. The peptides produced are not analogous to those produced during the natural complement cascades (Boulard, 1989). However, crude larval homogenates were more effective at degradation than the isolated enzymes (Baron, 1990). In addition, IgG is also degraded by hypodermin A, though the survival of the antigen binding heavy chain domains (Pruett, 1993) does allow the possibility of the preservation of antigen binding albeit with the loss of immune reactions dependent on Fc binding. There is little doubt that other proteins involved in immune interaction will also be degraded by one or all of the hypodermins. Although, the physiological significance of these observations has not yet been demonstrated, *in vivo* observations have shown a marked lack of inflammatory cell response around migrating larvae (Nelson and Weintraub, 1972). This lack of response correlates to low or weak lymphocyte proliferation responses (Baron and Weintraub 1987; Fisher *et al.*, 1991) and to the cleavage of C3 which inhibits the complement pathways and so also inhibits inflammatory and immune reactions (Boulard 1989; Baron, 1990). In addition, hypodermin A, but not C, has been shown to depress bovine lymphocyte mitogen responses (Chabaudie and Boulard, 1992, 1993). These observations combine to suggest that *Hypoderma* spp. larvae are able to survive in naive hosts because they depress the inflammatory and cellular response. In previously infected animals, the acquired cellular response is able to overcome the larval evasion mechanisms and inhibit migration and then kill the larvae soon after they enter the host (Gingrich, 1980). The rapid development and efficacy of the resistance response suggests that vaccination against *Hypoderma* spp. will be an effective control option.

## Lucilia cuprina

Resistance mechanisms or even the presence of resistance in other myiasis flies is less certain. Infections of *Lucilia cuprina* larvae do show a bias towards young sheep, consistent with the development of acquired resistance (Watts *et al.*, 1979). Repeatedly infecting a group of sheep results in a drop in the number of larvae recovered after about five infections in a proportion of the animals (Sandeman *et al.*, 1986). However, these findings are highly variable depending on the number of larvae at each infection and the interval between infections (Sandeman *et al.*, 1992). Thus, if resistance does develop, it would appear to depend on immune mechanisms which are very sensitive to the level and pattern of immune stimulation. The immune responses in which such variability is characteristic are hypersensitivity responses and especially IgE-mediated responses.

*Lucilia cuprina* larvae certainly cause inflammatory responses with skin inflammation and neutrophil influx being well described corollaries of infection (Broadmeadow *et al*, 1984; Bowles *et al.*, 1992). However, these responses occur in naive sheep and thus would appear to be mainly non-specific responses mediated by larval products and tissue damage. Hypersensitivity responses do develop in previously infected animals but these are usually intermediate or Arthus-like in timing with maximum skin weal responses between 2 and 4 hours after antigen exposure (O'Meara *et al.*, 1992; Sandeman *et al.*, 1992). Until recently, specific anti-*L. cuprina* IgE responses could not be measured because of the lack of anti-sheep IgE reagents and because the non-specific inflammatory response to the larval E/S products masked any attempt to carry out passive cutaneous anaphylaxis assays.

Despite these problems, the role of inflammatory responses has been analysed in sheep which are resistant or susceptible to *Lucilia cuprina* and to the predisposing condition, fleece rot, which is a bacterial dermatitis that renders the affected sheep highly attractive to oviposition by gravid flies. There is a clear difference in skin inflammatory response to an intradermal injection of *L. cuprina* E/S products between sheep bred for resistance (R) or susceptibility (S) to fleece rot and fly infection at Trangie Agricultural Research Centre (O'Meara *et al.*, 1992). This is true in naive and previously infected animals though the differences in naive animals are usually greater. This difference in inflammation correlates to faster and increased levels of protein leakage into the infection site (O'Meara *et al.*, 1992) and to higher levels of acute phase proteins such as C3 in resistant animals (O'Meara *et al.*, 1995). These results suggest that the resistance mechanism in these sheep may result from a fast inflammatory response causing the appearance of complement proteins, and serum enzyme inhibitors such as $\alpha 2$ macroglobulin on the skin surface soon after infection. These proteins could control both bacterial proliferation on the skin and possibly larval growth (Bowles *et al.*, 1990), though the latter was not observed in these experiments (O'Meara *et al.*, 1992). Recent studies have shown a greater number of IgE-positive cells (Colditz *et al.*, 1994) and mast cells (Nesa, unpublished) in the skin of sheep bred for resistance suggesting again the importance of inflammation in the resistance response. If inflammatory mechanisms are involved in resistance, then the impact of various hypersensitivity responses on the basic inflammation response may explain the variable nature of the development of resistance after repeated infection in unselected animals (Sandeman *et al*, 1992).

To add to the problem of defining the resistance mechanism, a study carried out to determine the heritability, phenotypic, genetic and environmental correlations of the E/S skin test fleece rot and flystrike resistance gave quite different results to those described above. An analysis of 1425 progeny from 123 sires has shown a positive genetic correlation of 0.3–0.4 between skin weal response and fleece rot susceptibility. Thus, the larger the weal response the greater the sensitivity to fleece rot (Raadsma and Sandeman, unpublished). This different result to that obtained by O'Meara *et al.*, (1992) might be explained by differences in genetic background between the flocks or by phenotypic differences.

In order to further examine this variation two single trait selection flocks have now been established for increased and decreased weal sizes after E/S antigen skin testing (Raadsma and Sandeman, unpublished). Breeding has now been under way for 2 years and a heritability estimate of 0.4 has been confirmed for the skin test. If it can be shown that the sheep which are bred as a result of using this test suffer no loss in commercial traits and are not more susceptible to other diseases, then this test offers the prospect of a rapid and cheap indirect selection marker for breeding flystrike resistance.

Immune responses to L. cuprina are like those to Hypoderma, slow to develop, with antibody responses reaching maximum titres only after four or five consecutive infections (Sandeman et al., 1985). Lymphocyte mitogenic responses have not been measured but a mainly neutrophilic cellular response does occur at the site of infection accompanied by eosinophils and T-cell infiltration with $CD4^+T$ helper cells, $\gamma\delta$ T cells and $T19^+$ cells all showing significant increases at the site (Bowles et al., 1994). However, at a secondary infection the same level of response occurred with only B-cell levels showing a significant increase in traffic through the local lymph node (Bowles et al., 1992, 1994), suggesting that the specific anamnestic response is limited possibly to a polyclonal B-cell response. Cytokine levels at the local lymph node and in the skin have also been analysed (Bowles et al., 1994; Elhay et al., 1994), with mRNA for the inflammatory cytokines IL-1-$\alpha$ and $\beta$, IL-6 and IL-8 all found in the local lymph node and those for IL-1-$\beta$, IL-8 and TNF-$\alpha$ were expressed in the skin. IL-2 and IFN-$\gamma$ mRNA levels increased as the infection progressed and as T cells migrated into the site. The antigens which stimulate these responses are many and varied (Skelly and Howells 1987; Seaton et al, 1992), but the enzymes released are similar to the hypodermins of Hypoderma spp. Trypsins and chymotrypsins are the major enzymes produced by Lucilia cuprina with three trypsins and one chymotrypsin described to date (Sandeman et al., 1990). The major antigens are also thought to include these enzymes (Seaton et al., 1992). The recent cloning of these enzymes (Casu et al., 1995) should enable their role in the infection and host response to be dissected.

The presence of enzymes in Lucilia which are very similar to those in Hypoderma larvae suggests that they may have similar activities against immune components. In fact, L. cuprina E/S products are known to be highly inhibitory to the stimulation of antibody responses (Kerlin and East, 1992), virtually abrogating an immunization response to the defined antigen ovalbumin when injected simultaneously into sheep. This effect probably explains the very slow and weak development of antibody responses during infections and may partly explain the limited anamnestic response (Bowles et al., 1994).

Other activities similar to those of Hypoderma larvae are the apparent cleavage of C3 (O'Meara et al., 1992) and IgG (Sandeman et al., 1995a), though the antigen-binding ability of the cleaved antibody is maintained despite this cleavage. Recently, another apparent evasion mechanism has been observed with the incubation of isolated sheep skin cells with L. cuprina E/S products. At low concentrations of products, MHC class II surface proteins are apparently cleaved from the cells while cell death occurs at higher concentrations (Ghosh, Bowles and Sandeman, unpublished). This reduction in class II expression may explain the suppression of antibody formation during infections and against a defined antigen in the presence of E/S products (Kerlin and East, 1992). If this is an enzyme effect then similar mechanisms may operate in *Hypoderma* and other myiasis infections. The actual impact of these evasion mechanisms on immunity to infections remains uncertain but they do suggest that the myiasis flies and, in particular, *Lucilia* larvae, have specific adaptations to a parasitic existence and are not just treating the host as moving carrion. They also suggest that myiasis flies may share quite similar biochemistry and that comparisons between *Lucilia and Hypoderma* may allow more rapid progress in analysing immune responses and developing novel vaccine strategies.

## Vaccination

Vaccination against myiasis flies has become a major aim of research into both *Hypoderma* and *Lucilia* infections. The existence of a clearly acquired resistance response in cattle against *Hypoderma* spp. has guided the development of a vaccine which attempts to mimic this response. Vaccines against *Lucilia* have not had a similar clear response to copy but have had to be developed by a more empirical approach. Thus, the major vaccine targets in *Lucilia cuprina* are novel antigens such as the gut peritrophic membrane, antigens identified by monoclonal and polyclonal antibodies and a set of E/S antigens recognized by local antibody probes.

### Hypoderma *vaccines*

In *Hypoderma*, the obvious vaccine antigens were the abundant enzymes, hypodermis A, B and C. Initial trials with crude E/S products or larval extracts demonstrated that resistance to a primary infection could be induced by immunization (Khan *et al.*, 1960; Magat and Boulard, 1970; Pruett and Barrett, 1985; Baron and Weintraub, 1986). However, most of the recent work has concentrated on the hypodermis with A, in particular, shown to stimulate varying levels of resistance depending on the adjuvant and protocol used (Pruett *et al.*, 1989; Chabaudie *et al.*, 1991; Baron and Colwell, 1991b).

As outlined earlier, the best effects seem to be obtained after immunization procedures designed to stimulate cellular responses (Baron and Colwell, 1991b), although in this experiment, a combination of hypodermis A, B and C was administered. Thus, it seems that a vaccine composed of at least hypodermin A plus a cell-stimulating adjuvant and/or possibly a cytokine such as IFN-γ should stimulate high levels of resistance (> 90%) to *Hypoderma lineatum*. The recent cloning of the hypodermins (Moire *et al.*, 1994) including the previously uncharacterized enzyme described as P2 (Schwinghammer *et al.*, 1988) which now appears to be a variant of hypodermin B (Moire *et al.*, 1994), will allow final development and testing of a vaccine. A vaccine which cross-reacted between *H. lineatum* and *H. bovis* would be desirable rather than producing two vaccines or one with up to eight related enzyme antigens.

Studies of cross-reactivity between these species have shown antibody recognition of similar epitopes on hypodermin A and C in both species but not hypodermin B (Pruett et al., 1990). However, immunization with hypodermin A gave the best levels of protection (Pruett et al., 1989) and the cross-reactivity of epitopes on this protein may allow development of a single antigen vaccine which controls both species. However, in a trial reported recently, hypodermin A had little effect on either *H. lineatum* or *H. bovis* survival (Chabaudie et al., 1991) when it was delivered subcutaneously with Freund's adjuvant, perhaps illustrating the importance of the method of administration and adjuvant selection to obtaining successful vaccine responses.

The commercial viability of a vaccine may not only depend on the efficacy of the final injection. Control of *Hypoderma* spp. with ivermectin has been shown to be highly effective (Dorchies et al., 1982) and its use for other internal parasites over a number of years may explain the apparent reduction in warble fly prevalence in the USA over the last few years (Scholl, 1993). This finding suggests that control of *Hypoderma* might be achieved over wide areas using existing control technologies. Such control programmes have, in fact, been previously attempted in Alberta, Canada and parts of Europe (Wilson, 1986) though only Great Britain has apparently controlled *Hypoderma* using this technique (Tarry, 1989). If the use of ivermectin improves the impact of such control programmes, then the development of a vaccine may not be commercially viable with the loss of its major market or may only be commercially viable for a limited period as an adjunct to the control programme.

## Lucilia *vaccines*

The only other anti-dipteran vaccine to reach a stage of development similar to those against *Hypderma* are vaccines against the sheep myiasis fly *Lucilia cuprina*. The larvae of *Lucilia* cause the formation of a skin lesion which can extend over large areas. The epidermis is sloughed off and the dermis exposed, in some cases, allowing the larvae to penetrate through the dermis to the underlying connective tissue (Sandeman et al., 1987). The infection proceeds rapidly as the larvae grow through three stages and two moults over 3 to 5 days. Affected sheep can die from infections of over 500 larvae from a combination of toxic effects (Broadmeadow et al., 1984). As discussed above, acquired resistance responses are difficult to stimulate against *L. cuprina* and naturally resistant sheep seem to be directly resistant to bacterial predisposing conditions rather than the larvae themselves. Research into vaccines has, as a result, had to be developed empirically with little assistance from natural immune mechanisms. Three major strategies have been followed. The first employed intuitive guesswork from a knowledge of the larval biology to define an initial target from which a set of novel antigens were isolated and tested in vaccine trials. The second approach used antibodies from animals immunized with various larval extracts to isolate larval antigens while the third used antibodies from naturally infected animals to define a set of natural antigens which were then tested in vaccine trials. All strategies have defined sets of antigens which give partial protection against infection.

The first or novel antigen approach yielded three major antigens from the peritrophic membrane of the larvae (East et al., 1993). The peritrophic membrane is a chitinous tube which surrounds the digesta and prevents contact between the gut contents over a certain molecular weight and the gut cells (Eisemann and Binnington, 1994). Three protein antigens (PM 44, PM 48 and PM 95 kDa) have been isolated from the peritrophic membrane and tested in vaccine trials (Tellam et al., 1994). Antiserum raised against these antigens produces an antibody-concentration-dependent inhibition of larval growth, both in vitro and in vivo (East et al., 1993). Larvae have been inhibited by up to 55% in vivo and if the antibody concentration is artificially increased in vitro, inhibition can reach levels of 90% with significant larval mortality (East et al., 1993). However, significant levels of mortality have not been observed in vivo and, if left on the skin, compensatory growth can occur, reducing the vaccine effect (Eisemann cited in Eisemann and Binnington, 1994). Electron microscope studies have shown that the antibodies bind to the peritrophic membrane and block the passage of nutrients to the larval midgut cells (Eisemann and Binnington, 1994). These antigens have now been cloned and, although they have complex structures of repeating domains which include six cysteine residues (Tellam et al., 1994), work is continuing to develop them for further trials.

The second approach to vaccine identification involved immunization of sheep with fractions of first instar E/S antigens and homogenate extracts separated by SDS–PAGE (Sandeman et al., 1996). Antisera collected from these sheep and adjuvant injected controls were tested in vitro for effects on larval growth. Antibodies were then isolated from sera with significant effects on growth, and used to isolate antigens from first instar homogenates. The antigens were used to vaccinate sheep against a challenge L. cuprina infection. In addition, a monoclonal antibody raised against midgut antigens, which also significantly affected larval growth in vitro (Fry et al., 1994), was used to isolate antigens for this vaccine trial. Three antigen preparations stimulated protective responses which resulted in up to 50% reduced larval recovery over controls: the monoclonal antigen; one of the SDS–PAGE fractions; and an ion-exchange fraction of E/S antigens with protease activity. No protection was obtained in sheep immunized with antigens isolated from control antibodies. The effects on larval numbers were combined with only low levels of growth inhibition suggesting that these responses are not mainly targeted against larval digestion as are the anti-peritrophic responses.

The third approach also used antibodies but these were isolated from the local lymph nodes of sheep experiencing normal infections. These antibodies are termed ASC probes and are secreted by B cells isolated from the local lymph nodes (Meeusen and Brandon, 1994). These antibodies identified four major bands on Western blots which were isolated and combined for vaccine trials (Bowles, 1996). Sheep immunized with these antigens had up to 65% fewer infections than control sheep and also had effects on larval growth and wound size in those infections which did develop. As suggested above, this total protection is quite different to the growth effects seen in peritrophic-antigen-vaccinated sheep and suggests a different mechanism. However, the second approach and these experiments are similar and similar antigens may have been isolated. Analysis of antibody levels in protected sheep did not reveal any correlation between antibody levels and protection. Thus, antibody does not seem to be directly involved but skin biopsies of protected sites have shown accumulations of T cells and MHC class II (probably Langerhans) cells, suggesting a cellular mechanism (Bowles, 1996).

These studies clearly illustrate the principle that an apparent lack of natural immunity need not be a deterrent to vaccine development, even using natural antigens. However, whether these antigens can be finally developed to a commercially viable vaccine is still to be determined. The sheep blowfly causes an acute pathology that can quickly cause major problems to the host. A vaccine will have to protect at a very high level to ensure that sheep are protected and that the farmer can reduce other control costs which currently include applications of persistent chemicals and a variety of animal husbandry procedures such as regular inspection of all susceptible animals. In addition, the current chemical control, cyromazine, gives high levels of protection for up to 12 weeks and there are no reports, as yet, of field resistance to this chemical.

Vaccination studies against other myiasis flies have been very limited. This is despite the importance of such infections in both human and animal health especially in less developed countries. Flies such as *Dermatobia hominis* cause large numbers of human and animal infections throughout South America and parts of North America. Studies have also suggested an acquired immune response is effective against this parasite (Sancho, 1988). This plus the current wealth of knowledge about *Hypoderma* infections should result in rapid progress towards a *Dermatobia* vaccine.

The screwworm, *Cochliomyia hominivorax*, also causes major problems in animals in South America, although it has been eradicated over North and much of Central America by the sterile insect release programme (Welch, 1990). As a result there is little interest in developing vaccines against this calliphorid though this situation may change if a vaccine is successfully developed against *Lucilia cuprina*.

*Oestrus ovis*, the nasal botfly, is a very common parasite of sheep worldwide but it is usually considered to be a relatively benign pathogen and control measures are not usually required. As a result the development of a vaccine is fairly unlikely despite the suggestion that vaccination against

*Oestrus ovis* in sheep seems a relatively simple process. Thus, Marchenko and Marchenko (1989) reported that only 0.4% of challenge larvae survived after three immunizations of lambs with an *O. ovis* homogenate.

The further development of vaccines against myiasis flies will probably depend on the success of the vaccines currently under development with *Hypoderma* representing the botfly group and *Lucilia* the skin infecting flies. If successful, these vaccines should stimulate interest in controlling similar species which may be less commercially significant but which have more deleterious effects in man and livestock in developing countries.

## CONCLUSIONS

The current state of knowledge of immune responses to dipteran parasites is highly variable depending on the species of parasite. Only two species have been studied in any detail and both of these cause myiasis infections. Work on haematophagous species has mainly concerned the analysis of hyper-sensitivity responses which cause debilitating acute and chronic dermal reactions. However, these studies have not so far led to reliable immune-based methods of controlling these pathologies and until more is known of the roles of cytokines and of pathway-specific, inhibitory molecules, there appears little prospect for the development of such treatments. The isolation of allergenic antigens for use in desensitization protocols is under way but the probability that a particular dipteran parasite only produces a single major allergen seems rather unlikely. Thus, desensitization will probably require a number of antigens to be isolated from each insect, produced in recombinant form and then used in specific combinations and/or amounts in individual patients.

Vaccination against haematophagous species has shown some promise with responses stimulated which affect feeding, fecundity and survival. However, these effects have rarely been such as to provide more than a partial protection to the host at least with the crude antigen mixtures used to date. The isolation of specific antigens is now required to determine whether the anti-parasite response can reach levels which either protect the host, or have a significant effect on dipteran breeding or at least reduce the probability of infection with vectored parasites. This last option may in fact be the best long-term strategy since it could reduce the main problem caused by many of these mosquito and fly species without leading to a strong selection pressure on the dipteran for resistance to the vaccine effects.

It is the area of immune evasion mechanisms that promise the most interesting findings over the next few years. To date both haematophagous and myiasis species have shown a range of mechanisms which apparently increase their chances of survival in the face of an active immune response. However, the mechanism of many of these anti-host responses is not known and their actual role in protecting the parasite while it is in contact with the host is usually not clear. Work on these parasite mechanisms may not only help our understanding of the parasite's survival strategies but also elucidate those host mechanisms which are effective and which are as a result inhibited by the parasite. This promising area of research and the pending trials of at least two vaccines against dipterans suggests an interesting and important period over the next few years, for all those involved and interested in the control of these parasites.

## REFERENCES

Abebe, M., Cupp, M.S., Ramberg, F.B. and Cupp, E.W. (1994) Anticoagulant activity in salivary gland extracts of black flies (Diptera: Simuliidae). *Journal of Medical Entomology* 31, 908–911.

Alger, N.E. and Cabrera, E.J. (1972) An increase in death rate of *Anopheles stephensi* fed on rabbits immunized with mosquito antigen. *Journal of Economic Entomology* 61, 165–168.

Anderson, G.S., Belton, P. and Kleider, N. (1991) *Culicoides obseletus* (Diptera: Ceratopogonidae) as a causal agent of Culicoides hypersensitivity (sweet itch) in British Columbia (Canada). *Journal of Medical Entomology* 28, 685–693.

Baldo, B.A. and Panzani, R.C. (1988) Detection of IgE antibodies to a wide range of insect species in subjects with suspected inhalant allergies to insects. *International Archives of Allergy and Applied Immunology* 85, 278–287.

Barbet, J.L., Bevier, D. and Greiner, E.C. (1990) Specific immunotherapy in the treatment of Culicoides hypersensitive horses: a double blind study. *Equine Veterinary Journal* 22, 232–235.

Baron, R.W. (1990) Cleavage of purified bovine complement component C3 in larval *Hypoderma lineatum* (Diptera: Oestridae) hypodermins. *Journal of Medical Entomology* 27, 899–904.

Baron, R.W. and Colwell, D.D. (1987) Enhancement of immune response to cattle grub infestations using bacterial toxins. Research Highlights – Lethbridge Research Station, Agriculture Canada, Ottawa, pp. 25–27.

Baron, R.W. and Colwell, D.D. (1991a) Mammalian immune responses to myiasis. *Parasitology Today* 7, 353–355.

Baron, R.W. and Colwell, D.D. (1991b) Enhanced resistance to cattle grub infestation (*Hypoderma lineatum* de Vill.) in calves immunized with purified hypodermin A, B and C plus monophosphoryl lipid A (MPL). *Veterinary Parasitology* 38, 185–198.

Baron, R.W. and Weintraub, J. (1986) Immunization of cattle against hypodermatosis [(*Hypoderma lineatum*) (de Vill.) and *H. bovis* (L.)] using *H. lineatum* antigens. *Veterinary Parasitology* 21, 43–50.

Baron, R.W. and Weintraub, J. (1987) Lymphocyte responsiveness in cattle previously infested and uninfested with *Hypoderma lineatum* (de Vill.) and *H. Bovis* (Diptera: Oestridae). *Veterinary Parasitology* 24, 285–296.

Barriga, O.O. (1993) A review on vaccination against protozoa and arthropods of veterinary importance. *Veterinary Parasitology* 55, 29–55.

Bean, G., Siefert, A., MacQueen, A. and Doube, B.M. (1987) Effect of insecticide treatment for control of buffalo fly on weight gains of steers in coastal central Queensland. *Australian Journal of Experimental Agriculture* 27, 329–334.

Beck, T., Moir, B. and Meppem, T. (1985) The costs of parasites to the Australian sheep industry. *Quarterly Review of the Rural Economy* 7, 336–343.

Billingsley, P.F. (1994) Vector–parasite interactions for vaccine development. *International Journal for Parasitology* 24, 53–58.

Bissonnette, E.Y., Rossignol, P.A. and Befus, A.D. (1993) Extracts of mosquito salivary gland inhibit tumour necrosis factor alpha release from mast cells. *Parasite Immunology* 15, 27–33.

Boulard, C. (1989) Degradation of bovine C3 by serine proteases from parasites, *Hypoderma lineatum* (Diptera: Oestridae). *Veterinary Immunology and Immunopathology* 20, 387–398.

Bowles, V.M., Feehan, J.P. and Sandeman, R.M. (1990) Sheep plasma protease inhibitors influencing protease activity and growth of *Lucilia cuprina in vitro*. *International Journal for Parasitology* 20, 169–174.

Bowles, V.M., Grey, S.T. and Brandon, M.R. (1992) Cellular immune responses in the skin of sheep infected with larvae of *Lucilia cuprina*, the sheep blowfly. *Veterinary Parasitology* 44, 151–162.

Bowles, V.M., Meeusen, E.N.T., Chandler, K.C., Verhagen, A., Nash, A.D. and Brandon, M.R. (1994) The immune response of sheep infected with larvae of the sheep blowfly, *Lucilia cuprina*, monitored via efferent lymph. *Veterinary Immunology and Immunopathology* 40, 341–352.

Bowles, V.M., Meeusen, E., Young, A.R., Nash, A.D. and Brandon, M.R. (1996) Vaccination of sheep against larvae of the sheep blowfly, *Lucilia cuprina*. *Vaccine* (in press).

Broadmeadow, M., Gibson, J.E., Dimmock, C.K., Thomas, R.J. and O'Sullivan, B.M. (1984) The pathogenesis of flystrike in sheep. *Wool Technology and Sheep Breeding* 32, 28–32.

Brown, A.H., Jr, Steelman, C.D., Johnson, Z.B., Rosenkrans, C.F., Jr. and Brasuell, T.M. (1992) Estimates of repeatability and heritability of horn fly resistance in cattle. *Journal of Animal Science* 70, 1375–1381.

Brummer-Korvenkontio, H. Lappalainen, P., Reunala, T. and Palusuo, T. (1994) Immunization of rabbits with mosquito bites: immunoblot analysis of IgG antimosquito antibodies in rabbits and man. *International Archives of Allergy and Applied Immunology* 93, 14–18.

Casu, R.E., Jarmey, J.M., Elvin, C.M. and Eisemann, C.H. (1994) Isolation of a trypsin-like serine protease gene family from the sheep blowfly *Lucilia cuprina*. *Insect Molecular Biology* 3, 159–170.

Chabaudie, N. and Boulard, C. (1992) Effect of hypodermin A, an enzyme secreted by *Hypoderma lineatum* (Insecta: Oestridae), on the bovine immune system. *Veterinary Immunology and Immunopathology* 31, 167–177.

Chabaudie, N. and Boulard, C. (1993) *In vitro* and *ex vivo* responses of bovine lymphocytes to hypodermin C, an enzyme secreted by *Hypoderma lineatum* (Insecta: Oestridae). *Veterinary Immunology and Immunopathology* 36, 153–162.

Chabaudie, N., Villejoubert, C. and Boulard, C. (1991) The response of cattle vaccinated with hypodermin A to a natural infestation of *Hypoderma bovis* and *Hypoderma lineatum*. *International Journal for Parasitology* 21, 859–862.

Champagne, D.E. (1994) The role of salivary vasodilators in bloodfeeding and parasite

transmission. *Parasitology Today* 10, 430–433.
Champagne, D.E., Smartt, C.T., Ribeiro, J.M.C. and James, A.A. (1995) The salivary gland-specific apyrase of the mosquito *Aedes aegypti* is a member of the 5' nucleotidase family. *Proceedings of the National Academy of Science* 92, 694–698.
Colditz, I.G., Lax, J., Mortimer, S.I., Clarke, R.A. and Beh, K.J. (1994) Cellular inflammatory responses in skin of sheep selected for resistance or susceptibility to fleece rot and flystrike. *Parasite Immunology* 16, 289–296.
Cogley, T.P. (1989) Effects of migrating *Gasterophilus intestinalis* larvae (Diptera: Gasterophilidae) on the mouth of the horse. *Veterinary Parasitology* 31, 317–331.
Connan, R.M. and Lloyd, S. (1988) Seasonal allergic dermatitis in sheep. *Veterinary Record* 123, 335–337.
Cross, M.L., Cupp, M.S., Cupp, E.W., Ramberg, F.B. and Enriquez, F.J. (1993a) Antibody responses of Balb/c mice to salivary antigens of haematophagous black flies (Diptera: Simuliidae). *Journal of Medical Entomology* 30, 725–734.
Cross, M.L., Cupp, M.S., Cupp, E.W., Galloway, A.L. and Enriquez, F.J. (1993b) Modulation of murine immunological responses by salivary gland extract of *Simulium vittatum* (Diptera: Simuliidae). *Journal of Medical Entomology* 30, 928–935.
Cross, M.L., Cupp, E.W. and Enriquez, E.J. (1994a) Differential modulation of murine cellular immune responses by salivary gland extract of *Aedes aegypti*. *American Journal of Tropical Medicine and Hygiene* 51, 690–696.
Cross, M.L., Cupp, E.W. and Enriquez, F.J. (1994b) Modulation of murine cellular immune responses and cytokines by salivary gland extract of the black fly, *Simulium vittatum*. *Tropical Medicine and Parasitology* 45, 119–124.
Dart, A.J., Hutchins, D.R. and Begg, A.P. (1987) Suppurative splenitis and peritonitis in a horse after gastric ulceration caused by the larvae of *Gasterophilus intestinalis*. *Australian Veterinary Journal* 64, 155–158.
Das, M.K., Mishra, A., Beuria, M.K. and Dash, A.P. (1991) Human natural antibodies to *Culex quinquefasciatus*: Age dependent occurrence. *Journal of the American Mosquito Control Association* 7, 319–321.
Desquesnes, M. (1990) Attempts to immunize rabbits against tsetse flies *Glossina fuscipes fuscipes* (Diptera: Glossinidae). *Revue d'Elevage et de Medecine Veterinaire des Pays Tropicaux* 43, 511–513.
Diven, D.G., Newton, R.C. and Ramsey, K.M. (1988) Heightened cutaneous reactions to mosquito bites in patients with acquired immune deficiency syndrome receiving zidovudine. *Archives of Internal Medicine* 149, 2296.
Dorchies, P., Franc, M. and de Lahitte, J.D. (1982) Antiparasitic treatment of cattle with Ivermectine. *Revue Medecine de Veterinaire* 133, 709–713.
Dougherty, C.T., Knapp, F.W., Burras, P.B., Willis, D.C. and Cornelius, P.L. (1994) Moderation of grazing behaviour of beef cattle by stable flies (*Stomoxys calcitrans*). *Applied Animal Behaviour Science* 4, 113–127.
East, I.J., Fitzgerald, C.J., Pearson, R.D., Donaldson, R.A., Vuocolo, T., Cadogan, L.C., Tellam, R.L. and Eisemann, C.H. (1993) *Lucilia cuprina*: inhibition of larval growth induced by immunization of host sheep with extracts of larval peritrophic membrane. *International Journal for Parasitology* 23, 221–229.
Eisemann, C.H. and Binnington, K.C. (1994) The peritrophic membrane: Its formation, structure, chemical composition and permeability in relation to vaccination against ectoparasitic arthropods. *International Journal for Parasitology* 24, 15–26.

Elhay, M.J., Hanrahan, C.F., Bowles, V.M., Seow, H.F., Andrews, A.E. and Nash, A.D. (1994) Cytokine mRNA expression in skin in response to ectoparasite infection. *Parasite Immunology* 16, 451–461.

Ellis, J.A., Shapiro, S.Z., ole Moi-Yoi, O. and Moloo, S.K. (1986) Lesions and saliva-specific antibody responses in rabbits with immediate and delayed hypersensitivity reactions to the bites of *Glossina morsitans centralis*. *Veterinary Pathology* 23, 661–667.

Fadok, V.A. and Greiner, E.C. (1990) Equine insect hypersensitivity: skin test and biopsy results correlated with clinical data. *Equine Veterinary Journal* 22, 236–240.

Fisher, W.F., Pruett, J.H., Howard, V.M. and Scholl, P.J. (1991) Antigen specific lymphocyte proliferative responses in vaccinated and *Hypoderma lineatum* infested calves. *Veterinary Parasitology* 40, 135–146.

Foster, A.P., Lees, P. and Cunningham, F.M. (1995) Platelet activating factor mimics antigen-induced cutaneous inflammatory responses in sweet itch horses. *Veterinary Immunology and Immunopathology* 44, 115–128.

Frazier, C.A. 1987 Diptera – mosquitoes and flies. In: Frazier, C.A. (ed.), *Insect Allergy*. Green, St Louis, pp. 136–200.

Fry, K.J., Seaton, D.S. and Sandeman, R.M. (1994) The production and characterization of monoclonal antibodies to *Lucilia cuprina* larval antigens. *International Journal for Parasitology* 24, 379–387.

Georgi, J.R. and Georgi, M.E. (1990) *Parasitology for Veterinarians*, 5th edn. W.B. Saunders Company, Philadelphia.

Gingrich, R.E. (1980) Differentiation of resistance in cattle to larval *Hypoderma lineatum*. *Veterinary Parasitology* 7, 243–254.

Gingrich, R.E. (1982) Acquired resistance to *Hypoderma lineatum*: comparative immune response of resistant and susceptible cattle. *Veterinary Parasitology* 9, 233–242.

Greiner, E.C., Fadok, V.A. and Rabin, E.B. (1990) Equine Culicoides hypersensitivity in Florida: biting midges aspirated from horses. *Medical and Veterinary Entomology* 4, 375–381.

Halldorsdottir, S., Lazary, S., Gunnarsson, E. and Larsen, H.J. (1991) Distribution of leucocyte antigens in Icelandic horses affected with summer eczema compared to non-affected horses. *Equine Veterinary Journal* 23, 300–302.

Hart, B.L. (1994) Behavioural defence against parasites: Interaction with parasite invasiveness. *Parasitology* 109, S139–S151.

Hatfield, P.R. (1988a) Anti-mosquito antibodies and their effects on feeding, fecundity and mortality of *Aedes aegypti*. *Medical and Veterinary Entomology* 2, 331–338.

Hatfield, P.R. (1988b) Detection and localisation of antibody ingested with a mosquito bloodmeal. *Medical and Veterinary Entomology* 2, 339–345.

Hoffman, D.R. (1986) Allergic reactions to biting insects. In: Levine, M.I. and Lockley, R.F. (eds), *Monograph on Insect Allergy*. American Academy of Allergy and Immunology, Milwaukee, USA, pp. 85–92.

Irby, W.S. and Apperson, C.S. (1989) Immunoblot analysis of digestion of human and rodent blood by *Aedes aegypti* (Diptera: Culicidae). *Journal of Medical Entomology* 26, 284–293.

Jones, C.J. and Lloyd, J.E. (1987) Hypersensitivity reactions and haematologic changes in sheep exposed to mosquito (Diptera: Culicidae) feeding. *Journal of Medical Entomology* 24, 71–76.

Kagen, S.L. (1986) Inhalant insect allergy. In: Levine, M.I. and Lockley, R.F. (eds), *Monograph on Insect Allergy*. American Academy of Allergy and Immunology, Milwaukee, USA, pp. 93–98.

Kay, B.H. and Kemp, D.H. (1994) Vaccines against arthropods. *American Journal of Tropical Medicine and Hygiene* 50 (Suppl. S), 87–96.

Kerlin, R.L. and Allingham, P.G. (1992) Acquired immune response of cattle exposed to buffalo fly (*Haematobia irritans exigua*). *Veterinary Parasitology* 43, 115–129.

Kerlin, R.L. and East, I.J. (1992) Potent immunosuppression by secretory/excretory products of larvae from the sheep blowfly *Lucilia cuprina. Parasite Immunology* 14, 595–604.

Khan, M.A., Connell, R. and le Q. Darcel, C. (1960) Immunization and parental chemotherapy for the control of cattle grubs *Hypoderma lineatum* (de Vill.) and *H. bovis* (L.) in cattle. *Canadian Journal for Comparative Medicine* 24, 177–180.

Kitani, H., Kotoh, N., Mitsonobu, F., Mifune, T., Okazaki, M. and Tanizaki, Y. (1991) Questionnaire survey on allergic reactions induced by black fly bites. *Okayama Igakkai Zasshi* 103, 1331–1336.

Krafsur, E.S., Whitten, C.J. and Novy, J.E. (1987) Screwworm eradication in North and Central America, *Parasitology Today* 3, 131–137.

Lal, A.A., Schriefer, M.E., Sacci, J.B., Goldman, I.F., Louis-Wileman, V., Collins, W.E. and Azad, A.F. (1994) Inhibition of malaria parasite development in mosquitoes by anti-mosquito-midgut antibodies. *Infection and Immunity* 62, 316–318.

Lazary, S., Marti, M., Szalai, G., Gaillard, C. and Gerber, H. (1994) Studies on the frequency and associations of equine leucocyte antigens on sarcoid and summer dermatitis. *Animal Genetics* 25 (Suppl. 1) 75–80.

Lecroisey, A., Giles, A., De Wolf, A. and Keil, B. (1978) Complete amino acid sequence of the collagenase from the insect *Hypoderma lineatum. Journal of Biological Chemistry* 262, 7546–7551.

Lecroisey, A., Tong, N.T. and Keil, B. (1983) Hypodermin B, a trypsin-related enzyme from the insect *Hypoderma lineatum*: Comparison with hypodermin A and *Hypoderma* collagenase, two serine proteases from the same source. *European Journal of Biochemistry* 134, 261–267.

Lecroisey, A., Giles, A., De Wolf, A. and Keil, B. (1987) Complete amino acid sequence of the collagenase from the insect *Hypoderma lineatum. Journal of Biological Chemistry* 262, 7546–7551.

Lehane, M.J. (1994) Digestive enzymes, haemolysins and symbionts in the search for vaccines against blood-sucking insects. *International Journal for Parasitology* 24, 27–32.

Lehrer, Z., Lehrer, M. and Verstraeten, C. (1988) Myiasis in sheep in Romania caused by *Wohlfartia magnifica* (Schiner) (Diptera: Sarcophagidae). *Annales de Medecine Veterinaire* 13, 475–481.

Lerner, E.A., Ribeiro, J.M.C., Nelson, R.J. and Lerner, M.R. (1991) Isolation of maxadilan, a potent vasodilatory peptide from the salivary glands of the Sand Fly *Lutzomyia longipalpis. Journal of Biological Chemistry* 266, 11234–11236.

Mackinnon, M.J., Meyer, K. and Hetzel, D.J.S. (1991) Genetic variation and covariation for growth, parasite resistance and heat tolerance in tropical cattle. *Livestock Production Science* 27, 105–122.

Magat, A. and Boulard, C. (1970) Essais de vaccination contre l'hypodermose bovine avec un vaccin contenant une collagenase brute extraite des larves de 1er stade d'*Hypoderma lineatum*. *Comptes Rendus l'Academie de Science*, Paris 270, 728–730.

Mant, M.J. and Parker, K.R. (1981) Two platelet aggregation inhibitors in tsetse (*Glossina*) saliva with studies of roles of thrombin and citrate in *in vitro* platelet aggregation. *British Journal of Haematology* 48, 601–608.

Marchenko, V.A. and Marchenko, V.P. (1989) Survival of the larvae of the sheep botfly, *Oestrus ovis* L. depending on the function of the immune system of the host's body. *Parazitologiia* 23, 129–133.

Marti, E., Gerber, H. and Lazary, S. (1992) On the genetic basis of equine allergic diseases: II. Insect bite dermal hypersensitivity. *Equine Veterinary Journal* 24, 113–117.

Mason, K.V. and Evans, A.G. (1991) Mosquito bite-caused eosinophilic dermatitis in cats. *Journal of the American Veterinary Medical Association* 198, 2086–2088.

Matha, V. and Weiser, J. (1988) Detection of antigens common to salivary glands and other tissues of tsetse fly, *Glossina palpalis palpalis* (Diptera: Glossinadae). *Folia Parasitologica* 35, 285–287.

Matha, V., Lukes, S. and Soldan, T. (1989) Passive transfer of humoral resistance against adults of the tsetse fly, *Glossina palpalis palpalis* (Diptera: Glossinidae), in rabbits. *Folia Parasitologica* 36, 375–377.

Meeusen, E.N.T. and Brandon, M.R. (1994) Antibody secreting cells as specific probes for antigen identification. *Journal of Immunological Methods* 172, 71–76.

Moire, N., Bigot, Y., Periquet, G. and Boulard, C. (1994) Sequencing and gene expression of Hypodermins A, B, and C in larval stages of *Hypoderma lineatum*. *Molecular and Biochemical Parasitology* 66, 233–240.

Mooring, M.S. and Hart, B.L. (1993) Animal grouping for protection from parasites: selfish herd and encounter dilution effects. *Behaviour* 123, 173–193.

Morrow, A.N., Quinn, P.J. and Baker, K.P. (1986) Allergic reactions in the horse: response to intradermal challenge with fractionated Culicoides. *Journal of Veterinary Medicine – Series B* 33, 508–517.

Murray, M.D. (1987) Effects of host grooming on louse populations. *Parasitology Today* 3, 276–278.

Nakagawa, T. and Gershwin, M.E. (1993) Immunotherapy of allergic diseases. *International Archives of Allergy and Immunology* 102, 117–120.

Nelson, W.A. (1988) Skin eruptions in ked infected sheep. *Veterinary Record* 122, 472.

Nelson, W.A. and Weintraub, J. (1972) *Hypoderma lineatum* (de Vill.) (Diptera: Oestridae): Invasion of the bovine skin by newly hatched larvae. *Journal of Parasitology* 58, 614–624.

O'Donnell, I.J., Green, P.E., Connell, J.A. and Hopkins, P.S. (1980) Immunoglobulin G antibodies to the antigens of *Lucilia cuprina* in the sera of fly struck sheep, *Australian Journal of Biological Science* 33, 27–34.

O'Meara, T.J., Nesa, M., Raadsma, H.W., Saville, D.G. and Sandeman, R.M. (1992) Variation in skin inflammatory responses between sheep bred for resistance of susceptibility to fleece rot and blowfly strike. *Research in Veterinary Science* 52, 205–210.

O'Meara, T.J., Nesa, M., Seaton, D.S. and Sandeman, R.M. (1995) A comparison of inflammatory exudates released from myiasis wounds on sheep bred for resistance or susceptibility to *Lucilia cuprina*. *Veterinary Parasitology* 56, 207–223.

Owen, J.B. and Axford, R.F.E. (1991) *Breeding for Disease Resistance in Farm Animals*. CAB International, Wallingford.

Parker, K.R. and Mant, M.J. (1979) Effects of tsetse salivary gland homogenate on coagulation and fibrinolysis. *Haemostasis* 42, 743–751.

Penneys, N.S., Nayar, J.K., Bernstein, H. and Knight, J.W. (1989) Chronic pruritic eruption in patients with acquired immune deficiency syndrome associated with increased titres to mosquito salivary gland antigens. *Journal of the American Academy of Dermatology* 21, 421–425.

Pruett, H.J., Jr (1993) Proteolytic cleavage of bovine IgG by hypodermin A, a serine protease of *Hypoderma lineatum* (Diptera: Oestridae). *Journal of Parasitology* 79, 829–833.

Pruett, J.H. and Barrett, C.C. (1984) Induction of intradermal skin reactions in the bovine by fractioned proteins of *Hypoderma lineatum*. *Veterinary Parasitology* 16, 137–146.

Pruett, J.H. and Barrett, C.C. (1985) Kinetic development of humoral anti-*Hypoderma lineatum* antibody activity in the serum of vaccinated and infested cattle. *Southwestern Entomologist* 10, 39–48.

Pruett, J.H., Barrett, C.C. and Fisher, W.F. (1987) Kinetic development of serum antibody to purified *Hypoderma lineatum* proteins in vaccinated and non-vaccinated cattle. *Southwestern Entomologist* 12, 79–88.

Pruett, J.H., Temeyer, K.B. and Burkett, B.K. (1988) Antigenicity and immunogenicity of *Hypoderma lineatum* soluble proteins in the bovine host. *Veterinary Parasitology* 29, 53–63.

Pruett, J.H., Fisher, W.F. and Temeyer, K.B. (1989) Evaluation of purified proteins of *Hypoderma lineatum* as candidate immunogens for a vaccine against bovine hypodermiasis. *Southwestern Entomologist* 14, 363–374.

Pruett, J.H., Scholl, P.J. and Temeyer, K.B. (1990) Shared epitopes between the soluble proteins of *Hypoderma lineatum* and *Hypoderma bovis* first instars. *Journal of Parasitology* 76, 881–888.

Pythal, C. and Rajan, A. (1993) Insect-inflicted nodular myositis on the trunk of an Indian elephant *Elephas maximus bengalensis* de Blainville. *Journal of Veterinary and Animal Science* 23, 55–57.

Ralley, W.E., Galloway, T.D. and Crow, G.H. (1993) Individual and group behaviour of pastured cattle in response to attack by biting flies. *Canadian Journal of Zoology* 71, 725–734.

Ramasamy, M.S. and Ramasamy, R. (1990) Effect of anti-mosquito antibodies on the infectivity of the rodent malaria parasite, *Plasmodium berghei*, to *Anopheles faruati*. *Medical and Veterinary Entomology* 4, 161–166.

Ramasamy, M.S., Ramasamy, R., Kay, B.H. and Kidson, C. (1988) Anti-mosquito antibodies decrease the reproductive capacity of *Aedes aegypti*. *Medical and Veterinary Entomology* 2, 87–93.

Ramasamy, M.S., Sands, M., Kay, B.H., Fanning, I.D., Lawrence, G.W. and Ramasamy, R. (1990) Anti-mosquito antibodies reduce the susceptibility of *Aedes aegypti* to arbovirus infection. *Medical and Veterinary Entomology* 4, 49–56.

Ramasamy, M.S., Sri-Krishnaraj, K.A., Wijekoone, S., Jesuthasan, L.S.B. and Ramasamy, R. (1992) Host immunity to mosquitoes: effect of anti-mosquito antibodies on *Anopheles tessellatus* and *Culex quinquefasciatus* (Diptera: Culicidae). *Medical and Veterinary Entomology* 29, 934–938.

Reunala, T., Lappalainen, P., Brummer-Korvenkontio, H., Coulie, P. and Palosuo, T. (1991) Cutaneous reactivity to mosquito bites: effect of Cetirizine and development of anti-mosquito antibodies. *Clinical and Experimental Allergy* 21, 617–622.

Reunala, T., Brummer-Korvenkontio, H. and Palosuo, T. (1994a) Are we really allergic to mosquito bites? *Annals of Medicine* 26, 301–306.

Reunala, T., Brummer-Korvenkontio, H., Palosuo, K., Miyaij, M., Ruizmaldonado, R., Love, A., Francois, G. and Palosuo, T. (1994b) Frequent occurrence of IgE and IgG4 antibodies against saliva of *Aedes communis* and *Aedes aegypti* mosquitoes in children. *International Archives of Allergy and Applied Immunology* 104, 366–377.

Ribeiro, J.M.C. (1987) Role of saliva in blood-feeding by arthropods. *Annual Review of Entomology* 32, 463–478.

Ribeiro, J.M.C. (1995) Blood-feeding arthropods: live syringes or invertebrate pharmacologists? *Infectious Agents and Disease* 4, 143–152.

Ribeiro, J.M.C., Nussenzveig, R.H. and Tortorella, G. (1994) Salivary vasodilators of *Aedes triseriatus* and *Anopheles gambiae* (Diptera: Culicidae). *Journal of Medical Entomology* 31, 747–753.

Sancho, E. (1988) *Dermatobia*, the neo-tropical warble fly. *Parasitology Today* 4, 242–246.

Sandeman, R.M. (1992) Biotechnology and the control of myiasis diseases. In: Yong, Y.K. (ed.) *Animal Parasite Control Utilizing Biotechnology*. CRC Press, Boca Raton, pp. 275–301.

Sandeman, R.M., Dowse, C.A. and Carnegie, P.R. (1985) Initial characterization of the sheep immune response to infections of *Lucilia cuprina*. *International Journal for Parasitology* 15, 181–185.

Sandeman, R.M., Bowles, V.M., Stacey, I.N. and Carnegie, P.R. (1986) Acquired resistance in sheep to infection with larvae of the blowfly, *Lucilia cuprina*. *International Journal for Parasitology* 16, 69–75.

Sandeman, R.M., Collins, B. and Carnegie, P.R. (1987) A scanning electron microscope study of *Lucilia cuprina* larvae and the development of blowfly strike on sheep. *International Journal for Parasitology* 17, 753–758.

Sandeman, R.M., Feehan, J.P., Chandler, R.A. and Bowles, V.M. (1990) Tryptic and chymotryptic proteases released by larvae of the blowfly, *Lucilia cuprina*. *International Journal for Parasitology* 20, 1019–1024.

Sandeman, R.M., Chandler, R.A., Collins, B.J. and O'Meara, T.J. (1992) Hypersensitivity responses and repeated infections with *Lucilia cuprina*, the sheep blowfly. *International Journal for Parasitology* 22, 1175–1177.

Sandeman, R.M., Chandler, R.A. and Seaton, D.S. (1995a) Antibody degradation in blowfly strike. *International Journal for Parasitology* 25, 621–628.

Sandeman, R.M., Raadsma, H.W. and Bowles, V.M. (1995b) Vaccination and host resistance in the control of *Lucilia cuprina*. In: *Proceedings of the Eighth International Congress of Parasitology, Turkey* (in press).

Sandeman, R.M., Aalard, J., Thompson, N.A. and Seaton, D.S. (1995c) Selection of antigens for vaccination against *Lucilia cuprina* in sheep using antisera inhibitory to larval growth *in vitro*. *Vaccine*, submitted for publication.

Sauer, J.R., McSwain, J.L. and Essenberg, R.C. (1994) Cell membrane receptors and regulation of cell function in ticks and blood sucking insects. *International Journal for Parasitology* 24, 33–52.

Schlein, Y. and Lewis, C.T. (1976) Lesions in haematophagous flies after feeding on rabbits immunised with fly tissues. *Physiological Entomology* 1, 55–59.

Scholl, P.J. (1993) Biology and control of cattle grubs. *Annual Review of Entomology* 38, 53–70.
Schwinghammer, K.A., Pruett, J.H. and Temeyer, K.B. (1988) Biochemical and immunochemical properties of HPLC peak 2, an ion-exchange fraction of common cattle grub (Diptera: Oestridae). *Journal of Economic Entomology* 81, 549–554.
Seaton, D.S, O'Meara, T.J., Chandler, R.A. and Sandeman, R.M. (1992) The sheep antibody response to repeated infection with *Lucilia cuprina*. *International Journal for Parasitology* 22, 1169–1174.
Skelly, P.J. and Howells, A.J. (1987) The humoral immune response of sheep to antigens from larvae of the sheep blowfly (*Lucilia cuprina*). *International Journal for Parasitology* 17, 1081–1087.
Snewin, V.A., Premawansa, S., Kapilananda, G.M.G., Ratnayaka, L., Udagama, P.V., Mattei, D.M., Khouri, E., Delgiudice, G., Peiris, J.S.M., Mendis, K.N. and David, P.H. (1995) Transmission blocking immunity in *Plasmodium vivax* malaria – antibodies raised against a peptide block parasite development in the mosquito vector. *Journal of Experimental Medicine* 181, 357–362.
Sri-Krishnaraj, K.A., Ramasamy, R. and Ramasamy, M.S. (1993) Fecundity of *Anopheles tessellatus* reduced by ingestion of murine anti-mosquito antibodies *Medical and Veterinary Entomology* 7, 66–68.
Steelman, C.D., Gbur, E.E., Tolley, G. and Brown, A.H. (1993) Individual variation within breeds of beef cattle in resistance to horn fly (Diptera: Muscidae). *Journal of Medical Entomology* 30, 414–420.
Tan, H.P., Lebek, L.K. and Nelsen-Cannarella, S.L. (1992) Regulatory role of cytokines in IgE-mediated allergy. *Journal of Leukocyte Biology* 52, 115–118.
Tarn, C.Y., Rosenkrans, C.F., Jr, Steelman, C.D., Brown, A.H. Jr and Johnson, Z.B. (1994) Plasma characteristics of beef cattle classified as resistant or susceptible to horn flies. *Journal of Animal Science* 72, 886–890.
Tarry, D.W. (1989) Warble flies are on their last legs. *Farmers Weekly* 111, 71.
Tellam, R., Casu, R., Elvin, C., Schorderet, S., East, I. and Eisemann, C. (1994) Characterization of candidate vaccine antigens from peritrophic membrane of *Lucilia cuprina* larvae. Abstracts International Congress of Parasitology 10–14 October, Izmir, Turkey, Volume 1, p. 99.
Theodus, C.M. and Titus, R.G. (1993) Salivary gland material from the sand fly *Lutzomyia longipalpis* has an inhibitory effect on macrophage function *in vitro*. *Parasite Immunology* 15, 481–487.
Tong, N.T., Imhoff, J.M., Lecroisey, A. and Keil, B. (1981) Hypodermin A, a trypsin-like neutral proteinase from the insect *Hypoderma lineatum*. *Biochemica et Biophysica Acta* 658, 209–219.
Ungar-Waron, H., Braverman, Y., Gluckman, A. and Trainin, Z. (1990) Immunogenicity and allergenicity of *Culicoides imicola* (Diptera: Ceratopogonidae) extracts. *Journal of Veterinary Medicine Series B* 37, 64–72.
Vaughan, J.A. and Azad, A.F. (1988) Passage of host immunoglobulin G from blood meal into haemolymph of selected mosquito species (Diptera: Culicidae). *Journal of Medical Entomology* 25, 472–474.
Watts, J.E., Murray, M.D. and Graham, N.P.H. (1979) The blowfly strike problem in New South Wales. *Australian Veterinary Journal* 55, 325–334.
Webster, K.A., Rankin, M., Goddard, N., Tarry, D.W. and Coles, G.C. (1992)

Immunological and feeding studies on antigens derived from the biting fly, *Stomoxys calcitrans*. *Veterinary Parasitology* 44, 143–150.

Weir, D.M. and Stewart, J. (1993) *Immunology*, 7th edn. Churchill Livingstone, Edinburgh, UK.

Welch, J.B. (1990) A detector dog for screwworms (Diptera: Calliphoridae). *Journal of Economic Entomology* 83, 1932–1934.

Wieman, G.A., Campbell, J.B., Deshazer, J.A. and Berry, I.L. (1992) Effects of stable flies (Diptera: Muscidae) and heat stress on weight gain and feed efficiency of feeder cattle. *Journal of Economic Entomology* 85, 1835–1842.

Willadsen, P., Riding, G.A., McKenna, R.V., Kemp, D.H., Tellam, R.L., Nielsen, J.N., Lahnstein, J., Cobon, G.S. and Gough, J.M. (1989) Immunologic control of a parasitic arthropod: identification of a protective antigen from *Boophilus microplus*. *Journal of Immunology* 143, 1346–1351.

Wilson, G.W.C. (1986) Control of warble fly in Great Britain and the European community. *Veterinary Record* 118, 653–656.

Wu, C.H. and Lan, L.J. (1989) Immunoblot analysis of allergens in crude mosquito extracts. *International Archives of Allergy and Applied Immunology* 90, 271–273.

Yeruham, I., Braverman, Y. and Orgad, U. (1993) Field observations in Israel on hypersensitivity in cattle sheep and donkeys caused by Culicoides. *Australian Veterinary Journal* 70, 348–352.

# 9

# Immunology of the Tick–Host Interface

## Stephen K. Wikel

*Department of Entomology, Oklahoma State University, Stillwater, Oklahoma 74078, USA*

### INTRODUCTION

The complex nature of host–arthropod immunological interactions becomes evident as you read this book. The immunology of the tick–host interface is the most extensively characterized at the cellular and molecular levels. Tick feeding induces a complex array of host immune responses and the tick has developed a sophisticated arsenal of immunomodulatory countermeasures (Wikel, 1982a, 1996; Wikel *et al.*, 1994). Tick-borne viruses, rickettsiae, bacteria and protozoa increase the complexity of these relationships. Immune responses arising from tick transmission of a disease causing agent differ from those occurring after needle inoculation of the same pathogen (Roehrig *et al.*, 1992; Golde *et al.*, 1993). In addition, many similar immune regulatory and effector mechanisms are stimulated by both ticks and the disease-causing agents they transmit (Playfair, 1993; Wikel, 1996). By characterizing tick–host immunological interactions, further understanding is gained about blood-feeding, digestion, acquisition and transmission of disease-causing agents, initiation of host infection with tick-borne pathogens, and tick control.

Host immunoregulatory and effector pathways stimulated by tick feeding involve antibodies, complement, cytokines, antigen-presenting cells, mast cells, basophils, eosinophils, B and T lymphocytes, and bioactive molecules (Willadsen, 1980; Wikel, 1982a, 1996; Brown, 1985; Brossard *et al.*, 1991). Acquired resistance to tick infestation is expressed as reduced engorgement

---

© CAB INTERNATIONAL 1996. From Wikel, S.K. (ed.) *The Immunology of Host–Ectoparasitic Arthropod Relationships.* CAB INTERNATIONAL, Wallingford.

weight, increased duration of feeding, impaired production of ova, decreased viability of ova, inhibition of moulting and death of feeding ticks (Wikel, 1982a, 1996). Host grooming behaviour is an important element in the expression of resistance to ticks (Bennett, 1969). Biologically active molecules associated with acquired resistance to ticks have established roles in the physiological basis of itch, which could stimulate host grooming (Alexander, 1986).

Acquired resistance to ticks has been most extensively studied using tick–laboratory animal and tick–bovine associations (Willadsen, 1980; Wikel, 1982a, 1996; deCastro et al., 1985; Newson and Chiera, 1989; Brossard et al., 1991; Ramachandra and Wikel, 1995). Laboratory animal–tick associations are characterized by more intense expression of acquired resistance than animal–tick relationships that occur naturally (Ribeiro, 1989). Laboratory animal studies allow for greater ease of analysis of immune responses to ticks due to the extensive amount of fundamental research that has been performed on the immune systems of these species and reagent availability. Results of studies with laboratory animals can be used to develop strategies for dissection of immune response pathways of natural hosts. Cattle and laboratory animal species develop similar immune responses to ticks (Wikel and Osburn, 1982; George et al., 1985; Wikel, 1982a, 1996).

Host genetic composition determines immune response capabilities. Both purebred and crossbred cattle of *Bos indicus* genetic composition are more resistant to tick infestation than purebred *Bos taurus* (deCastro and Newson, 1993). Cells derived from cattle of *Bos indicus* genetic composition develop more vigorous *in vitro* lymphocyte proliferative responses and greater macrophage elaboration of interleukin-1 (IL-1) upon exposure to tick salivary gland molecules than similar cells obtained from *Bos taurus* (Ramachandra and Wikel, 1995).

Characterization of the cellular and molecular immunological basis of acquired resistance and the multifaceted interactions occurring at the tick–host interface is steadily advancing. The significant gaps in understanding these interactions are being reduced and exciting new questions are emerging. This chapter examines various facets of the immunology of the tick–host interface and closes with the presentation of a comprehensive model of tick–host immune interactions.

## ACQUIRED RESISTANCE

The first report of acquired resistance to tick infestation came from Australia (Johnston and Bancroft, 1918). Resistance was attributed to substances introduced by the feeding tick, and resistant cattle developed cutaneous exudates at attachment sites. Experimental analysis of the basis for acquired resistance to tick feeding began with the observations by Trager (1939) that guinea pigs developed systemic immunity after one infestation with *Dermacentor variabilis* larvae. Histological examination of tick bite sites during initial and repeated infestations resulted in the determination that acquired resistance was due to an immunological response. Non-resistant guinea pigs did not develop a cellular infiltrate at tick feeding sites. Only a haemorrhagic pool surrounded tick mouthparts by the fourth day of feeding. Cutaneous responses at tick attachment sites on repeatedly infested, resistant, guinea pigs were dramatically different, consisting of epidermal hyperplasia, oedema and a granulocytic infiltrate.

Guinea pigs resistant to *D. andersoni* developed bite-site cutaneous reactions consisting of serous exudates and intra-epidermal vesicles containing an intense infiltrate of basophils accompanied by numerous eosinophils (Allen, 1973). Similar cutaneous responses to feeding were reported for other tick–host associations (Allen, 1989), including cattle expressing acquired resistance (Allen *et al.*, 1977). Histological reactions at tick bite-sites on previously infested hosts were cutaneous basophil hypersensitivity (CBH) responses (Allen, 1973). The CBH response is one type of delayed type hypersensitivity (Dvorak *et al.*, 1970). A fact of increasing importance in analysis of the immunological basis of acquired resistance was the discovery that delayed type hypersensitivity reactions are mediated by $T_H 1$ helper lymphocytes (Mossman and Coffman, 1989).

Degranulated mast cells and basophils were observed at sites of tick feeding (Brossard and Fivaz, 1982). More mast cells were degranulated at two hours after initiation of a second infestation than during the first exposure. Similarly, a greater number of degranulated basophils were observed on the fifth day of reinfestation.

Does the cellular infiltrate affect tick feeding? Basophils contribute to the altered tick biology associated with acquired resistance. Passive administration of anti-basophil antibodies reduced basophil infiltrates to bite-sites and reduced expression of guinea pig resistance to *Amblyomma americanum* infestation (Brown *et al.*, 1982). Passive administration of anti-eosinophil antibodies partially blocked the resistance response, indicating a role for eosinophils in acquired resistance. Basophil, mast cell and eosinophil functions are interrelated. Eosinophil-granule-derived molecules modulate basophils and mast cells, an example being the induction of histamine release by eosinophil major basic protein (Gleich *et al.*, 1995). Basophils and mast-cell-derived chemotactic factors for eosinophils include histamine, tetra-peptides of eosinophil chemotactic factor of anaphylaxis, products of the lipoxygenase pathway of arachidonic acid metabolism and platelet activating factor (Butterworth and Thorne, 1993).

What is the role of mast cells in anti-tick resistance? Acquired resistance was expressed during the third and fourth infestations of BALB/c mice with *D. variabilis*, and cutaneous reactions at bite-sites contained numerous mast cells and eosinophils (denHollander and Allen, 1985). Mast-cell-deficient mice, infested with *Haemaphysalis longicornis* larvae, did not acquire resistance to infestation, while congenic mast-cell-sufficient mice developed resistance (Matsuda *et al.*, 1985). Mast-cell-deficient mice acquired resistance to *D. variabilis* after two infestations (Steeves and Allen, 1990). Ultrastructural analysis revealed the accumulation of basophils, eosinophils and neutrophils at tick bite-sites on these mast-cell-deficient mice. The basophil response appeared to compensate for mast cell deficiency. A murine basophil response was found to be rare (Allen, 1989).

Tick feeding induced production of homocytotropic antibodies, which bound to Fc receptors on the surfaces of basophils and mast cells (Brossard and Girardin, 1979; Whelen and Wikel, 1993). When introduced into the host during feeding, homocytotropic antibodies complexed with tick salivary-gland-derived molecules and caused basophil and mast cell degranulation. Salivary gland antigens caused basophils to degranulate more readily during repeated infestations with *Ixodes ricinus* (Brossard *et al.*, 1982). Biologically reactive molecules released from these granules could affect tick feeding.

Histamine inhibited *D. andersoni* salivation and engorgement (Allen and Kemp, 1982; Paine *et al.*, 1983). The histamine content at tick attachment sites on animals expressing acquired resistance was elevated, an expected finding given the intense infiltration of basophils (Wikel, 1982b). Administration of type I (Brossard, 1982) or concurrent administration of type I and type II histamine receptor antagonists (Wikel, 1982b) reduced expression of anti-tick immunity. Ultrastructural examination of the digestive tracts of ticks feeding upon resistant host revealed the presence of intact basophils, intact eosinophils and granules of both cell types (Voss-McCowan, unpublished). Basophil and eosinophil granules were taken up by midgut cells, which showed signs of injury. Further investigations are needed to ascertain the importance of basophil-, mast-cell- and eosinophil-derived molecules upon expression of acquired resistance. Candidate molecules include: leukotrienes, prostaglandins, chondroitin sulphate, heparin, enzymes, eosinophil cationic protein, eosinophil peroxidase and eosinophil major basic protein. The physiological changes occurring within ticks feeding upon resistant hosts remains to be defined.

Interactions of the feeding tick with the immune system at the cutaneous interface are central to acquisition and expression of acquired resistance. Antigen-presenting cells involved in development of an immune response include mononuclear phagocytes, B lymphocytes and Langerhans cells (Harding *et al.*, 1988). Dendritic Langerhans cells found in the epidermis complex with antigen and transport antigen to the draining lymph nodes for interaction with specific T lymphocytes (Bos and Kapsenberg, 1993). Tick salivary gland antigens associate with Langerhans cells in the skin of resistant guinea pigs (Allen *et al.*, 1979). Using an adenosine triphosphatase staining technique for identification, Langerhans cells decreased at tick bite-sites during primary infestation and increased during the early phase of a second infestation, during which resistance was expressed (Nithiuthai and Allen, 1984a). Depletion of epidermal Langerhans cells impaired both the acquisition and expression of acquired resistance (Nithiuthai and Allen, 1984b). Langerhans cells effectively presented tick antigens *in vitro* to lymph node cells from sensitized guinea pigs, confirming their importance in acquired resistance (Nithiuthai and Allen, 1985).

Use of hybridization probes for identification of cell subpopulations and cytokines in histological sections provides new insights into cellular and molecular interactions occurring during acquired resistance. Skin sections obtained from BALB/c mice given three successive infestations with *I. ricinus* nymphs were characterized in regard to lymphocyte subpopulations, expression of intercellular adhesion molecule-1 (ICAM-1), interleukin-1-alpha (IL-1-$\alpha$) and tumour necrosis factor alpha (TNF-$\alpha$) (Mbow et al., 1994). CD4 T lymphocytes were more numerous than CD8 T cells during each infestation. B lymphocytes were not detected in the skin of either control or infested mice. Presence of ICAM-1 could contribute to the infiltration of cells into the feeding lesion. Positive staining for Ia antigens (class II major histocompatibility molecules) indicates the presence of cells capable of antigen presentation to T lymphocytes. CD4 $T_H1$ lymphocytes are effectors of delayed type hypersensitivity responses (Mossman and Coffman, 1989), an important component of acquired resistance (Wikel, 1982a, 1996). *In situ* hybridization was used to detect the pro-inflammatory cytokines IL-1 and TNF in the skin of infested mice. These cytokines have immunoregulatory roles (Durum and Oppenheim, 1993). As discussed later in this chapter, characterization of the cytokine response to infestation helps to define the immunological basis of acquired resistance.

Serum complement proteins have a role in acquired resistance. *In vivo* depletion of complement by administration of cobra venom factor resulted in blockage of expression of acquired resistance, but not acquisition of resistance to tick infestation (Wikel and Allen, 1977). Depletion of complement by cobra venom factor did not allow determination of the relative importance of alternative and classical pathways of complement activation. Guinea pigs totally deficient in classical complement pathway component C4 acquired and expressed anti-tick resistance in a manner similar to intact animals, suggesting a role for the alternative pathway of complement activation in acquired resistance (Wikel, 1979). Activation of the alternative and/or classical pathways of complement could contribute to tick bite-site lesion formation through generation of anaphylatoxins and chemotactic activities (Frank and Fries, 1989). The anaphylatoxins C3a and C5a cause degranulation of basophils and mast cells, resulting in release of eosinophil chemotactic factors (histamine, eosinophil chemotactic factor of anaphylaxis) and vasoactive molecules (Frank and Fries, 1989). C5a and soluble factors released by lymphocytes are chemotactic for guinea pig basophils (Ward et al., 1975). Tick salivary gland antigens, IgG and complement were immunolocalized at the dermal–epidermal junction of tick bite-sites on resistant animals (Allen *et al.*, 1979).

In addition to homocytotropic antibodies, circulating immunoglobulins that are tick reactive are induced by ixodid feeding (Willadsen, 1980; Wikel, 1982a, 1988, 1996; Brown, 1985; Wikel and Whelen, 1986). Circulating immunoglobulins can bind to introduced tick saliva molecules to form antigen–antibody complexes capable of fixing complement by the classical pathway. Biological reactivities of activated complement components C3 through the effector pathway are similar to those induced by alternative pathway activation. Additional biological activities would arise from activation of classical pathway components C1, C4 and C2. Antigen–antibody complexes are capable of binding to Fc receptors on a variety of cell types (Ravetch and Kinet, 1991). Evidence has not yet been provided to support a role for antibody-dependent cellular cytotoxicity in acquired resistance to ticks.

The role of antibodies in expression of acquired resistance was established with successful passive transfer studies. Trager (1939) transferred variable levels of resistance to *D. variabilis* larvae with serum from infested guinea pigs. Modest resistance and cutaneous basophil hypersensitivity to *Ixodes holocyclus* infestation were passively transferred with guinea pig serum (Bagnall, unpublished). Bovine resistance to infestation with *Boophilus microplus* was transferred from resistant host to susceptible calves by intravenous administration of plasma (40 ml kg$^{-1}$ recipient body weight) (Roberts and Kerr, 1976). Passive transfer of plasma from cattle expressing low resistance to infestation did not significantly affect engorgement of challenge ticks. Antibodies passively transferred resistance to *Rhipicephalus appendiculatus* (Rubaire-Akiki and Mutinga, 1980), *A. americanum* and *Rhipicephalus sanguineus* (Brown and Askenase, 1981). The mean weight of female *Rhipicephalus evertsi evertsi* was reduced after feeding on a rabbit which had been passively administered serum derived from a rabbit resistant to infestation with this species (Rechav *et al.*, 1991). Antibodies contribute to the expression of acquired resistance by both natural and laboratory animal hosts.

Rabbit acquired resistance to *I. ricinus* was passively transferred with serum (Brossard, 1977; Brossard and Girardin, 1979). Immunoblotting with sera capable of transferring resistance to infestation was used to identify possible protection inducing immunogens. A 25 kDa antigen reported to be of importance in acquisition of rabbit resistance to *I. ricinus* is a soluble protein found in extracts of integument of partially engorged females and engorged nymphs (Rutti and Brossard, 1989). More than 50% of infested rabbits developed antibodies to this molecule. Antibodies reactive with the 25 kDa molecule of *I. ricinus* complexed with a 20 kDa molecule in the integument of partially fed *R. appendiculatus* females (Rutti and Brossard, 1989). The manner in which the rabbit immune system is stimulated with integument molecules is unknown. Possibly, cross-reactive epitopes present in saliva are associated with molecules of different molecular masses. Host antibodies can complex with integument epitopes in the intact tick. Host serum IgG can pass across the tick gut into haemolymph (Ackerman *et al.*, 1981; Ben-Yakir, 1989; Wang and Nuttall, 1994).

The number of infesting ticks and the life cycle stage influence the antibody response (Fig. 9.1). Cattle given repeated low level infestations with *D. andersoni* adults developed precipitating antibodies to a salivary gland extract prepared from adults allowed to engorge for four days (Wikel and Osburn, 1982). Guinea pigs repeatedly infested with either *D. andersoni* adults or nymphs developed increasing titres of tick salivary gland specific antibodies through four exposures; however, titres decreased significantly 30 days after final infestation (Whelen *et al.*, 1986). Feeding of adults induced a more intense antibody response than nymphal engorgement. A positive correlation occurred between the level of serum gamma globulin and resistance of rabbits to *R. appendiculatus* larvae and nymphs (Rechav and Dauth, 1987).

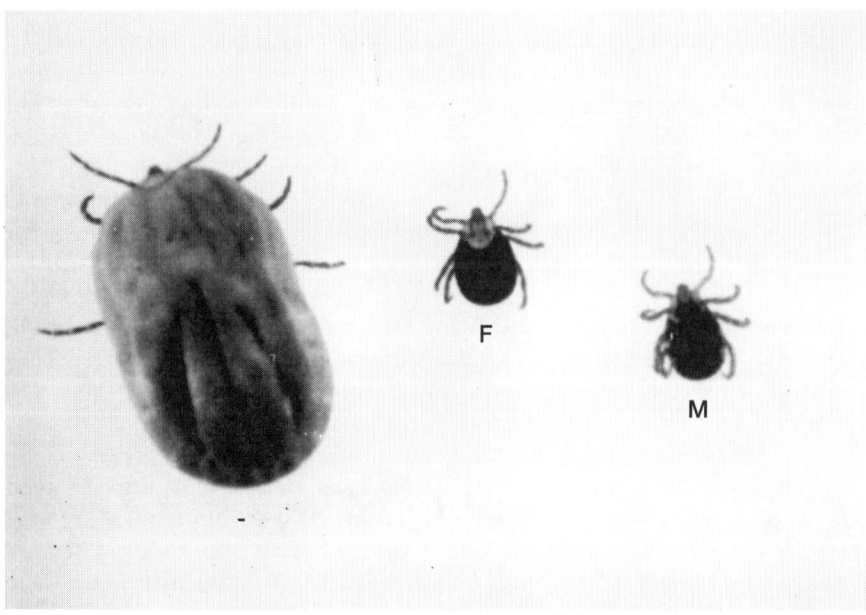

**Fig. 9.1.** *Dermacentor andersoni* unfed male (M), unfed female (F) and fully engorged female (E). The marked increase in the size of the engorged female represents a change from an unfed weight of approximately three milligrams to an approximate engorged weight of 1000 mg. Female of this species will engorge over a period of 8 to 17 days. An engorged female will produce as many as 7400 eggs. The feeding female introduces a changing array of saliva proteins into the host during the course of engorgement.

Antibodies induced by tick feeding complex with a variety of tick molecules; however, little is known about the biochemical nature of reactive molecules and their importance in acquisition and expression of resistance. Two allergens present in saliva of *B. microplus* were an esterase or carboxylic ester hydrolase (Willadsen and Williams, 1976) and a proteolytic enzyme inhibitor with independent sites for inhibition of chymotrypsin and trypsin (Willadsen and Riding, 1979). Guinea pig serum capable of transferring resistance to *A. americanum* infestation used to immunoprecipitate $^{35}$S and $^{125}$I labelled *A. americanum* salivary gland and attachment cement extracts reacted specifically with a 20 kDa protein/polypeptide (Brown *et al.*, 1984). Reactive epitope(s) associated with attachment cement might represent components of the cement and/or adsorbed salivary immunogens. Serum of guinea pigs immunized with *A. americanum* salivary gland extract and protected against challenge infestation contained antibodies that reacted with a 20 kDa protein/polypeptide (Brown *et al.*, 1984). Sera collected from immunized animals not resistant to challenge infestation did not immunoprecipitate the 20 kDa band.

Further analysis of *A. americanum* salivary gland antigens was performed by gel filtration, ion exchange chromatography and immunoaffinity chromatography with anti-tick antibodies (Brown and Askenase, 1986). Anti-tick IgG1 antibodies used for immunoaffinity purification were capable of passively transferring resistance and expression of cutaneous basophil hypersensitivity. Peaks obtained by gel filtration of salivary gland extracts were used to skin test naive and *A. americanum*-sensitive guinea pigs. Intradermal testing revealed specific reactivity in a fraction greater than 70 kDa molecular mass. Immunoprecipitation analysis of this fraction detected reactive bands of 20, 34 and 67 kDa. Anion exchange chromatography, using DEAE cellulose, resulted in 16 fractions of which 3 were skin-test reactive. Immunoprecipitation-reactive proteins/polypeptides of 20 and 67 kDa were identified. Immunoaffinity purification yielded additional reactive molecules of 33, 36 and 39 kDa. Normal guinea pig serum used for immunoaffinity fractionation of salivary gland extract resulted in identification of reactive molecules of 14, 27 and 55 kDa.

Serum collected from guinea pigs expressing acquired resistance to *A. americanum* larvae, nymphs or adults was used in immunoblotting to identify stage-specific and shared antigen of *A. americanum* ova, larvae, nymphs and female salivary gland extracts (Brown, 1988a). Antibodies reacted with numerous bands on electroeluted electrophoretograms separated by sodium dodecyl sulphate polyacrylamide gel electrophoresis (SDS–PAGE). Immunogens identified were unique to each life cycle stage, as well as shared among stages. Attention was directed toward 25 and 38 kDa proteins/polypeptides. The 25 kDa protein was reported to be the same as the 20 kDa molecule detected by immunoprecipitation (Brown, 1988a). Definitive proof of this association was not provided. Rabbits expressing acquired resistance to

A. americanum adults developed antibodies reactive with molecules in salivary gland extract of the same species, and proteins/polypeptides of 39, 40 and 41 kDa were reported to be linked to induction of host immunity (Brown, 1988b).

Sera derived from naturally immune and hyperimmune sheep and rabbits were analysed by immunoblotting with extracts of unfed and partially fed (12, 48, 72, 96 and 120 hour) female A. americanum (Jaworski et al., 1990). Antibodies contained in these sera consistently reacted with salivary gland polypeptides of 23, 33, 45, 58, 75 and 90 kDa. Antigens unique to tick gut were not detected. Amblyomma americanum salivary-gland-reactive antibodies bound 45 and 90 kDa polypeptides of salivary gland extracts of D. variabilis and I. dammini (Jaworski et al., 1990). Sera obtained from sheep infested five times were analysed for immunoblot reactivity with unfed salivary gland, three day fed salivary gland and gut extracts of the same tick species (Barriga et al., 1991). Immunogen recognition patterns changed during repeated infestations. Proteins/polypeptides that were predominantly recognized after the first exposure had molecular masses of 34, 36.5, 38 and 115 kDa. A 31.5 kDa molecule was reactive after the second infestation, while 29 and 35.5 kDa bands become evident after the third infestation of two sheep and the first infestation of one sheep. Recognition of the 29 and 35.5 kDa molecules occurred at the time anti-tick resistance appeared. Host antibody responses induced by infestation with this ixodid species are complex. Biological functions of reactive salivary gland molecules still need to be determined.

Rabbits infested three times with adult H. anatolicum anatolicum developed antibodies that complexed with 9 saliva and 17 salivary gland extract proteins/polypeptides (Gill et al., 1986). All 9 saliva and 12 of 17 salivary gland extract immunogens were glycoconjugates. Molecules of 96, 103 and 130 kDa molecular masses elicited immediate hypersensitivity reactions and the 103 and 130 kDa molecules elicited delayed hypersensitivity responses upon inoculation into the dermis of previously infested rabbits. The 130 kDa molecule had acid phosphatase activity and the 96 kDa molecule displayed both non-specific esterase and aminopeptidase activity.

Sera derived from guinea pigs resistant to R. appendiculatus was used to identify antigens by immunoblotting (Shapiro et al., 1986). Numerous antigens were detected in salivary glands. The most strongly reactive bands had molecular masses of 16, 20, 35, 58 and 94 kDa. The 16, 20 and 35 kDa antigens were also present in gut. The quantity and number of specific proteins changed during engorgement, including specific antigens. The 20 and 35 kDa proteins were reported to be expressed at a constant level by males and females. The 28 kDa protein was constantly expressed by male and the 88 kDa protein by female salivary glands. Some salivary gland antigens were detected only after initiation of feeding. Tick attachment cement was considered to be a source of antigens eliciting host responses (Shapiro et al., 1986).

Rabbit-acquired resistance to *R. appendiculatus* was linked to a 90 kDa salivary gland protein (Shapiro et al., 1987). Antisera to a 90 kDa immunogen in the salivary glands of *R. appendiculatus* reacted with similar size polypeptides in immunoblots of SDS–PAGE fractionated salivary glands of *D. variabilis* and *A. americanum* and 70 kDa polypeptides in attachment cement of *D. variabilis*, *A. americanum* and *R. sanguineus* (Jaworski et al., 1992). The 90 kDa polypeptide was conserved among ixodid genera and a component of attachment cement.

Salivary gland extracts of *D. andersoni* were analysed during a six to ten day feeding period (Gordon and Allen, 1987). Concentration and number of antigens varied during engorgement. Some variations correlated with skin test reactivity of tick-resistant guinea pigs. Antibodies derived from tick-resistant rabbits were used for immunoaffinity isolation of salivary gland extract immunogens (Gordon and Allen, 1987). Immunoaffinity purified molecules elicited immediate and delayed skin reactions from tick-resistant guinea pigs. Antigenic activity was associated with proteins/polypeptides in the 124 to 172 kDa molecular mass range. Both 31 and 82 kDa immunogens were associated with salivary glands of ticks with molecular masses of 10 mg or less. Importance of these immunogens in acquisition and expression of resistance remains to be established.

Protein composition and antigen content of the saliva of female *Amblyomma hebraeum* were analysed for ticks of different engorgement weights (Dharampaul et al., 1993). The highest saliva protein concentration occurred with engorgement weights of less than 100 mg. Larger ticks secreted a significantly greater fluid volume per minute, reducing saliva protein concentration. Expression of selected proteins was related to engorgement weight: 14 kDa for < 60 mg; 21 and 26 kDa for < 150 mg and 68 kDa at engorgement weights of > 100 mg. Sera of tick infested rabbits bound 13 antigens in the 23 to 200 kDa range. These antigens were detected only in the saliva of small ticks.

Antigenic determinants (epitopes) are shared among tick tissues and species. A 70 kDa antigen detected in the salivary glands of *Hyalomma truncatum*, *Hyalomma marginatum rufipes* and *R. evertsi evertsi* cross-reacted with epitopes in the digestive tracts of unfed adult *A. hebraeum*, *Rhipicephalus simus simus*, *R. evertsi evertsi*, *Rhipicentor nuttali* and *H. marginatum rufipes* (van Vuuren et al., 1992). The same 70 kDa antigen was localized by immunoblotting in the salivary glands of the argasid ticks, *Ornithodoros savignyi* and *Ornithodoros moubata*. Salivary gland extract proteins/polypeptides of several *Ornithodoros* species possess genus-specific epitopes (Wozniak et al., 1995).

What do these findings mean? A large number of tick salivary-gland-associated molecules can induce host antibody responses. Antibodies contribute to the expression of acquired anti-tick resistance. Tick salivary gland antigens change quantitatively and qualitatively during the course of feeding.

Cross-reactive epitopes are shared among phylogenetically diverse species. Genetic background of the host is a determining factor in antibody response patterns. Limited information is available on the chemical composition and function of tick-associated immunogens. Future studies should concentrate on determining molecular characteristics and biological roles of tick immunogens introduced during engorgement. Are antibody responses generated against pharmacologically active molecules introduced by the feeding tick? Obtaining this information will help elucidate the biochemical and cellular bases for acquisition and expression of tick resistance.

Antibodies are not the only effector elements of the immune system contributing to expression of acquired resistance. Rabbits given repeated experimental infestations with *I. ricinus* developed resistance that could not be correlated with levels of anti-tick antibodies (Bowessidjaou *et al.*, 1977). Expression of resistance was much stronger during a fourth infestation; however, anti-tick antibody titre was not elevated over that of the preceding infestation.

T lymphocytes are key elements in regulation and effector functions of the immune system, including antibody production, delayed type hypersensitivity and cytotoxic T cells (Green *et al.* 1983). Interaction of antigenic determinants with specific T-lymphocyte receptors activates complex cytokine signalling pathways, cell differentiation and activation (Janeway and Bottomly, 1994; Paul and Seder, 1994). In addition, the role of macrophages, Langerhans cells, other dendritic cells and cytokines elaborated by these cells orchestrate many aspects of innate and specific immunity (Kapsenberg *et al.*, 1990; Unanue, 1993; Stingl and Bergstresser, 1995; Trinchieri, 1995). Attention is now focused upon the role of T lymphocytes in acquired resistance.

Resistance of guinea pigs to infestation with *D. andersoni* was adoptively transferred more readily with viable lymphocytes than passively with serum derived from resistant hosts (Wikel and Allen, 1976). Salivary gland extracts prepared from female *D. andersoni*, which had engorged for four days, induced antigen-specific delayed type hypersensitivity skin reactions upon administration to tick-exposed hosts (Wikel *et al.*, 1978). Delayed type hypersensitivity reactions are mediated by $T_H1$-lymphocytes (Mosmann and Coffman, 1989). The same salivary gland extract stimulated *in vitro* blastogenesis of lymphocytes obtained from tick-infested guinea pigs starting at two to four days after termination of an initial infestation, reaching peak reactivity at 24 hours after initiation of a second infestation (Wikel *et al.*, 1978). *In vitro* lymphocyte responsiveness of tick-exposed hosts to the T-cell mitogen phytohaemagglutinin was reduced up to 42% (Wikel *et al.*, 1978).

Peripheral blood lymphocytes collected from *Bos taurus* that were given a third or fourth low level infestation with *D. andersoni* proliferated in an antigen-specific manner *in vitro* when cultured with salivary gland extract prepared from four day fed female *D. andersoni* (Wikel and Osburn, 1982). Tick-exposed cows and calves developed immediate and delayed skin reactivity upon intradermal administration of salivary gland extract. Peripheral blood lymphocytes obtained from purebred and crossbred *Bos indicus* calves infested with *A. americanum* proliferated *in vitro* in the presence of *A. americanum* and *Amblyomma cajennense* salivary gland extract (George et al., 1985). Tick-exposed calves developed immediate, five hour and delayed skin reactions after intradermal inoculation of *Amblyomma* salivary gland extracts (George et al., 1985).

Additional evidence for the participation of T lymphocytes in acquired immunity to ticks was provided by experiments involving administration of cyclosporin A to rabbits resistant to *I. ricinus* (Girardin and Brossard, 1989). Cyclosporin A treatment prior to, and during, a second infestation blocked development of immediate (type I) and delayed (type IV) cutaneous hypersensitivities induced by intradermal administration of *I. ricinus* salivary gland extract.

Sheep given four infestations with *R. evertsi evertsi* failed to develop peripheral blood lymphocyte *in vitro* transformation responses upon culture with salivary gland extract prepared from four to five day fed ticks (Neitz et al., 1993). Absence of lymphocyte reactivity was attributed to tick adaptation to a natural host. Blood was collected for analysis of lymphocyte reactivity on the fourth day after termination of each infestation. Peripheral blood lymphocyte responses to antigen were shown to be transient in nature (Wikel et al., 1978). The observed absence of tick-antigen-specific lymphocyte blastogenesis might simply reflect the limited number of time points sampled during the infestation regimen.

Rabbit immune responsiveness to high and low infestations with *I. ricinus* adults was evaluated (Schorderet and Brossard, 1993). The high infestation group received 25 pairs and the low infestation group was exposed to 5 pairs. Both groups were given two infestations prior to a third infestation, during which all rabbits were exposed to 15 pairs. Resistance was greater during the second exposure for the high infestation group. During the third infestation, the low tick burden group became more resistant than the high infestation group. Salivary gland and integumental antigens prepared from female *I. ricinus* stimulated blastogenesis of peripheral blood lymphocytes. *In vitro* lymphocyte blastogenesis response was of greater magnitude in the high infestation group at the end of the second infestation and higher in the low infestation group during the third exposure. Generally, salivary gland extract induced greater reactivity than integumental antigens (Schorderet and Brossard, 1993).

Interleukin-2 administration during primary exposure enhanced rabbit resistance to a second *I. ricinus* infestation (Schorderet and Brossard, 1994). More intense delayed cutaneous responses to salivary gland extract occurred in the interleukin-2 treated group. Interleukin-2 administration did not enhance anti-tick antibody production. *In vitro* proliferation of peripheral blood lymphocytes to salivary gland extract became positive during the first infestation with greatest responsiveness during the second exposure. The lack of significant differences between cells derived from interleukin-2 treated animals and controls was attributed to individual animal variations. Heightened delayed skin responsiveness and resistance might be due to enhanced production of memory cells in the presence of additional interleukin-2 during the first exposure.

BALB/c mice repeatedly infested with *I. ricinus* nymphs developed antibody and *in vitro* lymphocyte proliferative responses to a $15,000 \times g$ supernatant of whole nymphal homogenate (Borsky *et al.*, 1994). An anamnestic antibody response occurred with repeated exposures. *In vitro* spleen lymphocyte responsiveness to nymphal homogenate developed after primary exposure and peaked during initiation of the second infestation. This transient response remained below control values for samples prepared on days 28 and 35 of infestation. Transient pattern of *in vitro* lymphocyte reactivity was similar to that described for infestation with *D. andersoni* (Wikel *et al.*, 1978; Wikel and Osburn, 1982) and *I. ricinus* (Schorderet and Brossard, 1993).

T lymphocytes have important regulatory and effector roles in acquisition and expression of acquired resistance to tick infestation. Both $T_H 1$ and $T_H 2$ lymphocytes are involved in acquired resistance. Delayed type hypersensitivity component, manifested as cutaneous basophil hypersensitivity, is $T_H 1$-lymphocyte mediated. Anti-tick antibody responses involve participation of $T_H 2$ lymphocytes. Evidence has not been provided establishing a role for natural killer (NK) or $CD8^+$ cytotoxic T lymphocytes in acquired resistance to ticks. However, $CD8^+$ T lymphocytes infiltrated the skin of BALB/c mice undergoing repeated infestations with *I. ricinus* (Mbow *et al.*, 1994).

Specific tick immunogens reactive with T lymphocytes need to be identified. The temporal sequence of T-lymphocyte function during infestation should be determined. Consideration should be given to development of T-lymphocyte clones for identification of tick immunogens introduced into the host during infestation.

## PHARMACOLOGY OF TICK SALIVARY GLANDS AND TICK MODULATION OF HOST IMMUNE DEFENCES

The topic of arthropod modulation of host immune responses is covered in Chapter 5 and salivary gland pharmacology in Chapter 4 of this volume. Antihaemostatic and vasoactive properties of arthropod saliva have been reviewed (Ribeiro, 1987a; Champagne, 1994). This section provides a brief overview of this important topic so that a comprehensive picture can be developed regarding immunological interactions at the tick–host interface. Ticks have developed strategies for regulating complex interactions with host haemostatic, pain and inflammatory/immune responses. Ribeiro (1995) accurately described ticks as 'smart pharmacologists'.

Ticks are in continuous contact with their hosts for days while obtaining a bloodmeal. Close association with the host provides ample opportunity for haemostatic and immune defences to affect the tick. Successful feeding and pathogen transmission require the development of tick countermeasures to these defences.

Inhibition of host blood clotting is achieved by *I. dammini* (*I. scapularis*) through the action of salivary apyrase, which inhibits platelet aggregation by hydrolysing adenosine triphosphate (ATP) and adenosine diphosphate (ADP) released by injured cells to adenosine monophosphate (AMP) and orthophosphate (Ribeiro *et al.*, 1985). In addition, saliva inhibits platelet aggregation induced by collagen and platelet aggregation factor, possibly involving the action of apyrase (Ribeiro *et al.*, 1985). Tick salivary-gland-derived prostaglandin $E_2$ causes vasodilation, inhibits platelet aggregation and acts as an antagonist of vessel constrictors (Ribeiro *et al.*, 1985; Champagne, 1994). Saliva of *I. dammini* contains prostacyclin which blocks mast cell degranulation, causes vasodilation and inhibits platelet aggregation (Ribeiro *et al.*, 1988). Both the extrinsic and intrinsic pathways of coagulation are inhibited by salivary gland molecules of *D. andersoni* by acting on factors V and VII (Gordon and Allen, 1991). *Rhipicephalus appendiculatus* salivary gland extract inhibition of factor Xa or other components of the prothrombinase complex was attributed to a 65 kDa anticoagulant (Limo *et al.*, 1991).

Prostaglandin $E_2$ can potentiate pain caused by bradykinin, but salivary carboxypeptidase B of *I. dammini* saliva inactivates bradykinin (Ribeiro and Spielman, 1986; Ribeiro, 1995). Inhibition of pain at the tick feeding site reduces signals that would stimulate removal of the arthropod by host grooming.

Tick modulation of host immunity is directed against both innate and specific defences, providing an optimum environment for obtaining a bloodmeal and pathogen transmission (Wikel *et al.*, 1994; Wikel, 1996). Innate defences, capable of immediate reaction, include the alternative pathway of complement activation and natural killer cells. Specifically sensitized effector T lymphocytes and specific immunoglobulins may require days to develop due to clonal expansion. Tick salivary-gland-derived molecules modulate NK cell function, activation of the alternative complement pathway, anaphylatoxins, antibody production, T-lymphocyte proliferation, and cytokine elaboration by macrophages and T lymphocytes. All of these components of the immune response are relevant to host acquisition and expression of acquired resistance to ticks (Wikel 1982a, 1996; Wikel *et al.*, 1994).

*Ixodes dammini* saliva inhibits alternative complement pathway deposition of C3b and C5b upon activating surfaces and generation of the anaphylatoxin C3a (Ribeiro, 1987b). Anticomplement activity was attributed to a 49 kDa fraction obtained by gel permeation chromatography. The anaphylatoxin inhibitor is a saliva carboxypeptidase B (Ribeiro and Spielman,

1986; Ribeiro, 1995). Inhibition of alternative pathway activation prevents C3b-mediated uptake by phagocytic cells, generation of anaphylatoxins and membrane attack pathway assembly. Inactivation of anaphylatoxins reduces contraction of smooth muscle, release of vasoactive mediators from mast cells and basophils, and chemotaxis of neutrophils, eosinophils, basophils and monocytes through the action of C5a. Biological activities of complement appear to contribute to the formation of lesions associated with acquired resistance.

Guinea pigs infested with *D. andersoni* larvae were inhibited in their ability to generate a primary IgM response to a thymic-dependent antigen, as measured by plaque-forming cell assay (Wikel, 1985). Antibody production was reduced during both primary and secondary infestations, but returned to normal levels several days after termination of tick exposure. Rabbits immunized with bovine serum albumin (BSA) during infestation with *R. appendiculatus* produced mean anti-BSA titres of 1:64, while uninfested control rabbit titres were a mean of 1:2000 (Fivaz, 1989). Antibodies play an important role in acquired resistance to ticks. Down-regulation of the host humoral response has survival value for the feeding tick.

Tick infestation or salivary gland antigens modify *in vitro* responsiveness of lymphocytes to mitogens. Lymphocytes obtained from guinea pigs during the fifth day of both first and second infestations with *D. andersoni* were reduced in their ability to respond to the T-cell mitogens PHA and Con A, while responsiveness to the B-lymphocyte mitogen LPS was not affected (Wikel, 1982c). Suppression was greater during primary infestation. Possibly, some antibodies induced by the first five day exposure neutralized the immunosuppressant introduced during the second infestation. Rabbits infested with *I. ricinus* adults were reduced in their *in vitro* lymphocyte responses to Con A, a finding attributed to immunosuppression (Schorderet and Brossard, 1994). Concanavalin-A-induced lymphocyte proliferative responses of cattle infested with *Amblyomma variegatum* were depressed (Koney *et al.*, 1994). Purebred *Bos taurus* cows and calves infested with *D. andersoni* had peripheral blood lymphocyte *in vitro* responses to PHA reduced by up to 47% during third and fourth infestations (Wikel and Osburn, 1982).

Purebred *Bos taurus* infested with *B. microplus* had reduced *in vitro* lymphocyte responsiveness to PHA from the second through the fourth experimental infestations, when compared with similar cells from uninfested controls (Inokuma *et al.*, 1993). Suppression in this tick–host relationship was subsequently attributed to prostaglandin $E_2$ in *B. microplus* saliva (Inokuma *et al.*, 1994). Ability of *I. dammini* saliva to inhibit normal murine lymphocyte blastogenesis in the presence of PHA or ConA was attributed to a protein with a molecular mass of 5 kDa or greater (Urioste *et al.*, 1994). Ability of *D. andersoni* salivary glands to suppress normal lymphocyte *in vitro* responsiveness to Con A was associated with purified protein(s) having a molecular mass in the 36–43 kDa range (Bergman *et al.*, 1995). An increasing body of evidence is emerging to link host immunosuppression to tick salivary gland proteins. A role for prostaglandins in tick modulation of host immunity has yet to be definitively established.

Salivary gland extracts prepared daily from female *D. andersoni* during the course of engorgement modified normal mouse lymphocyte *in vitro* responsiveness to Con A or LPS (Ramachandra and Wikel, 1992). Salivary gland extract (SGE) prepared on day zero was not inhibitory to Con-A-induced proliferation; however, SGEs prepared on days one through nine inhibited responsiveness by 33.3 to 66.4%. Pre-incubation of lymphocytes with the same SGEs enhanced *in vitro* responsiveness to LPS from 6.3% for salivary glands from unfed ticks to 59.0% for day nine SGE (Ramachandra and Wikel, 1992). Peripheral blood lymphocytes obtained from uninfested purebred *Bos indicus* and *Bos taurus* had suppressed *in vitro* responsiveness to Con A and enhanced reactivity to LPS, when cultured with SGEs prepared daily from engorging female *D. andersoni* (Ramachandra and Wikel, 1995). Relative magnitude of suppression of Con A blastogenesis was similar for both *Bos indicus* and *Bos taurus*. However, Con-A-induced counts per minute of *Bos indicus* cells were a mean 34.5% greater than those of *Bos taurus*, while LPS-enhanced reactivity was 42.9% greater for *Bos indicus*. Cattle of *Bos indicus* genetic composition are considered to be innately resistant to tick infestation (deCastro and Newson, 1993). 'Innate resistance' might simply reflect enhanced immune response capabilities of *Bos indicus* when compared with *Bos taurus* breeds.

Cytokines regulate immune responses (Kroemer *et al.*, 1993). Pathogens have developed strategies for modulating the web of cytokine interactions to facilitate survival (Marrack and Kappler, 1994). The discovery that tick infestation modulates host cytokines was not unexpected. Salivary gland extracts prepared daily from female *D. andersoni* during engorgement significantly inhibited elaboration of the macrophage cytokines IL-1 on days zero through five and TNF on days zero through nine (Ramachandra and Wikel, 1992). The same SGEs slightly reduced lymphocyte production of IL-2 and IFN-$\gamma$. Interleukin-4 release was not inhibited by exposure of normal murine lymphocytes to SGEs (R.N. Ramachandra, unpublished observation). *Bos indicus* and *Bos taurus* macrophages were suppressed in their ability to elaborate IL-1 and TNF by the same SGEs (Ramachandra and Wikel, 1995). Tick salivary gland extracts appear to suppress macrophage and $T_H 1$-lymphocyte function. Modulation of the responsiveness of these cell populations provides a survival advantage for the feeding tick.

A salivary gland extract prepared from six-day-old engorged female *Dermacentor reticulatus* inhibited normal human NK cell function (Kubes *et al.*, 1994). Natural killer cells are an important first line of defence against viruses (Trinchieri, 1989). Impaired NK cell activity potentially increases host susceptibility to tick-borne disease-causing agents, particularly viruses.

Could tick mediated host immunosuppression be important in the transmission of disease-causing agents? Rabbit resistance to infestation with *D. andersoni* reduced the ability of a subsequent tick infestation to transmit *Francisella tularensis* (Bell *et al.*, 1979). Acquired resistance to tick feeding might neutralize or reduce the effectiveness of immunosuppressive molecules in tick salivary gland secretions, allowing greater expression of host innate and specific immune defences. Tick salivary gland factors facilitate transmission of Thogoto virus (Jones *et al.*, 1992). Salivary gland immunosuppressant(s) could prime the feeding site for establishment of infection. Blockade of tick-induced immunosuppression of the host by vaccination against salivary gland immunosuppressants could potentially reduce or inhibit successful transmission and establishment of any tick-borne pathogen.

## ACQUIRED RESISTANCE: A MODEL

Research findings to date are the foundation for the creation of a model of the immunological basis for development and expression of acquired resistance to ticks. Variations of selected aspects of this model will occur for different tick–host associations. Potential influence of tick-borne pathogens on tick–host immunological interactions has not been introduced into this model.

Tick feeding upon a previously unexposed animal stimulates multiple elements of the host immune system. Saliva introduced into the bite-site contains pharmacologically active molecules essential for coping with host haemostatic and immune defences. A changing array of secreted saliva proteins/polypeptides is deposited into the epidermis and dermis. In addition, immunogens appear to be adsorbed to, or are a component of, attachment cement. Tick saliva immunogens also appear to enter skin by percutaneous absorption. Immunogens are taken up by epidermal Langerhans cells, macrophages in the dermis or other antigen-presenting cells in skin and draining lymph nodes. Langerhans cells can migrate to the lymph nodes for presentation of tick-introduced molecules to antigen-specific B and T lymphocytes. Also, T-lymphocyte interactions with immunogens can occur in the skin. Dendritic cells might serve as antigen-presenting cells in the lymph node. Epidermal keratinocytes are capable of expressing class II major histocompatibility markers, secreting cytokines and presenting antigens. The role of keratinocytes, if any, in acquired resistance has not been determined.

Infestation stimulates innate defences with activation and release of cytokines by Langerhans cells, macrophages and other cells in the skin. Mast cell degranulation could occur due to salivary enzymes. Biologically active molecules released at the feeding site increase vascular permeability and contribute to the modest cellular influx arising during a first exposure. Activation of the alternative complement pathway would generate anaphylatoxins and other activated complement components that result in lesion development along with C5a-mediated leukocyte chemotaxis. Histamine and other bioactive molecules that arise at bite-sites during a primary infestation do not appear to alter tick engorgement.

The initial encounter with tick immunogens results in activation and clonal expansion of specific B- and T-lymphocyte clones, including development of memory cells. Both $T_H1$ and $T_H2$ lymphocytes are involved in acquired resistance. B lymphocytes are activated and differentiate into plasma cells producing homocytotropic or circulating tick-immunogen-specific immunoglobulins. Tick-specific, cell-mediated and antibody responses require days to weeks to develop (Table 9.1).

Continuous or repeat infestation brings feeding ticks into contact with immune effector elements and memory cells induced by initial exposure. Circulating antibodies and immunoglobulins bound to Fc receptors on mast cells, basophils or other leukocytes, along with sensitized T lymphocytes, can react with introduced tick immunogens. Mast cells resident in the dermis at the feeding site are degranulated through the direct action of saliva on the cell membranes or by tick immunogens complexing with homocytotropic antibodies. Lymphocytes arrive at the bite-site within hours of attachment. $T_H1$ lymphocytes orchestrate in part the basophil influx characteristic of the cutaneous hypersensitivity response to infestation of a sensitized animal. Tick-immunogen-reactive homocytotropic antibodies occupy basophil Fc receptors. Continuous introduction of tick immunogens results in antibody-mediated degranulation of basophils and mast cells with release of biologically

**Table 9.1.** Overview of events involved in acquisition of anti-tick immunity.

**Innate response** (immediate)

- Mast cell degranulation with release of vasoactive and chemotactic factors. Tick-specific homocytotropic antibodies not yet formed.
- Activation of Langerhans cells and macrophages with uptake of foreign material (note role in specific immunity).
- Activation of alternative pathway of complement with associated biological activities including anasphylatoxins, mobilization of leukocytes, chemoattraction, immune adherence and cell lysis.

**Development of specific adaptive immunity** (days to weeks)

- Saliva proteins/polypeptides change qualitatively and quantitatively during course of feeding.
- Antigen processing and presentation to T lymphocytes by Langerhans cells, macrophages, dendritic cells in lymph nodes (?) and/or B-lymphocytes (?). Release of immunoregulatory cytokines.
- Differentiation and clonal expansion of specific antigen-reactive B and T lymphocytes.
- Production of specific antibodies. Generation of regulatory and effector T lymphocytes.

active molecules. Basophil- and mast-cell-derived histamine, eosinophil chemotactic factor of anaphylaxis and leukotriene $B_4$ contribute to the influx of eosinophils at the bite-site.

Activation of complement through the alternative and/or classical pathways generates anaphylatoxins causing mast cell and basophil degranulation. C5a is chemotactic for the cells observed in tick feeding lesions on resistant hosts. The membrane attack complex of complement might contribute to feeding lesion development and tick rejection, an hypothesis that should be tested. Neutrophil accumulation would result in part due to the high molecular mass neutrophil chemotactic factor. Macrophages present at the bite-site likely clear saliva and cellular debris, as well as present antigen to immunocompetent cells.

$T_H 2$ lymphocytes are involved in the immunoregulatory events controlling antibody production. Circulating antibodies complex with saliva molecules and possibly bind to antigenic determinants within the feeding tick. Antigen–antibody complexes could activate complement by the classical pathway with the generation of an array of biological activities. These events likely contribute to feeding lesion development. Whether or not complement can be activated within the feeding tick remains to be determined. Complement might contribute to tick rejection directly, or in concert with antibodies. Homocytotropic antibody production is IL-4 regulated.

Mechanism(s) responsible for inhibition of engorgement, reduced production of ova and decreased viability are unknown. Ultrastructural evidence indicates that basophils, eosinophils and their granules in the tick gut are associated with degenerative changes in digestive tract cells and reduced tick viability. Histamine disrupts normal engorgement. Research is needed to determine the effects of prostaglandins, leukotrienes, eosinophil major protein, enzymes and other biologically active molecules on ticks. Mechanisms remain to be defined by which antibodies, complement and/or specifically sensitized T lymphocytes influence the feeding tick. Specific immunogens involved in these responses must be identified, isolated and characterized (Table 9.2).

Tick countermeasures arose against selected aspects of host immune defence pathways. Inhibition of the alternative pathway of complement, anaphylatoxins and NK cells reduces innate defences. Tick suppression of macrophage cytokine elaboration would impair both innate and specific immune responses dependent upon the normal function of these cells, which are central to host defences. Reduced antibody responses decrease the likelihood of neutralization of tick-introduced molecules and any direct impact of specific immunoglobulins on the vector. Impaired proliferative capacity of T lymphocytes could reduce both immunoregulatory and effector functions. Suppression of $T_H 1$-lymphocyte-derived cytokines, IL-2 and IFN-γ, could impair delayed type hypersensitivity reactions, T-cell proliferation, and activation of macrophages and NK cells (Table 9.3).

**Table 9.2.** Overview of events involved in expression of anti-tick immunity by a sensitized host.

**Response to tick feeding**

- Antigen processing and presentation by Langerhans cells, macrophages, dendritic cells in lymph nodes (?) and/or B lymphocytes. Release of cytokines.
- Antigen reactive $T_H1$ lymphocytes mediated development of cutaneous basophil hypersensitivity response. Possible lymphocyte-mediated degranulation of basophils.
- Preformed circulating tick-reactive immunoglobulins form antigen–antibody complexes with fixation of complement by the classical pathway. Direct action of antibodies and/or complement-associated biological activities contribute to feeding lesion formation and affect tick.
- Homocytotropic antibodies occupy Fc receptors on basophils and mast cells. Mediators released from these cells by tick immunogen interacting with homocytotropic antibodies during course of feeding.
- Alternative pathway of complement activation.

**Possible mechanisms by which acquired resistance could affect ticks**

- Direct action of antibodies and/or specifically sensitized T lymphocytes upon the tick with disruption of physiological processes.
- Basophil, mast cell and eosinophil-granule-associated mediator-induced changes through disruption of signals associated with feeding at salivary gland and/or digestive tract.
- Cytokines, enzyme acute phase proteins and/or other components released by the cellular infiltrate at the feeding site affect the tick.
- Direct action of complement components upon tick.

Acquired resistance is a state of balance between host defences and tick-mediated immunosuppression of the host. The balance cannot be tipped too far in either direction. The tick would not want the host immune system to be weakened to the point where the animal would succumb to opportunistic pathogens, thus depriving the tick of a bloodmeal. In turn, full expression of host immune competence might make obtaining a bloodmeal more difficult.

An understanding of the immunological basis of acquired resistance to ticks is emerging. However, many questions remain to be answered before the molecular and cellular basis of acquired immunity to ticks is fully understood. This is an area of research with vast basic and applied value.

# ACKNOWLEDGEMENTS

Research of the author is supported by the US Department of Agriculture, Oklahoma Center for Advancement of Science and Technology, Pfizer Animal Health, and Oklahoma Agricultural Experiment Station Project Number OKL02174.

**Table 9.3.** Immune responses associated with acquired resistance and tick-mediated immunomodulation of the host*.

| Host response | Tick countermeasure |
| --- | --- |
| Alternative complement pathway | Inhibitor of C3b deposition/convertase formation. |
| Anaphylatoxin B | Inactivation by salivary carboxypeptidase |
| Natural killer (NK) cell function | Reduced by salivary gland factor(s) |
| T-lymphocyte proliferation | Mitogen-induced *in vitro* responses reduced by salivary-gland-derived protein and possibly other factors |
| Macrophage elaboration of IL-1 and TNF factor(s) | Cytokine inhibition by salivary gland |
| $T_H 1$-lymphocyte cytokine elaboration of IL-2 and IFN-γ factor(s) | Cytokine inhibition by salivary gland |

\* See text for references to specific host responses to infestation and tick countermeasures.

# REFERENCES

Ackerman, S., Clare, F.B., McGill, T.W. and Sonenshine, D.E. (1981) Passage of host serum components, including antibody, across the digestive tract of *Dermacentor variabilis*. *Journal of Parasitology* 67, 737–740.

Alexander, J. O'D. (1986) The physiology of itch. *Parasitology Today* 2, 345–351.

Allen, J.R. (1973) Tick resistance: basophils in skin reactions of resistant guinea pigs. *International Journal for Parasitology* 3, 195–200.

Allen, J.R. (1989) Immunology of interactions between ticks and laboratory animals. *Experimental and Applied Acarology* 7, 5–13.

Allen, J.R. and Kemp, D.H. (1982) Observations on the behavior of *Dermacentor andersoni* larvae infesting normal and resistant guinea pigs. *Parasitology* 84, 195–204.

Allen, J.R., Doube, B.M. and Kemp, D.H. (1977) Histology of bovine skin reactions to *Ixodes holocyclus*, Neuman. *Canadian Journal of Comparative Medicine* 41, 26–35.

Allen, J.R., Khalil, H.M. and Graham, J.E. (1979) The location of tick salivary gland antigens, complement and immunoglobulin in the skin of guinea pigs infested with *Dermacentor andersoni* larvae. *Immunology* 38, 467–472.

Barriga, O.O., Andujar, F., Sahibi, H. and Andrzejewski, W.J. (1991) Antigens of *Amblyomma americanum* ticks recognized by repeatedly infested sheep. *Journal of Parasitology* 77, 710–716.

Bell, J.F., Stewart, S.J. and Wikel, S.K. (1979) Resistance to tick-borne *Francisella tularensis* by tick-sensitized rabbits: allergic klendusity. *American Journal of Tropical Medicine and Hygiene* 28, 876–880.

Bennett, G.F. (1969) *Boophilus microplus* (Acari: Ixodidae): experimental infestations

on cattle restrained from grooming. *Experimental Parasitology* 26, 323–328.
Ben-Yakir, D. (1989) Quantitative studies of host immunoglobulin G in the hemolymph of ticks (Acari). *Journal of Medical Entomology* 26, 243–246.
Bergman, D.K., Ramachandra, R.N. and Wikel, S.K. (1995) *Dermacentor andersoni*: salivary gland proteins suppressing T-lymphocyte responses to concanavalin A *in vitro*. *Experimental Parasitology* 81, 262–271.
Borsky, I., Hermanek, J., Uhlir, J. and Dusbabek, F. (1994) Humoral and cellular immune response of BALB/c mice to repeated infestations with *Ixodes ricinus* nymphs. *International Journal for Parasitology* 24, 127–132.
Bos, J.D. and Kapsenberg, M.L. (1993) The skin immune system: progress in cutaneous biology. *Immunology Today* 14, 75–78.
Bowessidjaou, J., Brossard, M. and Aeschlimann, A. (1977) Effects and duration of resistance acquired by rabbits on feeding and egg laying in *Ixodes ricinus*. *Experientia* 33, 528–530.
Brossard, M. (1977) Rabbits infested with the adults of *Ixodes ricinus* L.: passive transfer of resistance with immune serum. *Bulletin de la Societé Pathologie Exotique* 70, 289–294.
Brossard, M. (1982) Rabbits infested with adult *Ixodes ricinus* L. effects of mepyramine on acquired resistance. *Experientia* 38, 702–704.
Brossard, M. and Fivaz, V. (1982) *Ixodes ricinus* L.: mast cells, basophils and eosinophils in the sequence of cellular events in the skin of infested and reinfested rabbits. *Parasitology* 85, 583–592.
Brossard, M. and Girardin, P. (1979) Passive transfer of resistance in rabbits infested with adult *Ixodes ricinus* L.: humoral factors influence feeding and egg laying. *Experientia* 35, 1395–1396.
Brossard, M., Monneron, J.P. and Papatheodorou, V. (1982) Progressive sensitization of circulating basophils against *Ixodes ricinus* L. antigens during repeated infestations of rabbits. *Parasite Immunology* 4, 355–361.
Brossard, M., Rutti, B. and Haug, T. (1991) Immunological relationships between host and ixodid ticks. In: Toft, C.A., Aeschliman, A. and Bolic, L. (eds), *Parasite–Host Associations: Coexistence or Conflict*. Oxford University Press, Oxford, pp. 177–200.
Brown, S.J. (1985) Immunology of acquired resistance to ticks. *Parasitology Today* 1, 165–171.
Brown, S.J. (1988a) Western blot analysis of *Amblyomma americanum*-derived stage specific and shared antigens using serum from guinea pigs expressing resistance. *Veterinary Parasitology* 28, 163–171.
Brown, S.J. (1988b) Characterization of tick antigens inducing host immune resistance. II. Description of rabbit acquired immunity to *Amblyomma americanum* ticks and identification of potential tick antigens by western blot analysis. *Veterinary Parasitology* 28, 245–259.
Brown, S.J. and Askenase, P.W. (1981) Cutaneous basophil responses and immune resistance of guinea pigs to ticks: passive transfer with peritoneal exudate cells or serum. *Journal of Immunology* 127, 2164–2167.
Brown, S.J. and Askenase, P.W. (1986) *Amblyomma americanum*: physiochemical isolation of a protein derived from tick salivary gland that is capable of inducing immune resistance in guinea pigs. *Experimental Parasitology* 62, 40–50.
Brown, S.J., Galli, S.J., Gleich, G.J. and Askenase, P.W. (1982) Ablation of immunity to *Amblyomma americanum* by anti-basophil serum: cooperation between basophils

and eosinophils in expression of immunity to ectoparasites (ticks) in guinea pigs. *Journal of Immunology* 129, 790–796.

Brown, S.J., Shapiro, S.Z. and Askenase, P.W. (1984) Characterization of tick antigens inducing host immune resistance. I. Immunization of guinea pigs with *Amblyomma americanum*-derived salivary gland extracts and identification of an important salivary gland protein antigen with guinea pig anti-tick antibodies. *Journal of Immunology* 133, 3319–3325.

Butterworth, A.E. and Thorne, K.J.I. (1993) Eosinophils and parasitic diseases. In: Smith, H. and Cook, R.M. (eds), *Immunopharmacology of Eosinophils*. Academic Press, London, pp. 119–150.

Champagne, D.E. (1994) The role of salivary vasodilators in bloodfeeding and parasite transmission. *Parasitology Today* 10, 430–433.

deCastro, J.J. and Newson, R.M. (1993) Host resistance in cattle tick control. *Parasitology Today* 9, 13–17.

deCastro, J.J., Cunningham, M.P., Dolan, T.T., Dransfield, R.D., Newson, R.M. and Young, A.S. (1985) Effects on cattle of artificial infestations with the tick *Rhipicephalus appendiculatus*. *Parasitology* 90, 21–33.

denHollander, N. and Allen, J.R. (1985) *Dermacentor variabilis*: acquired resistance to ticks in BALB/c mice. *Experimental Parasitology* 59, 118–129.

Dharampaul, S., Kaufman, W.R. and Belosevic, M. (1993) Differential recognition of saliva antigens from the ixodid tick *Amblyomma hebraeum* (Acari: Ixodidae) by sera from infested and immunized rabbits. *Journal of Medical Entomology* 30, 262–266.

Durum, S.K. and Oppenheim, J.J. (1993) Proinflammatory cytokines and immunity. In: Paul, W.E. (ed.), *Fundamental Immunology*, 3rd edn. Raven Press, New York, pp. 801–835.

Dvorak, H.F., Dvorak, A.M., Simpson, B.A., Richardson, H.B., Leskowitz, S. and Karnovsky, M.J. (1970) Cutaneous basophil hypersensitivity. II. A light and electron microscopic description. *Journal of Experimental Medicine* 132, 558–582.

Fivaz, B.H. (1989) Immune suppression induced by the brown ear tick *Rhipicephalus appendiculatus* Neuman, 1901. *Journal of Parasitology* 75, 946–952.

Frank, M.M. and Fries, L.F. (1989) Complement. In: Paul, W.E. (ed.), *Fundamental Immunology*. Raven Press, New York, pp. 679–701.

George, J.E., Osburn, R.L. and Wikel, S.K. (1985) Acquisition and expression of resistance by *Bos indicus* and *Bos indicus* × *Bos taurus* calves to *Amblyomma americanum* infestation. *Journal of Parasitology* 71, 174–182.

Gill, H.S., Boid, R. and Ross, C.A. (1986) Isolation and characterization of salivary antigens from *Hyalomma anatolicum anatolicum*. *Parasite Immunology* 8, 11–25.

Girardin, P. and Brossard, M. (1989) Effects of cyclosporin A on humoral immunity to ticks and on cutaneous immediate and delayed hypersensitivity reactions to *Ixodes ricinus* L. salivary gland antigens in re-infested rabbits. *Parasitology Research* 75, 657–662.

Gleich, G.J., Kita, H. and Adolphson, C.R. (1995) Eosinophils. In: Frank, M.M., Austen, K.F., Claman, H.N. and Unanue, E.R. (eds), *Samter's Immunologic Diseases*. Little, Brown and Company, Boston, pp. 205–245.

Golde, W.T., Burkot, T.R., Sviat, S., Keen, M.G., Mayer, L.W., Johnson, B.J.B. and Piesman, J. (1993) The major histocompatibility complex-restricted response of recombinant inbred strains of mice to natural tick transmission of *Borrelia*

burgdorferi. *Journal of Experimental Medicine* 177, 9–17.
Gordon, J.R. and Allen, J.R. (1987) Isolation and characterization of salivary antigens from the female tick, *Dermacentor andersoni*. *Parasite Immunology* 9, 337–352.
Gordon, J.R. and Allen, J.R. (1991) Factor V and VII anticoagulant activities in the salivary glands of feeding *Dermacentor andersoni* ticks. *Journal of Parasitology* 77, 167–170.
Green, D.R., Flood, P.M. and Gershon, R.K. (1983) Immunoregulatory T-cell pathways. *Annual Review of Immunology* 1, 439–463.
Harding, C.V., Leyva-Cobian, F. and Unanue, E.R. (1988) Mechanisms of antigen processing. *Immunological Reviews* 106, 78–92.
Inokuma, H., Kerlin, R.L., Kemp, D.H. and Willadsen, P. (1993) Effects of cattle tick (*Boophilus microplus*) infestation on the bovine immune system. *Veterinary Parasitology* 47, 107–118.
Inokuma, H., Kemp, D.H. and Willadsen, P. (1994) Prostaglandin E2 production by the cattle tick (*Boophilus microplus*) into feeding sites and its effect on the response of bovine mononuclear cells to mitogen. *Veterinary Parasitology* 53, 293–299.
Janeway, C.A. and Bottomly, K. (1994) Signals and signs for lymphocyte responses. *Cell* 76, 275–285.
Jaworski, D.C., Muller, M.T., Simmen, F.A. and Needham, G.R. (1990) *Amblyomma americanum*: identification of tick salivary gland antigens from unfed and early feeding females with comparisons to *Ixodes dammini* and *Dermacentor variabilis*. *Experimental Parasitology* 70, 217–226.
Jaworski, D.C., Rosell, R., Coons, L.B. and Needham, G.R. (1992) Tick (Acari: Ixodidae) attachment cement and salivary gland cells contain similar immunoreactive polypeptides. *Journal of Medical Entomology* 29, 305–309.
Johnston, T.H. and Bancroft, M.J. (1918) A tick-resistant condition in cattle. *Proceedings of the Royal Society of Queensland* 30, 219–317.
Jones, L.D., Kaufman, W.R. and Nuttall, P.A. (1992) Modification of the skin feeding site by tick saliva mediates virus transmission. *Experientia* 48, 779–782.
Kapsenberg, M.L., Teunisseu, M.B.M. and Bos, J.D. (1990) Langerhans cells: a unique subpopulation of antigen presenting dendritic cells. In: Bos, J.D. (ed.), *Skin Immune System (SIS)*. CRC Press, Boca Raton, Florida, pp. 109–124.
Koney, E.B.M., Morrow, A.N., Heron, I., Ambrose, N.C. and Scott, G.R. (1994) Lymphocyte proliferative responses and the occurrence of dermatophilosis in cattle naturally infested with *Amblyomma variegatum*. *Veterinary Parasitology* 55, 245–256.
Kroemer, G., deAlboran, I.M., Gonzalo, J.A. and Martinez-A, C. (1993) Immunoregulation by cytokines. *Critical Reviews in Immunology* 13, 163–191.
Kubes, M., Fuchsberger, N., Labuda, M., Zuffova, E. and Nuttall, P.A. (1994) Salivary gland extracts of partially fed *Dermacentor reticulatus* ticks decrease natural killer cell activity *in vitro*. *Immunology* 82, 113–116.
Limo, M.K., Voigt, W.P., Tumbo-Oeri, A.G. and ole-MoiYoi, O.K. (1991) Purification and characterization of an anticoagulant from the salivary glands of the ixodid tick *Rhipicephalus appendiculatus*. *Experimental Parasitology* 72, 418–429.
Marrack, P. and Kappler, J. (1994) Subversion of the immune system by pathogens. *Cell* 76, 323–332.
Matsuda, H., Fukui, K., Kiso, Y. and Kitamura, Y. (1985) Inability of genetically mast

cell deficient w/w$^v$ mice to acquire resistance against larval *Haemaphysalis longicornis* ticks. *Journal of Parasitology* 71, 443–448.

Mbow, M.L., Rutti, B. and Brossard, M. (1994) Infiltration of CD4$^+$, CD8$^+$ T-cells, and expression of ICAM-1, Ia antigens, IL-1alpha and TNF-alpha in the skin lesion of BALB/c mice undergoing repeated infestations with nymphal *Ixodes ricinus* ticks. *Immunology* 82, 596–602.

Mossman, T.R. and Coffman, R.L. (1989) TH1 and TH2 cells: different patterns of lymphokine secretion lead to different functional properties. *Annual Review of Immunology* 7, 145–173.

Neitz, A.W.H., Gothe, R., Pawlas, S. and Groeneveld, H.T. (1993) Investigations into lymphocyte transformation and histamine release by basophils in sheep repeatedly infested with *Rhipicephalus evertsi evertsi* ticks. *Experimental and Applied Acarology* 17, 551–559.

Newson, R.M. and Chiera, J.W. (1989) Development of resistance in calves to nymphs of *Rhipicephalus appendiculatus* (Acarina: Ixodidae) during test feeds. *Experimental and Applied Acarology* 6, 19–27.

Nithiuthai, S. and Allen, J.R. (1984a) Significant changes in epidermal Langerhans cells of guinea-pigs infested with ticks (*Dermacentor andersoni*). *Immunology* 51, 133–141.

Nithiuthai, S. and Allen, J.R. (1984b) Effects of ultraviolet irradiation on the acquisition and expression of tick resistance in guinea-pigs. *Immunology* 51, 153–159.

Nithiuthai, S. and Allen, J.R. (1985) Langerhans cells present tick antigens to lymph node cells from tick sensitized guinea pigs. *Immunology* 55, 157–163.

Paine, S.H., Kemp, D.H. and Allen, J.R. (1983) In vitro feeding of *Dermacentor andersoni* (Stiles): effects of histamine and other mediators. *Parasitology* 86, 419–428.

Paul, W.E. and Seder, R.A. (1994) Lymphocyte responses and cytokines. *Cell* 76, 241–251.

Playfair, J.H.L. (1993) Overview: parasitism and immunity. In: Lachmann, P.J., Peters, K., Rosen, F.S. and Walport, M.J. (eds), *Clinical Aspects of Immunology*, 5th edn. Blackwell Scientific Publications, Oxford, pp. 1439–1454.

Ramachandra, R.N. and Wikel, S.K. (1992) Modulation of host immune response by ticks (Acari: Ixodidae): effect of salivary gland extracts on host macrophages and lymphocyte cytokine production. *Journal of Medical Entomology* 29, 818–826.

Ramachandra, R.N. and Wikel, S.K. (1995) Effects of *Dermacentor andersoni* (Acari: Ixodidae) salivary gland extracts on *Bos indicus* and *Bos taurus* lymphocytes and macrophages: *in vitro* cytokine elaboration and lymphocyte blastogenesis. *Journal of Medical Entomology* 32, 338–345.

Ravetch, J.V. and Kinet, J.-P. (1991) Fc receptors. *Annual Review of Immunology* 9, 457–492.

Rechav, Y. and Dauth, J. (1987) Development of resistance in rabbits to immature stages of the ixodid tick *Rhipicephalus appendiculatus*. *Medical and Veterinary Entomology* 1, 177–183.

Rechav, Y., Clarke, F.C., Els, D.A. and Dauth, J. (1991) Development of resistance in laboratory animals to adults of the tick *Rhipicephalus evertsi evertsi*. *Medical and Veterinary Entomology* 5, 29–34.

Ribeiro, J.M.C. (1987a) Role of saliva in blood-feeding by arthropods. *Annual Review of Entomology* 32, 463–478.

Ribeiro, J.M.C. (1987b) *Ixodes dammini*: salivary anti-complement activity. *Experimental Parasitology* 64, 347–353.
Ribeiro, J.M.C. (1989) Role of saliva in tick/host interactions. *Experimental and Applied Acarology* 7, 15–20.
Ribeiro, J.M.C. (1995) How ticks make a living. *Parasitology Today* 11, 91–93.
Ribeiro, J.M.C. and Spielman, A. (1986) *Ixodes dammini*: salivary anaphylatoxin inactivating activity. *Experimental Parasitology* 62, 292–297.
Ribeiro, J.M.C., Makoul, G.T., Levine, J., Robinson, D.R. and Spielman, A. (1985) Antihemostatic, antiinflammatory and immunosuppressive properties of the saliva of a tick, *Ixodes dammini*. *Journal of Experimental Medicine* 161, 332–344.
Ribeiro, J.M.C., Makoul, G.T. and Robinson, D.R. (1988) *Ixodes dammini*: evidence for salivary prostacyclin secretion. *Journal of Parasitology* 74, 1068–1069.
Roberts, J.A. and Kerr, J.D. (1976) *Boophilus microplus*: passive transfer of resistance in cattle. *Journal of Parasitology* 62, 485–489.
Roehrig, J.T., Piesman, J., Hunt, A.R., Keen, M.G., Happ, C.M. and Johnson, B.J.B. (1992) The hamster immune response to tick-transmitted *Borrelia burgdorferi* differs from the response to needle inoculated, cultured organisms. *Journal of Immunology* 149, 3648–3653.
Rubaire-Akiki, C.M. and Mutinga, M.J. (1980) Immunological reactions associated with rabbit resistance to *Rhipicephalus appendiculatus* (Neumann) infestations. *Bulletin of Animal Health and Production in Africa* 28, 49–59.
Rutti, B. and Brossard, M. (1989) Repetitive detection by immunoblotting of an integumental 25-kDa antigen in *Ixodes ricinus* and a corresponding 20-kDa antigen in *Rhipicephalus appendiculatus* with the sera of pluriinfested mice and rabbits. *Parasitology Research* 75, 325–329.
Schorderet, S. and Brossard, M. (1993) Changes in immunity to *Ixodes ricinus* by rabbits infested at different levels. *Medical and Veterinary Entomology* 7, 186–192.
Schorderet, S. and Brossard, M. (1994) Effects of human recombinant interleukin-2 on resistance, and on the humoral and cellular response of rabbits infested with adult *Ixodes ricinus* ticks. *Veterinary Parasitology* 54, 375–387.
Shapiro, S.Z., Voigt, W.P. and Fujisaki, K. (1986) Tick antigens recognized by serum from a guinea pig resistant to infestation with the tick *Rhipicephalus appendiculatus*. *Journal of Parasitology* 72, 454–463.
Shapiro, S.Z., Buscher, G. and Dobbelaere, D.A.E. (1987) Acquired resistance to *Rhipicephalus appendiculatus* (Acari: Ixodidae): identification of an antigen eliciting resistance in rabbits. *Journal of Medical Entomology* 24, 147–154.
Steeves, E.B.T. and Allen, J.R. (1990) Basophils in skin reactions of mast cell-deficient mice infested with *Dermacentor variabilis*. *International Journal for Parasitology* 20, 655–667.
Stingl, G. and Bergstresser, P.R. (1995) Dendritic cells: a major story unfolds. *Immunology Today* 16, 330–333.
Trager, W. (1939) Acquired immunity to ticks. *Journal of Parasitology* 25, 57–81.
Trinchieri, G. (1989) Biology of natural killer cells. *Advances in Immunology* 47, 187–376.
Trinchieri, G. (1995) Interleukin-12: a proinflammatory cytokine with immunoregulatory functions that bridge innate resistance and antigen-specific adaptive immunity. *Annual Review of Immunology* 13, 251–276.
Unanue, E.R. (1993) Macrophages, antigen-presenting cells, and the phenomena of

antigen handling and presentation. In: Paul, W.E. (ed.), *Fundamental Immunology*, 3rd edn. Raven Press, New York, pp. 111–144.

Urioste, S., Hall, L.R., Telford, III S.R. and Titus, R.G. (1994) Saliva of the Lyme disease vector, *Ixodes dammini*, blocks cell activation by a non-prostaglandin $E_2$-dependent mechanism. *Journal of Experimental Medicine* 180, 1077–1085.

vanVuuren, A.M.J., Crause, J.C., Verschoor, J.A., Spickett, A.M. and Neitz, A.W.H. (1992) The identification of a shared immunogen present in the salivary glands and gut of ixodid and argasid ticks. *Experimental and Applied Acarology* 15, 205–210.

Wang, H. and Nuttall, P.A. (1994) Excretion of host immunoglobulin in tick saliva and detection of IgG-binding proteins in tick haemolymph and salivary glands. *Parasitology* 109, 525–530.

Ward, P.A., Dvorak, H.F., Cohen, S., Yoshida, T., Data, R. and Selvaggio, S.S. (1975) Chemotaxis of basophils by lymphocyte dependent and lymphocyte independent mechanisms. *Journal of Immunology* 114, 1523–1527.

Whelen, A.C. and Wikel, S.K. (1993) Acquired resistance of guinea pigs to *Dermacentor andersoni* mediated by humoral factors. *Journal of Parasitology* 79, 908–912.

Whelen, A.C., Richardson, L.K. and Wikel, S.K. (1986) DOT-ELISA assessment of guinea pig antibody responses to repeated *Dermacentor andersoni* infestations. *Journal of Parasitology* 72, 155–162.

Wikel, S.K. (1979) Acquired resistance to ticks. Expression of resistance by C4-deficient guinea pigs. *American Journal of Tropical Medicine and Hygiene* 28, 586–590.

Wikel, S.K. (1982a) Immune responses to arthropods and their products. *Annual Review of Entomology* 27, 21–48.

Wikel, S.K. (1982b) Histamine content of tick attachment sites and the effects of $H_1$ and $H_2$ histamine antagonists on the expression of resistance. *Annals of Tropical Medicine and Parasitology* 76, 179–185.

Wikel, S.K. (1982c) Influence of *Dermacentor andersoni* infestation on lymphocyte responsiveness to mitogens. *Annals of Tropical Medicine and Parasitology* 76, 627–632.

Wikel, S.K. (1985) Effects of tick infestation on the plaque-forming cell response to a thymic dependent antigen. *Annals of Tropical Medicine and Parasitology* 79, 195–198.

Wikel, S.K. (1988) Immunological control of hematophagous arthropod vectors: utilization of novel antigens. *Veterinary Parasitology* 29, 235–264.

Wikel, S.K. (1996) Host immunity to ticks. *Annual Review of Entomology* 41, 1–22.

Wikel, S.K. and Allen, J.R. (1976) Acquired resistance to ticks. I. Passive transfer of resistance. *Immunology* 30, 311–316.

Wikel, S.K. and Allen, J.R. (1977) Acquired resistance to ticks. III. Cobra venom factor and the resistance response. *Immunology* 32, 457–465.

Wikel, S.K. and Whelen, A.C. (1986) Ixodid–host immune interaction. Identification and characterization of relevant antigens and tick-induced host immunosuppression. *Veterinary Parasitology* 20, 149–174.

Wikel, S.K. and Osburn, R.L. (1982) Immune responsiveness of the bovine to repeated low-level infestations with *Dermacentor andersoni*. *Annals of Tropical Medicine and Parasitology* 76, 405–414.

Wikel, S.K., Graham, J.E. and Allen, J.R. (1978) Acquired resistance to ticks. IV. Skin reactivity and *in vitro* lymphocyte responsiveness to salivary gland antigen. *Immunology* 34, 257–263.

Wikel, S.K., Ramachandra, R.N. and Bergman, D.K. (1994) Tick-induced modulation of the host immune response. *International Journal for Parasitology* 24, 59–66.

Willadsen, P. (1980) Immunity to ticks. *Advances in Parasitology* 18, 293–313.

Willadsen, P. and Riding, G.A. (1979) Characterization of a proteolytic enzyme inhibitor with allergenic activity. Multiple functions of a parasite-derived protein. *Biochemical Journal* 177, 41–47.

Willadsen, P. and Williams, P.G. (1976) Isolation and partial characterization of an antigen from the cattle tick, *Boophilus microplus*. *Immunochemistry* 13, 591–597.

Wozniak, E.J., Butler, J.F. and Zan, S.G. (1995) Evidence of common and genus-specific epitopes on *Ornithodoros* spp. tick (Acari: Argasidae) salivary proteins. *Journal of Medical Entomology* 32, 484–489.

# 10

# Immunology of Scabies

## Larry G. Arlian

*Department of Biological Sciences and Department of Microbiology and Immunology, Wright State University, Dayton, Ohio 45435, USA*

### INTRODUCTION

*Sarcoptes scabiei* infest humans and many other species of wild and domestic mammals (Fain, 1968). The mites reside in the burrows they make in the lower stratum corneum of the skin. Mite varieties from different host species exhibit few morphological differences, but they are physiologically different. This is evidenced by the fact that most varieties of scabies mites that naturally infest different host species are host specific and cannot establish permanent infestations on foreign hosts (Arlian *et al.*, 1984a, b, 1988a). Scabies mites are the source of antigens that elicit immunologic reactions by the host. Burrowing mites secrete saliva and possibly other products such as moulting enzymes, and deposit exuviae, ova, faecal and excretory material in the burrow. The expelled products and degrading dead mites contain antigenic molecules that reach the various effector components of the immune system. This chapter will review both the key historic and current research information that has provided a limited characterization and understanding of the immune response mechanism. In some instances, details of the methods or host characteristics are given when these are important to the interpretation of the results of specific studies and/or the broader picture. Though limited in scope, early studies have provided some important insights that have been confirmed and expanded by more recent studies.

© CAB INTERNATIONAL 1996. From Wikel, S.K. (ed.) *The Immunology of Host–Ectoparasitic Arthropod Relationships*. CAB INTERNATIONAL, Wallingford.

## ANTIGENS

Historically, immunologic studies of scabies infestations have been limited because of the lack of sufficient mites to produce the antigenic extracts or to induce the experimental infestations needed for controlled studies. This problem is mentioned many times by the authors of various published papers. This problem was overcome when we successfully established a colony of *S. scabiei* var. *canis* (canine scabies) on laboratory rabbits, developed a method of harvesting large quantities of mites from crusts and demonstrated that various strains of scabies mites are the sources of common antigens that can be used as references in immunological studies (Arlian *et al.*, 1984b, 1985, 1996a). Enough var. *canis* mites (milligram quantities) can be collected to prepare moderate volumes of extracts, and to experimentally infest additional hosts. To make extracts the collected mites are dried, ground, defatted in anhydrous diethyl ether and extracted in glass-distilled water 1:20, w/v (Arlian *et al.*, 1985). The extract supernatant containing soluble protein is sterile filtered (0.2 µm) into vials and then used in various immune and biochemical assays. Antiserum to the extract is produced by immunization (Morgan and Arlian, 1994). Very limited quantities of similar extracts have also been produced by other researchers by collecting individual scabies mites as they migrated from the skin scales and crusts of humans, pigs, dogs and foxes (Prakken and van Vloten, 1949; Sheahan, 1975a; Falk and Bolle, 1980b; Dobson, 1981; Wooten and Gaafar, 1984; Davis and Moon, 1990b; Bornstein and Zakrisson, 1993a).

Using crossed immunoelectrophoresis (CIE), we found that New Zealand white rabbits infested with *S. scabiei* var. *canis* produced antibodies to 12 different antigens (Arlian *et al.*, 1985; Morgan and Arlian, 1994). In contrast, rabbits immunized with an extract made from scabies mite bodies produced antibodies to 20 different antigenic molecules (Morgan and Arlian, 1994). Western blotting revealed that serum from rabbits infested or immunized with *S. scabiei* var. *canis* contained antibodies specific for more than 25 proteins resolved by reducing and non-reducing sodium dodecyl sulphate polyacrylamide gel electrophoresis (SDS–PAGE) of primarily whole body extracts.

## CROSS-REACTIONS BETWEEN *S. SCABIEI* AND OTHER MITES

Indirect evidence raised the possibility of cross-reactivity between scabies mites and the allergy-causing house dust mites *Dermatophagoides pteronyssinus* (DP) and *D. farinae* (DF). Falk and Bolle (1980a) found that some scabies patients who were RAST positive to DP and had high levels of serum IgE showed reduced concentrations of specific antibodies to DP and total IgE after treatment for scabies infestations. Also, protective immunity against scabies mites was induced by vaccination with an extract of DF and DP (50/50 mix) (Arlian et al., 1995).

Several studies directly showed that antigens introduced into the host by *S. scabiei* or molecules in extracts made from these mites cross-reacted with immunogens in extracts produced from the house dust mites, DF and DP. Falk et al., (1981) found using heterologous and tandem CIE that DF and *S. scabiei* extracts contained four and three common antigens, respectively. In more detailed investigations using CIE, Arlian et al. (1988b, 1991) found that antisera from scabies-infested rabbits contained antibodies directed at six and seven antigens in extracts of DF and DP, respectively. Likewise, CIE reactions of antisera produced to DF and DP extracts precipitated proteins (antigens) in extracts produced from *S. scabiei* var. *canis* mites. Furthermore, 8 and 15 proteins/peptides in extracts of var. *canis* resolved by SDS–PAGE, bound IgE present in the serum of DP- and DF-sensitive patients with no history of scabies infestation.

These studies clearly demonstrate immunologic cross-reactivity between scabies and house dust mites. This cross-reactivity creates difficulties in understanding the immune response to scabies in humans since the prevalence of dust mite sensitivity is high and dust mites are common in homes in humid geographical areas (Arlian et al., 1992). It also makes the task of isolating relevant antigens for development of a serologic diagnostic test or vaccine for scabies difficult.

## IMMUNE REACTIONS TO SCABIES ANTIGENS

Historically, the host's immunological responses to antigens from scabies mites have been characterized as developing through five distinct phases. These phases were based on the responses of guinea pigs to flea bites and on human and rabbit responses to mosquito bites (Benjamin et al., 1960; McKeil and West, 1961; Davis and Moon, 1990a, b; Cabrera et al., 1993). These phases included: (i) induction; (ii) delayed hypersensitivity; (iii) delayed and immediate hypersensitivity; (iv) immediate hypersensitivity; and (v) desensitization. However, based on the reported descriptions of the cell infiltrate in biopsy specimens of scabietic lesions and the results of intradermal skin tests, it appears that a late phase allergic reaction also occurs. This reaction is usually a temporal extension of the immediate reaction and for that reason has been called the late phase allergic reaction (Lemanske and Kaliner, 1993). Both immediate and late phase reactions result from the interaction of an antigen with IgE antibodies while delayed reactions do not. Therefore, neither immediate nor late phase reactions should occur in scabietic patients that are not predisposed to the allergic reactions (e.g. IgE-mediated) caused by the offending substances from scabies mites. Non-atopic individuals should express only delayed hypersensitivity reactions. The widely varied descriptions in the literature of this histopathology in scabies lesions better fits this classification.

Progression through phases and the presumed reaction mechanisms as they are currently understood are described below. As we gain a greater understanding of the immune response mechanism to scabies, these will undoubtedly be modified.

**Phase I. Induction.** There is no response to a skin test with scabies mite extract. Clinical signs of the scabies infestation are not evident, but antigen-presenting cells have recognized and begun processing antigenic material. Humans experiencing their first scabies infestation may not exhibit clinical signs for 4 to 8 weeks depending on the severity of the infestation.

**Phase II. Delayed hypersensitivity.** The clinical signs of redness, swelling and itching manifest in 24–48 hours at the site of the infestation or antigen exposure (e.g. skin test) in sensitized hosts. This probably represents the time for antigen presentation in the local lymph nodes, T-cell activation and clonal expansion, migration of the antigen-specific T cells to the dermal lesion, interaction of these cells with released cytokines and other cells (possibly keratinocytes, Langerhans cells, macrophages and B- cells) and subsequent infiltration of inflammatory cells. The cellular infiltrate contains $CD4^+$ and $CD8^+$ T cells, other lymphocytes and inflammatory cells. Delayed hypersensitivity reactions depend on antigen-specific receptors on the surfaces of $CD4^+$ T cells. Immunoglobulin and complement deposition have been demonstrated during the delayed response as discussed later.

**Phase III. Immediate, delayed, and late phase hypersensitivity.**
Antigen challenge (e.g. skin test or mite burrowing) can result in immediate, delayed and late phase reactions. Skin tests may exhibit immediate reactions in 10–20 minutes that manifest as wheal and flare (swelling and redness) responses in sensitized individuals. Immediate reactions are local allergic responses triggered by the binding of offending antigen to the IgE that is already bound to receptors on the surfaces of mast cells in the dermis. Immediate reactions usually evolve into delayed hypersensitivity and late phase reactions in sensitized allergic individuals. As infestations progress in naive hosts, the capability to express delayed hypersensitivity reactions by skin testing develops earlier than the capability to express immediate hypersensitivity/late phase reactions (Davis and Moon, 1990a). Skin tests may give delayed hypersensitivity reactions and no immediate reaction.

Histologically, late phase reactions are characterized by a mixed cellular infiltrate containing eosinophils, neutrophils and $CD4^+$ T cells, and no immunoglobulin or complement with or without fibrin deposition in the lesion (Lemanske and Kaliner, 1993). Early (30 minutes) late phase reactions are characterized by a cell infiltrate that is predominantly composed of eosinophils (90%). Mature late phase reactions (3–24 hours) are characterized by high percentages of neutrophils (e.g. 82%) in the infiltrate with about equal percentages of eosinophils and mononuclear cells (Lemanske and Kaliner, 1993).

**Phase IV. Immediate hypersensitivity.** Skin tests result only in immediate reactions that are manifested within 10–20 minutes in sensitized individuals. There are no delayed reactions later at the injection sites.

**Phase V. Desensitization.** Immunity has been induced so that no hypersensitivity occurs on exposure to the antigen. Theoretically, chronic exposure to scabies antigens may result in desensitization by blocking or down-regulating T-cell receptors as occurs in immunocompetent patients who develop allergic contact dermatitis (Polak and Rinck, 1978; Gaspari *et al.*, 1988). This may occur in crusted scabies and may explain why these patients experience little pruritus and thus allow the mite infestation to expand without seeking treatment.

## IMMEDIATE HYPERSENSITIVITY

Elevated concentrations of circulating IgE and/or the demonstration of immediate hypersensitivity indicate an allergic reaction to antigens released into the host by a parasite. To clarify the terminology used here, antigens that induce IgE production and allergic reactions are termed allergens. The clinical signs of allergic reactions to allergens from a parasite may be manifested in many ways and depend on the parasite's location on or in the host's body, the severity of the infection or infestation and the host's sensitivity to the offending allergen. The allergic response is caused by IgE-type immunoglobulins that are directed at the immunogenic epitopes on the allergenic molecule. The binding of allergens to specific IgE on the surfaces of mast cells and basophils trigger the release of pharmacological agents that affect particular cells and tissues and give rise to the typical allergic clinical signs. In extreme cases, total body or systemic anaphylaxis can result in 10–20 minutes following exposure to the offending allergen.

Systemic anaphylaxis to scabies infestations (antigens) in humans and other animals has not been reported while anaphylaxis to other arthropods (e.g. bee stings) is common. However, there is evidence that in some humans, antigens from scabies mites induce IgE production and IgE-mediated reactions. An immediate hypersensitivity response to antigens from scabies mites may result following sensitization. Likewise, immediate-type reactions have been demonstrated for pigs previously sensitized to scabies antigens. The following is a summary of the key studies that demonstrated immediate hypersensitivity. Details concerning extracts and patients are included since they are important to the interpretation of the results.

Early evidence for IgE-mediated reactions to scabies infestations in humans was provided by skin test studies conducted during the 1930s and 1940s when scabies in humans in the developed world was prevalent

(Kitchevatz and Pochtich, 1934; Heilesen, 1946; Prakken and Van Vloten, 1949). Heilesen (1946) performed intracutaneous skin test studies using an extract made from scabies mites picked from the crusts of a patient with crusted scabies. About 300 (1.2 mg wet weight) and 483 (1.6 mg wet weight) adult mites, respectively, were crushed, defatted with ether and the soluble protein was extracted in 6 ml of saline. Intracutaneous tests were performed with 0.1 ml of a 1:5000 or 1:3000 (w/v) extract. Using 2+ (++) skin reactions as positive (twice the size of the negative control), 3 of 38 patients with scabies showed positive reactions. The three skin test positive patients had had scabies symptoms (itching and eruptions) for 2 months, 3 months and of unknown duration, respectively. One patient previously had scabies once and the second four times previously. Although eight additional patients were doubtfully positive (+ or ± reactions) five of them had been infested with scabies previously. Ten of the skin test negative patients had scabies one to five times previously with symptoms of the current infestations ranging from 4 days to 9 months. Therefore, neither their histories nor the duration of their previous infestations seemed to relate to their allergic reactions. These data suggest that some individuals may exhibit an allergic reaction to scabies while others may not, but that this is only one aspect of a complex immune response. In fact, many patients with scabies do not present with allergic reactions. These results confirmed an earlier report that intracutaneous skin tests in patients with scabies were often positive when using saline extracts made from crusts of a crusted scabies patient (Kitchevatz and Pochtich, 1934).

In another study, Prakken and Van Vloten (1949) prepared a crude scabies extract from the skin scales taken from a 65-year-old male with crusted scabies. Sixteen grams of skin scales were ground, defatted then extracted in 100 ml of physiological saline. The skin scales contained ova, excreta and all life stages of mites. [As a reference, crusts from pigs may contain up to 18,000 mites per gram (Dobson, 1981).] Following intracutaneous injections of a $1:2^{12}$ w/v dilution of the saline scabies extract, 18 of 24 patients with active scabies and 9 of 16 patients previously treated for scabies and no longer exhibiting symptoms, showed immediate hypersensitivity reactions. However, 6 of the 18 patients with immediate hypersensitivity reactions to scabies extract also reacted to a control extract prepared from the skin scales from a patient with erythroderma Wilson-Brocq but without scabies. The extract induced no reactions in 18 individuals who had never had scabies. A more concentrated scabies extract gave positive intracutaneous reactions in both the scabies cases and in the normal controls who had never had scabies. Interestingly, no delayed hypersensitivity reactions were observed 24–48 hours later; however, not all patients were examined. In the same study the presence of circulating antibodies (now known to be IgE) was demonstrated by the Prausnitz–Küstner (P–K) passive transfer test. Serum from a 65-year-old patient with crusted scabies was transferred by intracutaneous injection to 12 normal persons with negative intradermal skin reactivity to the same crude crust extract ($1:2^{12}$ w/v). Following serum transfer, 8 of the 12 patients then gave positive intracutaneous skin reactions when challenged with the crude crust extract. Interestingly, similar P–K transfer experiments were negative with four normal subjects that received serum from a patient with 'ordinary' (non-crusted) scabies even though that patient showed a strong skin test reaction to the crude crust extract. The significance of this is unclear, although it would indicate that a significant level of circulating IgE to scabies was not produced in this particular case of 'ordinary' (non-crusted) scabies.

Falk and Bolle (1980b) skin tested, by both prick and intracutaneous methods, 12 patients with previous histories of scabies and 3 persons with skin sensitivity to the house dust mite, DP, but who had never had scabies. The concentration of total serum IgE was elevated in only 2 of the 12 patients with a history of scabies. The extract was made from 200 (0.8 mg) crushed, defatted female mites extracted in 2 ml of saline. The final concentration was 1:2500 (w/v). Intracutaneous injections were performed with approximately 20 µl of the extract. Five of the seven patients who had scabies during the previous 2–7 months had positive intracutaneous skin responses while skin prick tests were negative in all cases. The fact that the prick tests were negative while the intracutaneous tests were positive illustrated the differences in sensitivity of the two skin test methods and the difficulty of interpreting results of seemingly similar diagnostic clinical tests. The remaining five patients, who had been free of scabies for more than a year (15–24 months) were skin test negative (intracutaneous). In the same study, two individuals who had never had scabies were passively sensitized (P–K test) by the intracutaneous injection of 0.1 ml of serum from five of the patients with scabies. Two of the scabietic donor patients were atopic (RAST class 4 to DP) and three had no history of atopy. Passive P–K transfer induced positive skin test reactions in both test cases. Although sample size was small and the responses could be due to cross-reactivity between scabies mites and DP, these results again were evidence that some individuals with scabies produce scabies-mite-specific IgE. Also, IgE-mediated immediate hypersensitivity specific for scabies antigens can be induced following passive serum transfer. However, the IgE-type sensitivity may wane rapidly since no individual that had been cured of scabies for more than a year gave an immediate hypersensitivity skin reaction. It is possible that none of the five individuals in this group of patients was predisposed to atopy for antigens from scabies mites.

Several other studies, some conducted in different laboratories (clinics), have determined the total IgE or specific IgE levels in the serum of patients with scabies. Falk and Bolle (1980a) found that of 135 patients with scabies, 35% were RAST positive to the allergy-causing dust mite, DP, and 45% had elevated total serum IgE concentrations. In another study, 39 of 94 patients with scabies were found to have elevated total serum IgE levels at the time of diagnosis (Falk, 1980). Because of the high prevalence of elevated IgE specific for the dust mite DP, or the elevated total IgE concentrations among patients with scabies, it was suggested that scabies infestations were associated with atopic disease (possibly because of dust mite sensitivity) and that atopy may predispose individuals to scabies infestation (Falk, 1980; Falk and Bolle, 1980a). In contrast to the finding of Falk and colleagues, Hancock and Ward (1974) determined that the total serum IgE concentrations of 100 patients with scabies were within normal limits: 64 of these patients had been itching for more than 3 weeks.

Rantanen *et al.*, (1981) determined both the total and scabies-specific IgE levels for 16 patients with scabies that had pruritis and rashes lasting from 1 to 6 months. One patient had crusted scabies and 11 had widespread scabies. An extract (1000 mites per 0.5 ml saline) made from *S. scabiei* var. *suis* mites collected from the ears of pigs was used for radioallergosorbent testing (RAST) of five patients. Two of the five patients who were RAST tested to scabies had IgE antibodies specific for *S. scabiei* var. *suis* antigens. Of the 16 patients, 8 had elevated total serum IgE levels. Although the sample size was small, the percentage of patients with IgE specific for var. *suis* was consistent with the incidence of positive skin tests in the studies that used extracts containing var. *hominis*, and that suggested that other scabies strains can be used as a reference for testing.

Based on the results of all of these studies it appears that in humans, atopy is not always a predisposing factor to scabies infestation. It is clear that some infested individuals are predisposed to allergic reactions to allergens from these mites but at least an equal percentage of individuals are not.

Cutaneous hypersensitivity was investigated in pigs experimentally infested with *S. scabiei* var. *suis* (Davis and Moon, 1990b). Pigs were experimentally infested with 100 (low dose) or 1000 (high dose) mites and then skin tested weekly for 9 weeks. Injection sites were observed after 15 minutes and 24 hours for evidence of immediate and delayed hypersensitivity reactions, respectively. An extract was made from 50 female mites and contained 0.32 µg extractable protein per mite. Each skin test injection contained 5 µg of protein (equivalent to 15.5 mites). Over the course of the study, weekly skin results revealed that the onset of delayed hypersensitivity reactions developed sooner than the onset of immediate sensitivity reactions. Immediate sensitivity followed the onset of delayed hypersensitivity reactions by approximately 2.5 weeks for pigs infested with both the low and high doses of mites. For low-dose pigs, delayed reactions were not evident until week three but all low dose pigs exhibited delayed responses by week six. In low-dose pigs immediate hypersensitivity responses were not evident until 4 weeks post-infestation and they were concurrent with delayed responses that, in most cases, had developed earlier. Several low-dose pigs did not give immediate reactions until 5 and 7 weeks post-initial infestation. The rate of sensitization was a function of their level of infestation. Pigs infested with the high dose (1000 mites) developed delayed and immediate hypersensitivity sooner than did the low-dose pigs. Some high-dose pigs showed delayed reactions at 1 week and all showed them by 4 weeks. Concurrent immediate and delayed hypersensitivity were evident by week three in some high-dose pigs.

A study by Sheahan (1975b) compared the skin test (intradermal) reactions of scabietic pigs who were reared on conventional diets with those on low iron diets. Pigs on conventional diets were tested at 3, 7, 17 and 20 weeks following the initial scabies infestation. Pigs on the low iron diet were skin tested 8 weeks following infestation. Well-developed immediate reactions (15 minutes) were observed in pigs skin tested at 7 weeks but not at 3 weeks post-infestation. Delayed reactions at the injection sites were evident at 3 weeks. A dense lymphocyte and eosinophil infiltrate was present in the dermis by 24 hours. The perivascular infiltrate consisted of 73% lymphocytes and 27% eosinophils and neutrophils. Iron-deprived infested pigs gave similar immediate and delayed reactions but gave less intense responses compared to pigs on conventional diets. No pigs developed immediate reactions without delayed reactions. Both immediate and delayed reactions, coupled with the presence of eosiniphils in the tissue, suggest that the delayed reactions included a late phase response. This would explain Sheahan's conclusion that the delayed reactions resembled local anaphylactic responses (Sheahan, 1974).

These studies with pigs show that in naive hosts with progressing infestations, an intradermal inoculation with var. *suis* extract first induces cell-mediated (delayed hypersensitivity) reactions, followed later by the onset of allergic (immediate) reactions. Therefore, the capability for delayed reactions develops earlier in the infestation than does the capability of immediate reactions. These pig studies also show that the level of exposure (parasitic load) as well as the diet influence the rate at which sensitivity develops and the resulting intensity of the reaction. Finally, in contrast to humans, these limited studies also suggest that most pigs develop immediate reactions and therefore produce IgE to molecules from *S. scabiei* var. *suis*. The onset of immediate reactions was coupled to, but followed, the onset of delayed reactions.

## CELLULAR INFILTRATES IN THE EPIDERMIS AND DERMIS OF HUMANS AND OTHER ANIMALS WITH SCABIES (DELAYED AND LATE PHASE REACTIONS)

It is clear that delayed and likely late phase reactions can be components of the immune and inflammatory responses to scabies. However, the temporal progression of the infiltrating cells in the dermis of humans with scabies has not been characterized so it is not possible to differentiate or to understand these progressive reactions. Several studies have characterized the cellular infiltrates in punch biopsy specimens from patients with nodular, papular, vesicular and crusted scabies infested for usually undetermined durations. The results of these studies are highly varied which makes understanding the cell-mediated response and the immune response mechanism difficult. Several recent studies of the temporal progression of the cellular infiltrate in the scabietic lesions of naive and sensitized animals characterized the immunohistopathology associated with delayed and late phase reactions in non-primate hosts.

In humans, in general, the cell-mediated reaction to scabies antigens is characterized by a mixed cellular infiltrate consisting of lymphocytes (T cells, B cells, plasma cells) histiocytes (macrophages), eosinophils and neutrophils in the deep dermis, superficial dermis (dermal/epidermal junction) and perivascular areas. The infiltrate is initially sparse and becomes dense as the duration and intensity of the infestation increases. The infiltrate develops in the perivascular area and then spreads into the dermis and epidermis. It is apparent that the reported relative percentages of each cell type in the infiltrate vary greatly between studies and among patients within a study. Differences in the reported composition of the infiltrate probably reflect the different types of hypersensitivities (e.g. induction, delayed, late phase and desensitization) between patients, the varying durations (progression) of the infestations, sites in the lesion where the infiltrate was examined (e.g. epidermis, epidermal/dermal junction, deep dermis, perivascular, etc) and differences in the ages of the patients (e.g. older, immunocompromized). For example, the cell infiltrates in chronic scabies lesions (of long duration) appear to consist mostly of mononuclear cells (probably T cells), macrophages and plasma cells, but this generalization requires confirmation by well-controlled studies. Earlier stages of cell-mediated reactions contain more neutrophils. The infiltrates of late phase reactions contain eosinophils and are mediated by IgE while the delayed hypersensitivity responses are not.

In early studies of the histopathology of scabietic lesions, Konstantinov and Stanoeva (1973) found that infiltrates of old nodular lesions continued numerous histiocytes and plasma cells but the infiltrates in younger lesions consisted almost exclusively of lymphocytes. In contrast, Hejazi and Mehregan (1975) found that for 14 patients with scabies, the predominant cells in the dermal infiltrates were eosinophils and lymphocytes. The ages of these patients were 18 to 81 and the durations of their lesions varied from 2 weeks to 1 year, but specific histopathology associated with these age factors was not given. However, Head *et al.* (1990) reported that the cell infiltrate throughout the dermis in a skin biopsy specimen from the abdomen of an 87-year-old patient with scabies contained predominantly lymphocytes, histiocytes and eosinophils.

Ackerman (1977) examined the histopathology of punch biopsy specimens of the skin from patients with papular and papulovesicular (n=45), nodular (n=17) and crusted (n=4) scabies. Papular scabies lesions contained mixed cell infiltrates consisting of histiocytes, lymphocytes and numerous eosinophils both superficially and in the deep perivascular areas. In contrast, nodular scabies lesions showed dense infiltrates of lymphocytes, histiocytes, plasma cells, eosinophils and atypical mononuclear cells. Crusted scabies biopsies were characterized by dense lympho-histiocytic infiltrates with numerous plasma cells. This study clearly illustrates the variable histopathology that may result in the different clinical forms and progressive stages of scabies. Individual variations in the histopathology between patients were not reported but these would have been interesting. Likewise, since eosinophils were reported in the infiltrates of some forms of scabies, correlations with allergic histories and with the possibility of late phase reactions would have been interesting.

In a study involving biopsy specimens from 19 patients with scabies, Falk and Eide (1981) found that their infiltrates consisted mainly of lymphocytes, with fewer histiocytes, and varying numbers of neutrophils and eosinophils. Eosinophil numbers varied from zero to many between patients. The densities of eosinophils in the tissue infiltrates and in the blood correlated directly with the severity of the rash. Eosinophils were not correlated with an atopic disease history but this would not preclude allergic reactions to scabies.

A later study of seven patients by Falk and Matre (1982) determined that lymphocyte density was greater in older lesions compared to younger ones and the composition of the cell infiltrates were similar regardless of scabies infestation severity. The infiltrates contained mostly lymphocytes, fewer histiocytes, some neutrophils and occasionally a few eosinophils. T lymphocytes were the predominant cells in the dermal infiltrates as determined by using indirect immunofluorescence and by demonstrating their abilities to form rosettes with sheep erythrocytes. In two cases macrophages were more numerous than T lymphocytes; otherwise few macrophages and B lymphocytes were detected.

Reunala et al. (1984) evaluated biopsy specimens from nodular or papulovesicular skin lesions of seven patients who demonstrated clinical features of scabies for 1 to 2.5 months. Four of the patients had nodular scabies. The biopsies were characterized by α-naphthyl acetate esterase (ANAE) staining and immunoperoxidase labelling. T lymphocytes were the dominant cells in the epidermal, dermal and perivascular infiltrates of all of the patients. The density of T lymphocytes decreased progressively from the deep dermis (64% of the mononuclear cells) to the superficial dermis (16%). In contrast, macrophages were prevalent but the density varied from the deep dermis (10%) to the epidermis (47%) of all mononuclear cells. B lymphocytes comprised about 25% of all of the mononuclear cells in the deep dermis, 33% in the superficial dermis and 38% in the epidermis. Plasma cells were detected in the dermal infiltrate of only one case of nodular scabies and these cells constituted 10% of all mononuclear cells. IgE, IgG and IgA-positive plasma cells occurred in four, two and one specimens, respectively. Only one nodular specimen had countable numbers of eosinophils and neutrophils were infrequent.

Results of a study using planimetric analysis combined with [$^3$H]thymidine labelling indices suggested that scabies mites activate local spontaneous blastogenesis of mononuclear cells (presumably T cells) in the perivascular areas of the dermis of scabietic lesions (Van Neste, 1982). Many mononuclear cells in this infiltrate were engaged in the cell proliferation cycle and were synthesizing DNA.

Van Neste and Lachapelle (1981) found that the infiltrates in biopsy specimens from three patients with crusted scabies contained many lymphohistiocytic cells and C3 deposits in the dermal–epidermal junction. The ratio of B cells to T cells in the peripheral blood of these patients was increased.

The histopathological characteristics of scabies lesions in the epidermis and dermis of other mammals are generally similar to those reported for humans except that the infiltrate in pigs with scabies consistently contains eosinophils. This finding is consistent with the observation that in pigs both delayed and allergic reactions occurred together. Sheahan (1975a) reported that the cell infiltrates in the epidermis of 16 naturally infested and 14 experimentally infested pigs contained mostly neutrophils and eosinophils near the mite burrows. Experimentally and naturally infested pigs had similar dermal and epidermal infiltrates after 7 weeks of infestation. Neutrophils and eosinophils were present in the epidermis, while diffuse infiltrates consisting mostly of eosinophils and mononuclear cells were common in the dermis. In addition, numerous mast cells were present around small vessels in the dermis of chronic, crusted lesions. Experimentally infested pigs exhibited similar epidermal and dermal infiltrates after 1 and 7 weeks of infestation but with an increased density at 7 weeks. Experimentally infested pigs also had significantly elevated serum gamma globulin concentrations 7 weeks after infestation.

Morsy and Gaafar (1989) sequentially monitored and quantitated the density of immunoglobulin-secreting cells that infiltrated the cutaneous tissues of pigs that were experimentally infested with *S. scabiei* var. *suis*. In pigs infested once at either 6 or 30 days of age, the densities of IgG-, IgM- and IgA-secreting cells increased while the infestation developed and then decreased significantly during recovery. For a multiple infestation experiment, after the initial infestation, 6-day-old pigs developed cell densities similar to the pigs infested only once with similar significant cell decreases upon disease regression. However, following four subsequent infestations (at 63, 77, 91 and 105 days post-initial infestation) these same pigs showed only slight increases in the density of the various secreting cells over the densities observed during recovery from the first infestation. Cell densities during subsequent infestations were less than one-third of those observed during the first infestation. In newborn pigs, IgG-secreting cells were the most numerous followed by IgM- and then IgA-secreting cells. IgM-secreting cells were the most numerous secreting cells in older pigs. These results suggest that the infiltrate contained antibody-secreting plasma cells.

Using monoclonal antibodies directed at specific cell surface molecules, we further delineated the cell types in the infiltrates in the lesions of naive and sensitized dogs infested with scabies (Stemmer and Arlian, unpublished data). The dogs were immune to reinfestation following their first infestations (Arlian et al., 1996b). Monoclonal and polyclonal antibodies to the canine leukocyte antigens were used to determine the density of cells in the infiltrates that expressed specific surface markers. Sequential biopsy specimens of skin from scabietic lesions were excised at specific times during an initial first (sensitization) infestation, during recovery following the successful treatment for scabies and during a second (challenge) infestation.

During sensitization, cells in the epidermis and dermis expressing class II MHC, CD1a, CD3, CD4, CD8, CD11c and CD45A antigens showed monophasic or biphasic density increases. The densities of cells expressing these antigens dropped to low levels during the successful treatment and recovery periods. The infiltrates of reinfested hosts again showed progressively increasing densities of cells expressing these antigens and then a decline as the infestation spontaneously cured due to adaptive immunity. Following reinfestation the densities of cells expressing these surface antigens again increased. No cells expressing CD21 antigens were evident in the epidermal infiltrate during sensitization and challenge. In contrast to the epidermis, some cells expressing CD21 were apparent in the deep dermis. Density of the infiltrates increased earlier during the challenge phase than during the sensitization phase of the infestation.

By comparing the relative densities and rates of progression of cells with specific markers in the epidermal and dermal infiltrates, several important conclusions or inferences concerning the immune response mechanism were made. Few B lymphocytes migrated into the scabietic lesions. Langerhans cells were actively involved in the immune response. The infiltrates contained numerous T lymphocytes of both the $CD4^+$ and $CD8^+$ T-cell subtypes. There was a strong representation of memory T cells in the infiltrates.

## SERUM ANTIBODY

Several studies have determined total serum antibody isotype concentrations during infestation and following successful treatment for human scabies. Concentrations of circulating antibody isotypes specific for *S. scabiei* var. *hominis* antigens have not been determined in humans. However, circulating antibody isotypes recognizing var. *canis* and var. *suis* antigens have been determined in a few cases for humans and in other animals with scabies.

These studies provide indirect evidence that scabies infestations are associated with a humoral immunological response.

In one of the first studies of the humoral response to scabies, Hancock and Ward (1974) found that total serum IgE levels were normal in 100 patients with scabies: 64 and 36 of the patients had been itching for either more than or less than 3 weeks, respectively. Serum IgA levels for both groups were significantly lower than the normal mean value, and it was suggested that this may be correlated with low secretory IgA in skin secretions that may predispose individuals to scabies infestation. There was no correlation between the severity of scabies infestation and immunoglobulin isotype variation or degree of sensitization. IgG and IgM were significantly higher than controls with no difference between the two groups.

A series of studies by Falk and colleagues indirectly indicated that scabies induced a circulating IgE response in many patients in addition to the IgM and IgG responses already reported by Hancock and Ward. Falk (1980) determined total serum antibody isotype levels for 154 patients with scabies (ages 11–76 years). These patients had been exhibiting pruritis and skin rashes for 1 week to 12 months before diagnosis and none had had scabies previously. Of 94 patients 39 had increased total serum IgE concentrations. The study also confirmed the IgA, IgG and IgM results of Hancock and Ward (1974). Serum IgA concentrations for all 154 patients were lower during infestation than at 6 weeks and 9 months following successful treatment. In contrast, serum IgG and IgM concentrations during infestation were significantly higher than at 6 weeks and 9 months following successful treatment which indicated the humoral response had waned. Serum complement C3 and C4 concentrations were the same both during and following treatment.

In another study Falk and Bolle (1980b) found that of 135 patients with scabies, 45% had elevated total serum IgE concentrations while only 23% of the 120 patients evaluated by questionnaire had atopic disease. In a subsequent study of 120 patients with scabies, Falk (1981) found that total serum IgE concentrations of 80 non-atopic and 29 of 40 atopic patients were significantly lower 12 months after successful treatment for scabies than during their infestations. In a related study, Falk and Eide (1981) found that the severity of skin rashes was directly correlated with total serum IgE concentrations in patients with scabies. Patients with severe or moderate rashes had significantly greater IgE levels than patients with moderate or mild rashes.

Like previous studies, Nassef et al. (1991) also found that total serum IgA was significantly decreased and total serum IgG and IgM were significantly elevated in 40 scabietic patients who presented at an outpatient dermatology clinic. However, mean total serum IgE, C3 and C4 in these patients were not significantly different from those of a control group of non-scabietic individuals.

Recently Morsy et al. (1993) determined serum immunoglobulin and complement (C3 and C4) concentrations in 16 patients with typical scabies. Mean total IgG, IgM and IgE concentrations were significantly greater for the patients with active scabies as compared to the levels observed 3 months after treatment and to the controls. IgA levels were significantly decreased before treatment but increased to levels not significantly different from controls after treatment. C3 and C4 levels were not significantly different from controls and did not change significantly after treatment.

Finally, Van Neste and Salmon (1978) determined serum C1q binding activity in 29 patients with scabies and found increased circulating immune complexes in the treated patients. Interestingly, there were no differences in the levels of circulating complexes between untreated scabies patients and controls. Salo et al. (1981) also found no circulating immune complexes in an examination of 18 patients with scabies.

Two studies of HLA typing found significantly increased expression of HLA-A antigens (human leukocyte antigen) in scabietic patients (Falk and Thorsby, 1981; Morsy et al., 1990).

Because sufficient amounts of S. scabiei var. hominis material from humans cannot be obtained, S. scabiei from other animals have been used in immunological studies. Three studies have detected scabies-specific IgE antibodies in the sera of scabietic patients by using an extract prepared from scabies mites from other animals (Rantanen et al., 1981; Dahl et al., 1985; Arlian et al., 1996a). Results of these studies provide direct evidence that antigens of S. scabiei can induce IgE antibodies in individuals with and without an allergic history. Rantanen et al. (1981) examined scabies-specific IgE in 16 patients with scabies, 50% of whom had elevated total serum IgE levels. Two (both with elevated total IgE) of five sera examined had IgE-antibody-specific for S. scabiei var. suis. Only 1 patient of the 16 had a history of atopy. The extract of S. scabiei var. suis for these immune assays was made from ground mites harvested from infested pig ears. The extract had a soluble protein content of 176 μg ml$^{-1}$.

Dahl et al. (1985) found that 6 of 20 patients infested with S. scabiei var. hominis had a RAST score of at least two indicating the presence of circulating IgE specific for S. scabiei var. suis. The porcine-specific IgE levels were not significant 2 to 4 weeks after treatment. The extract used was prepared by washing 2000 mites (not ground) for 2 hours in 4 ml of 0.9% NaCl. The protein content was about 130 μg ml$^{-1}$. This extract was unique in that it contained surface antigens (washed whole bodies), while Rantanen et al., (1981) used an extract that contained internal proteins (crushed mites) as well. Likewise, Arlian et al., (1996a) and Bornstein and Zakrisson (1993a, b) have shown that soluble proteins in extracts produced from one scabies variety can be detected using serum antibodies from hosts infested with a different variety. For example, serum antibodies built to either var. canis or var. hominis will recognize antigenic deteminants on soluble proteins in extracts made from var. suis and var. canis (Arlian et al., 1996a). Likewise, antibodies produced to var. *canis* will recognize soluble proteins in an extract made from the bodies of var. *suis*. Both homologous and heterologous CIE analyses of extracts of var. *canis* and var. *suis*, reacted with anti-var. *canis* serum, gave similar antigen–antibody precipitin profiles. This indicated the extracts contained proteins of similar electrophoretic mobility. Serum from a patient with scabies had circulating IgE that bound to antigens in the extracts of both var. *canis* and var. *suis*. Tandem CIE reactions of var. *canis* and var. *suis* extracts with anti-var. *canis* serum resulted in antigen–antibody precipitin peaks with both fused and partially fused legs. This was evidence that the two different mite extracts contained identical (homologous) antigenic molecules and some different molecules that possessed both shared and unique epitopes (partial homology). Using an extract of *S. scabiei* var. *vulpes*, serum antibodies were detectable 5 to 7 weeks after experimental infestation of pigs with *S. scabiei* var. *suis* (Bornstein and Zakrisson, 1993b). The results of these studies clearly showed that various animal and human varieties of scabies mites produce some common antigens/epitopes and that scabies mites of one variety can be used for immunologic studies of different hosts.

Although many of these studies did not directly measure scabies-specific serum antibodies (IgA, IgE, IgG and IgM), collectively the data from the various human studies consistently suggested:

**1.** Scabies antigens induce an IgM and IgG response in most cases and these are independent of an IgE response.
**2.** There may or may not be an IgE (allergic) response to scabies.
**3.** Low serum IgA is associated with scabies infestation but whether or not this correlates with low secretory IgA in skin secretions and predisposition to scabies, as suggested, remains to be determined.
**4.** Circulating IgG or IgM antibody–antigen complexes have not been found and these may not form or concentrations may be low due to deposition in scabietic lesions and other secondary sites.
**5.** Serum C3 and C4 concentrations are not increased during infestation.

## IMMUNOGLOBULIN AND COMPLEMENT DEPOSITS IN SKIN LESIONS

Several studies have detected immunoglobulin and complement deposits in skin lesions of patients with scabies. Van Neste and Lachapelle (1981) reported band-like deposits of C3 in the dermal–epidermal junction of three patients with crusted scabies. Likewise, Frentz *et al*. (1977) found that biopsy specimens from 2 of 11 patients with scabies had C3 and/or IgM deposits along the basal membranes in scabietic lesions; 4 of the 11 patients also had IgE deposits in the vessel walls of their scabietic lesions.

Using direct immunofluorescence, Hoefling and Schroeter (1980) determined the presence of IgM, IgA and C3 in the scabietic lesions of four consecutive patients. Three patients had C3 and one had IgM at the dermal–epidermal junction while three had IgM and C3 in dermal vessels. Two patients had IgG and one had IgA deposits in the stratum corneum. In one patient, C3 deposition was found on a mite in the burrow.

Salo et al. (1981) studied the skin biopsies taken from 18 patients who had had scabies for an average of 3 months. Six patients had nodular, one had crusted and the remaining had papulovesicular scabies. Complement C3 deposits were found in 13 of the 18 patients, 6 of whom had nodular scabies. Twelve and three patient had C3 deposits in the vessel walls and the basement membranes, respectively. Three patients had IgM and two had IgA deposits in the vessel walls. No circulating immune complexes were found but IgM antibodies (RF type) were detected in the sera of 5 of 15 patients.

The presence of C3 deposits indicates that the complement system functions during scabies infestation. From these data it is unclear what role the classical and/or alternative complement cascades play in the immune response to scabies mites. Immune-complex-mediated reactions (classical pathway) involve formation of circulating antigen-antibody complexes which deposit in tissues such as skin. Once deposited, these complexes activate complement which contributes to the inflammatory response (e.g. attraction of neutrophils and eosinophils). The absence of circulating immune complexes and elevated levels of C3 and C4 is puzzling if the classical pathway functions unless there is a rapid clearance of the circulating complexes and C3. Presence of C3 deposits and antibodies in the tissues near the scabietic lesions suggests the alternative pathway is activated during scabies infestations. Finding circulating antigen–antibody complexes after treatment for scabies (Van Neste and Salmon, 1978) suggests the classical pathway is activated later.

## PROTECTIVE IMMUNITY

Resistance (reduced mite burden) to reinfestation with scabies following an initial infestation has been demonstrated for both humans and other animals. Mellanby (1944) observed that humans showed reduced mite burdens during second scabies infestations when compared to the initial infestation.

Bornstein and Zakrisson (1993a) experimentally infested six dogs with a low dose of *S. scabiei* var. *vulpes* from foxes. Seven weeks after spontaneous clearing the dogs were reinfested. Serum IgG (specific for var. *vulpes*) titres for the dogs were similar during both the primary and secondary infestations. During the primary infestations IgG titres peaked at 5–7 weeks then waned as the infestations spontaneously cleared. However, reinfested dogs that showed clinical symptoms of sarcoptic mange developed more rapid (1–2 weeks) and higher circulating IgG levels during their second infestations compared to their first infestations. Reinfested dogs that showed little or no clinical signs of sarcoptic mange had significantly lower antibody titres compared to the dogs with pronounced clinical symptoms. Although not indicated by the authors, these results suggest that protective immunity may be associated with a down-regulated humoral response.

Studies by Arlian et al. (1994a, b, 1996b) showed that reinfested rabbits and dogs also expressed resistance to scabies mites and developed much lower mite levels than were developed during their previous scabies infestation. Rabbits experimentally infested with S. scabiei var. canis developed elevated circulating scabies-specific total antibody titres as measured by ELISA and CIE during their primary (sensitizing) infestations and again during their second and third infestations that were 20 (challenge I) and 65 (challenge II) days, respectively, following successful treatment of their previous infestations. The antibody titres followed the classical pattern for primary and secondary infestations with faster and higher titres developing during the second and third infestations. Concentrations of serum complement C3 paralleled the changes in circulating antibodies. This finding for rabbits is in contrast to the finding that serum C3 and C4 levels were normal in humans infested with scabies (Falk, 1980; Nassef et al., 1991; Morsy et al., 1993). Both antibody and C3 levels dropped by more than 60% during the 65 days following each successful treatment. Of significance was that after the first infestation 65% of the hosts exhibited protective immunity to reinfestation with scabies. The hosts with acquired immunity had significantly lower antibody titres during both challenges compared to the hosts that had no protective immunity. The sequential skin cell infiltrates in the vicinity of the scabies mites were determined in parallel to monitoring serum antibody. Both the resistant and non-resistant hosts exhibited similar cellular infiltrates that consisted primarily of neutrophils, plasma cells, macrophages and mononuclear cells. However, upon attempted reinfestation the resistant hosts developed faster and more dense neutrophil infiltrates than did the non-resistant hosts.

As discussed earlier, extracts of house dust mites and scabies mites contain many cross-reacting antigens (Arlian et al. 1988b, 1991). This raised the possibility that relevant antigens in a dust mite extract might be used to immunize against scabies. Therefore, a study was conducted in which rabbit hosts were first immunized with a mixed house dust mite extract (50:50 DF/DP) (Arlian et al., 1995). Attempts were then made to infest the immunized hosts with S. scabiei var. canis: 71% of the immunized hosts exhibited induced protective immunity to infestation with S. scabiei var. canis. As in the earlier study the resistant hosts exhibited lower circulating antibody levels and generally stronger cell-mediated responses compared to non-resistant hosts.

In a third study, the induction of protective immunity was investigated using a host and its natural parasite (Arlian et al., 1996b). We demonstrated that 88% of dogs previously infested with *S. scabiei* var. *canis* and then cured expressed complete protective immunity to reinfestation with scabies. Eight dogs that were experimentally infested developed progressive heavy scabies infestations that required treatment for cure. All of the dogs were successfully treated for scabies at 57 or 64 days following the initial infestation. Twenty-seven days following treatment all dogs were free of scabies lesions and skin scrapes were negative. Six weeks following treatment the eight dogs were again infested as were two control dogs that were not previously infested. The two control dogs exhibited progressing infestations and positive skin scrapes for the 64-day study period. Seven of the eight reinfested dogs spontaneously cured and weekly skin scrapes were negative by 64 days. In four skin scrapes only one mite was recovered from the eighth dog indicating it had acquired resistance. Clinical symptoms during the challenge peaked at 24 days then waned in parallel with the disappearance of the mites. Sequentially collected biopsies of skin containing scabies lesions showed that during both the initial infestation and subsequent challenge, dogs developed mixed cellular infiltrates that contained mononuclear cells, neutrophils, plasma cells and mast cells. The densities of the infiltrates developed more rapidly during reinfestation than for the sensitizing first infestation. During the challenge mononuclear cells and neutrophils were more numerous but during sensitization mononuclear and mast cells were more numerous. The circulating scabies-specific antibody responses were also more rapid during the challenge. The cell-mediated and circulating antibody responses waned in parallel with the spontaneously declining mite numbers and the clearing of the infestations. The role played by each infiltrating cell type in the resistance mechanism remains to be determined. The prominence of neutrophils in the infiltrates suggests that the oxidative burst by these cells may be important in eliminating the mites. The initial infestation induced strong circulating scabies-specific total Ig(Fc), IgG1, IgG2 and IgM responses by day 16 that returned to normal levels following successful treatment for scabies. A similar but more rapid (detected at 8 days) and less intense response developed during the reinfestation. The weaker circulating antibody responses exhibited by resistant hosts were consistent with those reported in our rabbit studies.

The data from our three studies suggest that: (i) primary infestation induces protective immunity; (ii) vaccination against scabies is possible; (iii) the cell-mediated response is the primary mechanism responsible for protective immunity; and (iv) the $T_H1$ system (cell mediated) is more up-regulated in protected hosts while the $T_H2$ system (humoral) is more up-regulated in non-resistant hosts. The mechanisms for regulation of the $T_H1$ and $T_H2$ systems are not known at this time.

## TRANSFER OF ANTIBODIES IN COLOSTRUM

Antibodies to scabies in infested hosts may be passed through the colostrum to suckling young. Using ELISA and an antigenic extract prepared from *S. scabiei* var. *vulpes*, Bornstein and Zakrisson (1993b) found that the sera of neonatal pigs born to a sow chronically infested with *S. scabiei* var. *suis* contained scabies-specific antibodies that decreased to relatively low levels when the piglets were about 3 weeks old.

## EPIDERMAL KERATINOCYTES

Keratinocytes are capable of producing and secreting a broad array of imunomodulatory cytokines that include IL-1 IL-1RA, IL-3, IL-6, IL-7, IL-8, IL-10, IL-12, GM-CSF, M-CSF, TNF-$\alpha$, TGF-$\alpha$ and $\beta$, PDGF, bFGF, prostaglandins, leukotrienes, *cis*-urocannic acid and MCGF (Stingl *et al.*, 1993). Normally constitutive production is very low but productions may be up-regulated by specific stimuli (both physical and chemical). These mediators may also affect the function of other cells such as Langerhans cells. Keratinocytes may functionally interact with lymphocytes ($CD4^+$ and $CD8^+$). Therefore, keratinocytes may be important components of the immunological unit (skin-associated lymphoid tissue – SALT) that responds to scabies mites.

In preliminary studies we have begun to investigate the role of keratinocytes in scabies infestations using human skin organotypic culture (human skin equivalents). Reconstructed epidermis was obtained by culturing keratinocytes seeded on to a dermal substrate and then raised to the air–medium interface. First, collagen matrices (dermal substrate) were prepared by combining fibroblasts (from the explants of human cadaver skin) with rat tail collagen (Bell *et al.*, 1981, 1983). Dissociated human epidermal cells from a second biopsy specimen were applied directly to the surface of the contracted collagen matrix after it was raised to the surface of the culture medium that was supplemented with epidermal growth factor and hydrocortisone. A bilayered human skin equivalent formed with a fully cornified epidermis and a well organized basement membrane with associated hemidesmosomes. The differentiation of these human skin equivalents is morphologically and biochemically similar to human skin (Harriger and Hull, 1992). Large numbers of all life stages of *Sarcoptes scabiei* var. *canis* that were placed on the surface of the skin equivalents burrowed into the stratum corneum. ELISA assays of the culture medium revealed that the burrowed scabies stimulated production and release of IL-1-$\beta$ from the keratinocytes in the epidermis. Scabies mites plus the addition of IL-6 ($10 \, \text{ng ml}^{-1}$) or HGF ($3 \, \text{ng ml}^{-1}$) to the culture medium stimulated the release of even greater amounts of IL-1-$\beta$. Adding IL-6 or HGF alone to the culture medium (without scabies infestations) did not stimulate release of IL-1-$\beta$ by the skin equivalents.

## MODEL FOR IMMEDIATE, DELAYED HYPERSENSITIVITY AND LATE PHASE REACTIONS

The published data clearly demonstrate that antigens from scabies mites produce delayed hypersensitivity reactions in both atopic (allergic) and non-atopic humans. Presumably these individuals are predisposed to react to antigens from scabies but as with most diseases the sensitivity and severity of the reaction vary between individuals. The hallmarks of delayed hypersensitivity reactions are that they are mediated by antigen-specific receptors on the surface of T cells (in particular $CD4^+$), develop gradually (hours), are not mediated by IgE, lack tissue specificity, require T cells and their products, are antibody independent, and are characterized by cell infiltrates dominated by mononuclear cells.

In addition, reports clearly indicate that scabies antigens will induce IgE-mediated allergic reactions (immediate and late phase) in some individuals but not in others. If an individual is predisposed to an IgE response, development of the IgE response (both circulating IgE and skin reactivity) is a function of the duration and severity of the infestation. The data show that individuals may give a positive skin test but not exhibit detectable levels of circulating IgE-type antibodies which is similar to reports for other allergies (e.g. dust mites). Immediate hypersensitivity reactions to the skin injection of antigens appears within 10–20 minutes, neutrophils and eosinophils migrate quickly into the affected tissue and mononuclear cells appear late in the reaction. Late phase allergic reactions, in most cases (66–85%), are extensions of immediate reactions (Lemanske and Kaliner, 1993). The clinical signs and histopathology of allergic reactions in patients with scabies are consistent with those of the late phase reaction. Skin testing may result in isolated immediate hypersensitivity, isolated late phase and dual reactions which are consistent with reactions to other allergens.

The clinical signs of scabies are not evident for a month or more the first time a patient becomes infested with scabies. Reinfestation while the host is still sensitized results in clinical signs within 48 hours. The response times for both initial and subsequent exposures are likely a function of the time it takes for the antigen once released into the stratum corneum to reach effector cells in the lower epidermis (e.g. Langerhans cells and keratinocytes) and/or dermis (e.g. macrophages and dermal dendritic cells) in order to initiate an immune response. The details of the delayed hypersensitivity response mechanism are yet unknown. However, it is assumed that the response conforms to the generally understood mechanisms for delayed hypersensitivity T-cell immune reactions. Most of the limited information from the research on scabies mites, or on the immunologically related house dust mites, suggests this is true.

Presumably antigen-presenting cells (APCs) in the epidermis (Langerhans cells) and/or dermis (macrophages) internalize an antigen, process it, then re-express it at the cell surface in association with class II MHC molecules. There is evidence that activities of keratinocytes are directly or indirectly up-regulated by scabies antigens (Arlian, unpublished). In scabies infestations cytokines from up-regulated keratinocytes may play a role in activating APCs by mechanisms reported by Heufler et al. (1987), Witmer-Pack et al. (1987), and Trefzer et al. (1993). These activated APCs migrate through the lymph vessels to regional lymph nodes and possibly through the circulation to the spleen where they interact with T lymphocytes in the medullary region of the node or white pulp of the spleen. This interaction triggers proliferation and differentiation of epitope-specific T cells and B cells. Sensitized T cells in lymphoid organs are activated to release antigen-specific T-cell factor (TCF) that passes into the circulation and then becomes extravascular in dermal tissue. TCF may bind to receptors on mast cells and other cells (Langerhans cells, macrophages, keratinocytes), sensitize them and up-regulate their function. The offending antigen may bind to antigen-specific T-cell factors (e.g. IgE) on the surface of mast cells, macrophages and other cells that causes the cells to release serotonin, IL-1, and other chemical mediators. Keratinocytes may be activated to release mediators as well. These mediators bind to receptors on local endothelial cells that respond and open gaps between the junctions and up-regulate the expressions of ICAM-1 (intercellular adhesion molecule-1), VCAM-1 (vascular cell adhesion molecule) and ELAM (E-selectin) which in turn recruit and produce adhesion of lymphocytes (T and B cells), neutrophils, eosinophils and basophils. Specific-antigen-sensitized T cells ($CD4^+$) return from the lymphoid tissue and pass through these gaps into the dermis near scabies mites and infiltrate the area. The vast majority of T cells infiltrating the dermis probably express the cutaneous lymphocytes-associated antigen (CLA), a skin specific homing receptor (Picker et al., 1990). There is evidence in scabies infestations that local lymphocyte blastogenesis may also occur in the perivascular area of the dermis (Van Neste, 1992). In the dermis, these migrated or blastogenic T cells interact again with APCs (macrophages/Langerhans cells) exhibiting antigen–class II MHC complexes on the surface that activate T-cell subsets ($T_H1$ and $T_H2$) to produce and release various cytokines. These cytokines regulate the balance between the $T_H1$ and $T_H2$ subsets and they are chemotactic for non-specific inflammatory cells such as neutrophils, eosinophils, B cells, plasma cells, some T cells (e.g. $CD8^+$), NK cells and others. Activated $T_H2$ cells secrete cytokines (IL-4, IL-5, IL-6, IL-10) that promote B-cell responses and down-regulate $T_H1$ activity (IL-4, IL-10) (Ezzell, 1993). Activated $T_H1$ cells secrete cytokines that up-regulate a cell-mediated response (IL-2) and down-regulate $T_H2$ responses ($IFN_\gamma$) (Ezzell, 1993). The stimulus that accounts for the balance in expression of $T_H1$ and $T_H2$ cells remains to be identified. The balance between the activated $T_H1$ and $T_H2$ cells may be responsible for the immune-based resistance to scabies that has been observed in sensitized hosts (Arlian et al., 1994b, 1995). Hosts resistant to scabies exhibit stronger cell-mediated and weaker humoral responses than hosts that express no resistance. Research with CD4$^+$ cells specific for immunologically related house dust mites also provide indirect evidence that this balance may be important (Shearer and Huston, 1993). *In vitro* clones of CD4$^+$ T cells specific for house dust mite allergens from atopic persons secreted IL-4 but not IFN-γ and induced IgE production. In contrast, clones from non-atopic persons secreted IFN-γ but little IL-4 and suppressed *in vitro* IgE production. Upon interaction with $T_H2$ helper cells and antigen, plasma cells proliferate and differentiate in the cortical regions of the lymph node and white pulp of the spleen and produce antibodies. Antibodies released in the secondary lymphoid tissue, or released after the plasma cells migrate into the dermal lesion, account for the circulating and tissue-deposited antibodies, respectively, that have been associated with scabies infestations. Complement activation, particularly in the tissues, seems to be involved in scabies infestations but it is not clear what route of activation may be involved (classical or alternative pathway). Future research is required to verify many aspects of this general model and to provide additional details of the immune response mechanisms.

## ACKNOWLEDGEMENTS

Some research reported in this chapter was supported by grant RO1 AI 17252 from the National Institutes of Allergy and Infectious Disease, National Institutes of Health, Bethesda, Maryland.

Special thanks to Jacqui Dippold, Chris Rapp, DiAnn Vyszenski-Moher and Marge Morgan for assistance during preparation of this chapter.

## REFERENCES

Ackerman, A.B. (1977) Histopathology of human scabies. In: Orkin, M., Maibach, H.I., Parish, L.C. and Schwartzman, R.M. (eds), *International Conference on Scabies, 1976 Scabies and Pediculosis*. Lippincot, Philadelphia, pp. 88–95.

Arlian, L.G., Runyan, R.A., Achar, S. and Estes, S.A. (1984a) Survival and infestivity of *Sarcoptes scabiei* var. *canis* and var. *hominis*. *Journal of the American Academy of Dermatology* 2, 210–215.

Arlian, L.G., Runyan, R.A. and Estes, S.A. (1984b) Cross infestivity of *Sarcoptes scabiei*. *Journal of the American Academy of Dermatology* 10, 979–986.

Arlian, L.G., Runyan, R.A., Sorlie, L.B., Vyszenski-Moher, D.L. and Estes, S.A. (1985) Characterization of *Sarcoptes scabiei* var. *canis* (Acari: Sarcoptidae) antigens and induced antibodies in rabbits. *Journal of Medical Entomology* 22, 321–323.

Arlian, L.G., Vyszenski-Moher, D.L. and Cordova, D. (1988a) Host specificity of *S. scabiei* var. *canis* (Acari: Sarcoptidae) and the role of host odor. *Entomological Society of America* 25(1), 52–56.

Arlian, L.G., Vyszenski-Moher, D.L. and Gilmore, A.M. (1988b) Cross-antigenicity between *Sarcoptes scabiei* and the house dust mite, *Dermatophagoides farinae* (Acari: Sarcoptidae and Pyroglyphidae). *Journal of Medical Entomology* 25, 240–247.

Arlian, L.G., Vyszenski-Moher, D.L., Ahmed, S.G. and Estes, S.A. (1991) Cross-antigenicity between the scabies mite, *Sarcoptes scabiei*, and the house dust mite, *Dermatophagoides pteronyssinus*. *Journal of Investigative Dermatology* 96(3), 350–354.

Arlian, L.G., Bernstein, D., Bernstein, I.L., Friedman, S., Grant, A., Lieberman, P., Lopez, M., Metzer, J., Platts-Mills, T., Schatz, M., Spector, S., Wasserman, S.I. and Zeiger, R.S. (1992) Prevalence of dust mites in the homes of people with asthma living in eight different geographical areas of the United States. *Journal of Allergy and Clinical Immunology* 90, 292–300.

Arlian, L.G., Morgan, M.S., Vyszenski-Moher, D.L. and Stemmer, B.L. (1994a) *Sarcoptes scabiei*: the circulating antibody response and induced immunity to scabies. *Experimental Parasitology* 78, 37–50.

Arlian, L.G., Rapp, C.M., Vyszenski-Moher, D.L. and Morgan, M.S. (1994b) *Sarcoptes scabiei*: histopathological changes associated with acquisition and expression of host immunity to scabies. *Experimental Parasitology* 78, 51–63.

Arlian, L.G., Rapp, C.M. and Morgan, M.S. (1995) Resistance and immune response in scabies-infested hosts immunized with *Dermatophagoides* mites. *American Journal of Tropical Medicine and Hygiene* 52, 539–545.

Arlian, L.G., Morgan, M.S. and Arends, J.J. (1996a) Immunologic cross-reactivity among various strains of *Sarcoptes scabiei*. *Journal of Parasitology* 82, 66–72.

Arlian, L.G., Morgan, M.S., Rapp, C.M. and Vyszenski-Moher, D.L. (1996b) The development of protective immunity in canine scabies. *Veterinary Parasitology* 62, 133–140.

Bell, E., Ehrlich, H.P., Sher, S., Merrill, C., Sarber, B., Hull, B., Nakatsuji, T., Church, D. and Buttle, D. (1981) Development and use of a living skin equivalent. *Plastic and Reconstructive Surgery* 67, 386–392.

Bell, E., Sher, S., Hull, B., Merrill, C., Chamson, A., Asselineau, D., Dubertret, L., Lapiere, L., Nusgens, B. and Neveux, Y. (1983) The reconstruction of living skin. *Journal of Investigative Dermatology* 81(Suppl. 1), 2s–10s.

Benjamini, E., Feingold, B.F. and Kartman, L. (1960) Allergy to flea bites. III. The experimental induction of flea bite sensitivity in guinea pigs by exposure to flea bites and by antigen prepared from whole flea extracts of *Ctenocephalides felis felis*. *Experimental Parasitology* 10, 214–222.

Bornstein, S. and Zakrisson, G. (1993a) Humoral antibody response to experimental *Sarcoptes scabiei* var. *vulpes* infection in the dog. *Veterinary Dermatology* 4, 107–110.

Bornstein, S. and Zakrisson, G. (1993b) Clinical picture and antibody response in pigs infected by *Sarcoptes scabiei* var. *suis*. *Veterinary Dermatology* 4, 123–131.

Cabrera, R., Agar, A. and Dahl, M.V. (1993) The immunology of scabies. *Seminars in Dermatology* 12, 15–21.

Dahl, J.C., Schwartz, B., Graudal, C., Christophersen, J. and Henrisken, S.A. (1985) Serum IgE antibodies to the scabies mite. *International Journal of Dermatology* 24, 313–315.

Davis, D.P. and Moon, R.D. (1990a) Dynamics of swine mange: a critical review of the literature. *Journal of Medical Entomology* 27, 727–737.

Davis, D. and Moon, R. (1990b) Density of the itch mite *Sarcoptes scabiei* (Acari:

Sarcoptidae) and temporal development of cutaneous hypersensitivity in swine mange. *Veterinary Parasitology* 36, 285–293.

Dobson, K.J. (1981) External parasites. In: Leman, A.D. *et al.* (eds) *Diseases of Swine*. Iowa State University Press, Ames, Iowa, pp. 579–589.

Ezzell, C. (1993) AIDS' unlucky strike: when T cells would rather switch than fight. *The Journal of NIH Research* 5, 59–64.

Fain, A. (1968) Etude de la variabilite de *Sarcoptes scabiei* avec une revision des sarcoptidae. *Acta Zoologica et Pathologica Antverprensia*, Antwerp.

Falk, E.S. (1980) Serum immunoglobulin values in patients with scabies. *British Journal of Dermatology* 102, 57–61.

Falk, E.S. (1981) Serum IgE before and after treatment for scabies. *Allergy* 36, 167–174.

Falk, E.S. and Bolle, R. (1980a) IgE antibodies to house dust mite in patients with scabies. *British Journal of Dermatology* 102, 283–288.

Falk, E.S. and Bolle, R. (1980b) *In vivo* demonstration of specific immunological hypersensitivity to scabies mite. *British Journal of Dermatology* 103, 367–373.

Falk, E.S. and Eide, T.J. (1981) Histologic and clinical findings in human scabies. *International Journal of Dermatology* 20, 600–605.

Falk, E.S. and Matre, R. (1982) *In situ* characterization of cell infiltrates in the dermis of human scabies. *American Journal of Dermatopathology* 4, 9–15.

Falk, E.S. and Thorsby, E. (1981) HLA antigens in patients with scabies. *British Journal of Dermatology* 104, 317–320.

Falk, E.S., Dale, S., Bolle, R. and Haneberg, B. (1981) Antigens common to scabies and house dust mites. *Allergy* 36, 233–238.

Frentz, G., Veien, N.K. and Eriksen, K. (1977) Immunofluorescence studies in scabies. *Journal of Cutaneous Pathology* 4, 191–193.

Gaspari, A.A., Jenkins, M.K. and Katz, S.I. (1988) Class II MHC-bearing keratinocytes induce antigen-specific unresponsiveness in hapten-specific TH1 clones. *Journal of Immunology* 141, 2216.

Hancock, B.W. and Ward, M. (1974) Serum immunoglobulin in scabies. *The Journal of Investigative Dermatology* 63, 482–484.

Harriger, M.D. and Hull, B.E. (1992) Cornification and basement membrane formation in a bilayered human skin equivalent maintained at an air–liquid interface. *Journal of Burn Care and Rehabilitation* 13, 187–193.

Head, E.S., MacDonald, E.M., Ewert, A. and Apisarnthanarax, P. (1990) *Sarcoptes scabiei* in histopathologic sections of skin in human scabies. *Archives of Dermatology* 126, 1475–1477.

Heilesen, B. (1946) *Studies on Acarus scabiei and Scabies*. Rosenkilde and Bagger, Copenhagen.

Hejazi, N. and Mehregan, A.H. (1975) Scabies–historical study of inflammatory lesions. *Archives of Dermatology* 111, 37–39.

Heufler, C., Koch, F. and Schuler, G. (1987) Granulocyte–macrophage colony-stimulating factor and interleukin-1 mediate the maturation of murine epidermal Langerhans cells into potent immunostimulatory dendritic cells. *Journal of Experimental Medicine* 167, 700.

Hoefling, K.K. and Schroeter, A.L. (1980) Dermatoimmunopathology of scabies. *American Academy of Dermatology* 3, 237–240.

Kitchevatz, M. and Pochtich, Z. (1934) Reaction allergiques cutanees dans la gale. *Bulletin Soc Fr Dermatol Syphilogr* 41, 989.

Konstantinov, D. and Stanoeva, L. (1973) Persistent scabious nodules, *Dermatologica* 147, 321–327.
Lemanske, R.F. and Kaliner, M.A. (1993) Late phase allergic reactions. In: Middleton, E., Reed, C.E., Ellis, E.F., Adkinson, N.F., Yunginger, J.W. and Busse, W.W. (eds), *Allergy Principles and Practice* 4th edn. Mosby, St Louis, pp. 321–361.
McKiel, J.A. and West, A.S. (1961) Effects of repeated exposure of hypersensitive humans and laboratory rabbits to mosquito antigens. *Canadian Journal of Zoology* 39, 597–603.
Mellanby, K. (1944) The development of symptoms, parasitic infection and immunity in human scabies. *Parasitology* 35, 197–206.
Morgan, M.S. and Arlian, L.G. (1994) Serum antibody profiles of *Sarcoptes scabiei* infested or immunized rabbits. *Folia Parasitologica* 41, 223–227.
Morsy, G.H. and Gaafar, S.M. (1989) Responses of immunoglobulin-secreting cells in the skin of pigs during *Sarcoptes scabiei* infestation. *Veterinary Parasitology* 33, 165–175.
Morsy, T.A., Romia, S.A., Al-Ganayni, G.A., Abu-Zakham, A.A., Al-Shazly, A.M. and Rezk, R.A. (1990) Histocompatibility (HLA) antigens in Egyptians with two parasitic skin diseases (scabies and leishmaniasis). *Journal of Egyptian Society of Parasitology* 20, 565–572.
Morsy, T.A., Kenawi, M.Z., Zohdy, H.A., Abdalla, K.F. and Fakahany, A.F.E. (1993) Serum immunoglobulin and complement values in scabietic patients. *Journal of Egyptian Society of Parasitology* 23, 221–228.
Nassef, N.E., Makled, K.M., Elzayat, E.A. and Sanad, M.M. (1991) Humoral and cell mediated immune responses in scabietic patients. *Journal of Egyptian Society of Parasitology* 21, 765–770.
Picker, L.J., Terstappen, L.W.M.M., Rostt, L.S., Streeter, P.R., Stein, H. and Butcher, E.C. (1990) Differential expression of homin-associated adhesion molecules by T cell subsets in man. *Journal of Immunology* 145, 3247.
Polak, L. and Rinck, C. (1978) Mechanism of desensitization in DNCB-contact sensitive guinea pigs. *Journal of Investigative Dermatology* 70, 98.
Prakken, J.R. and Van Vloten, T.J. (1949) Allergy in scabies. *Citation* 99, 124–131.
Rantanen, T., Björkstén, F., Reunala, T. and Salo, O.P. (1981) Serum IgE antibodies to scabies mite. *Acta Dermatovener* 61, 358–360.
Reunala, T., Ranki, A., Rantanen, T. and Salo, O.P. (1984) Inflammatory cells in skin lesions of scabies. *Clinical and Experimental Dermatology* 9, 70–77.
Salo, O.P., Renunala, T., Kalimo, K. and Rantanen, T. (1981) Immunoglobulin and complement deposits in the skin and circulating immune complexes in scabies. *Acta Dermatovener* 62, 73–76.
Sheahan, B.J. (1974) Experimental *Sarcoptes scabiei* infection in pigs: clinical signs and significance of infection. *The Veterinary Record* 94, 202–209.
Sheahan, B.J. (1975a) Pathology of *Sarcoptes scabiei* infection in pigs. I. Naturally occurring and experimentally induced lesions. *Journal of Comparative Pathology* 85, 87–95.
Sheahan, B.J. (1975b) Pathology of *Sarcoptes scabiei* infection in pigs. II. Histological, histochemical and ultrastructural changes at skin test sites. *Journal of Comparative Pathology* 85, 97–110.
Shearer, W.T. and Huston, D.P. (1993) The immune system, an overview. In: Middleton, E., Reed, C.E., Ellis, E.F., Adkinson, N.F., Yunginger, J.W. and Busse,

W.W. (eds), *Allergy Principles and Practice*, 4th edn. Mosby, St Louis, pp. 3–22.

Stingl, G., Hauser, C. and Wolff, K. (1993) The epidermis: an immunologic microenvironment. In: Fitzpatrick, T.B., Eisen, A.Z., Wolff, K., Freedberg, I.M. and Austen, K.F. (eds), *Dermatology in General Medicine*, 4th edn. McGraw-Hill, Inc., New York, pp. 172–197.

Trefzer, U., Brockhaus, M., Lötscher, H., Parlow, F., Budnik, A., Grewe, M., Christoph, H., Schöpf E., Luger, T.A. and Krutmann, J. (1993) The 55-kd tumor necrosis factor receptor on human keratinocytes is regulated by tumor necrosis factor alpha and by ultraviolet B radiation. *Journal of Clinical Investigations* 92, 462–470.

Van Neste, D. (1982) Immuno-allergological aspects of scabies: a comparative study of spontaneous blastogenesis in the dermal infiltrates of common and hyperkeratotic scabies, allergic contact dermatitis and irritant dermatitis. *Archives of Dermatological Research* 274, 159–167.

Van Neste, D. and Lachapelle, J.M. (1981) Host–parasite relationships in hyperkeratotic (Norwegian) scabies: pathological and immunological findings. *British Journal of Dermatology* 105, 667–678.

Van Neste, D. and Salmon, J. (1978) Circulating antigen antibody complexes in scabies. *Dermatologica* 157, 221–224.

Witmer-Pack, M.D., Olivier, W., Valinsky, J., Schuler, G. and Steinman, R.M. (1987) Granulocyte/macrophage colony-stimulating factor is essential for the viability and function of cultured murine epidermal Langerhans cells. *Journal of Experimental Medicine* 166, 1484.

Wooten, E.L. and Gaafar, S.M. (1984) Detection of serum antibodies to sarcoptic mange mite antigens by the passive hemagglutination assay in pigs infested with *Sarcoptes scabiei* var. *suis*. *Veterinary Parasitology* 15, 309–316.

# 11

# Immune Responses to Mange Mites and Chiggers

## William J. Wrenn

*Department of Biology, University of North Dakota, Grand Forks, North Dakota 58202, USA*

## INTRODUCTION

Mange is a dermatitis of mammals caused by mites. The more important families include burrowing skin mites in the Demodicidae, Psorergatidae and Sarcoptidae; superficial skin mites in the Psoroptidae; and fur mites in the Myobiidae, Myocoptidae, Listrophoridae and Cheyletidae. All stages of these mites occur on the host and they are transferred between hosts by direct body contact. Only a few species in some of these families have been studied from an immunologic aspect and most of these infest animals other than humans. Genera of mange-causing mites included in this review include species in *Psoroptes*, *Otodectes*, *Demodex*, *Myocoptes* and *Myobia*. Sarcoptic mange mites are included elsewhere in this publication. Chiggers are the parasitic hexapod larvae of mites in the family Trombiculidae. The larvae do not feed on blood but take in epidermal components which have been disintegrated by enzymes introduced with the saliva; postlarvae are free-living in the substrate where they feed on small arthropods or their eggs. Chiggers are of medical importance because some species are vectors of the rickettsia that causes scrub typhus and the saliva injected with the bite of others elicits an allergic reaction and dermatitis by the host. Only a few are of veterinary importance, for example, those that cause lesions on turkeys in commercial flocks in the southern United States. The northern fowl mite, *Ornithonyssus sylviarum*, family Macronyssidae, is common in the northern temperate regions of the world and feeds on the blood of numerous species of wild and domestic birds. It is also an important poultry pest.

© CAB INTERNATIONAL 1996. From Wikel, S.K. (ed.) *The Immunology of Host–Ectoparasitic Arthropod Relationships*. CAB INTERNATIONAL, Wallingford.

# PSOROPTIC MANGE MITES

Psoroptic mange mites are obligatory, non-burrowing mites that pierce the host skin with the chelicerae and cause inflammation and oedema at the site of infestation (Sweatman, 1958a). They feed on tissue fluids but some species ingest whole blood (DeLoach and Wright, 1981). Domestic sheep and cattle are primarily affected but other hosts including rabbits, goats, horses, donkeys, mules, bighorn sheep, white-tailed deer and wapiti are also parasitized and earlier workers believed that each species of host harboured a different species of mite. Sweatman (1958a) differentiated the following five species of psoroptic mange mites based on host species, site of infestation on the host, and the length of the outer opisthosomal seta ($l_4$) of the male: *Psoroptes cuniculi* (Delafond), *P. cervinus* Ward, *P. natalensis* Hirst, *P. equi* (Hering) and *P. ovis* (Hering). Boyce *et al.* (1990) concluded that morphometric analyses using several characters was more useful for identification of the mites than either the host species or site of infestation on the host. Wright *et al.* (1983) reported that reciprocal cross-matings of *P. cuniculi* from rabbits and *P. ovis* from bovids revealed that these two species were not reproductively isolated. Furthermore, it has been demonstrated that some forms of *Psoroptes* from different hosts are generally antigenically similar (Kirkwood, 1986; Boyce and Brown, 1991).

Only two species, *P. ovis* and *P. cuniculi*, are of major veterinary importance. Psoroptic mange or sheep scab or cattle scab is caused by infestations of *P. ovis*, which prefers densely haired or woolled areas, and initial lesions usually occur on the trunk. Feeding activity of the mite causes inflammation with a resulting heavy crusting and alopecia. The mites move to the periphery of a denuded area and in this way spread over the body. Because the disease is so pruritic, the infested animal can lose condition and die (Miller, 1984; Kirkwood, 1986). Infestations of the ear canker mite of rabbits, *P. cuniculi*, cause otitis in goats, horses, sheep and rabbits. Affected rabbits shake their heads, and rub and scratch the ears; a crust appears in the pinna and a ceruminous otitis forms in the ear canal (Yunker, 1973; Miller, 1984).

Culbertson (1935) used the tube ring preciptin test to demonstrate precipitating antibodies to whole mite extract in rabbits infested with *P. cuniculi*. He speculated that sloughing of affected skin was an Arthus reaction and that the slough was a dried deposit of serum from the infested area. Fox *et al.* (1967) confirmed these results with the agar-gel immunodiffusion test and precipitation bands revealed at least two antigens.

Fisher (1972) used agar-gel methods to detect precipitating antibodies in the sera of sheep infested with *P. ovis* and sera of rabbits parasitized by *P. cuniculi*. Fisher and Wilson (1977) demonstrated precipitating antibodies in the sera of *P. ovis*-infested cattle and that these antibodies could be detected by extracts of either *P. ovis* or *P. cuniculi*. Fisher (1972) found at least three antigens in whole body extracts of *P. ovis* and not less than four in *P. cuniculi* and Fisher and Wilson (1977) noted that common antigens were shared by these mites. Weisbroth *et al.* (1972) used immunoelectrophoresis to detect seven distinct antigen–antibody reactions with sera from *P. cuniculi*-infested rabbits.

deVos *et al.* (1980) found antibody responses to *P. ovis* antigens in sera from five Nelson Desert Bighorn sheep, *Ovis canadensis nelsoni*, which did not have apparent lesions or clinical signs of psoroptic mange. Three other sheep from the same area in the Black Mountains in northwestern Arizona were also positive for *P. ovis*, based both on the presence of mites in ear scrapings and on serology, while precipitating antibodies were not detected in four other sheep.

Wright and DeLoach (1980) reported that *P. cuniculi* ingested red blood cells of the rabbit host during feeding and, noting this, Wikel (1982) concluded that elements of the host immune system might have an effect on *P. cuniculi*. Wright and DeLoach (1981) found that *P. ovis* also ingested erythrocytes when feeding on cattle and DeLoach and Wright (1981) further observed that *P. cuniculi*, *P. ovis* and a species of *Psoroptes* from mountain sheep in New Mexico also ingested blood.

Guillot (1981a) noted that *P. ovis* populations persisted and increased on stanchioned cattle during the summer months, which normally were considered as unfavourable periods for mite development. Cattle that were infested with mites, but not stanchioned in summer, were able to groom themselves, suggesting that this behaviour contributed to a decline in mite numbers as well as a decline in the extent of the mange. Guillot and Wright (1981) studied the level of infestation of rabbits by *P. cuniculi* as it related to season, age of host, apparent degree of susceptibility of parents' resistance and the host's ability to groom. The only factor they observed that affected infestation levels was grooming by the host. These workers did not demonstrate the presence or absence of antibodies in the experimental rabbits and they noted that no apparent immunity developed in the rabbits as a result of infestation. They also found that infestation of rabbits by *P. ovis* was traumatic and often resulted in death if the mite infestation was allowed to increase; infestations by *P. ovis* on rabbits were successful only if the hosts were collared. Guillot (1981b) observed that cattle, after a two to five month latent period or recovery from scabies subsequent to treatment in a toxaphene dip, were as susceptible to *P. ovis* as cattle infested for the first time. He also noted that self-grooming behaviour was the only factor which consistently limited the number of *P. ovis* and the extent of body scab.

Wikel (1982), who studied the skin reactivity of *P. cuniculi*-infested rabbits at different durations and intensities, noted immediate, 3 to 5 hour and 24 hour skin reactions in parasitized hosts whereas clean animals had only an immediate reaction; he suggested that immediate reactivity of naive rabbits was an indication of a vasoactive substance in the antigen. Wikel (1982) also observed that the lymph node cells of *P. cuniculi*-infested rabbits, in the presence of mite extract, experienced antigen-specific *in vitro* proliferation, whereas cells from naive controls were not reactive. Wikel (1982) concluded that results of skin test studies suggested cytotropic antibody, precipitating antibody and cell-mediated immune responsiveness to *P. cuniculi* infestation.

Fisher (1983a) used enzyme-linked immunosorbent assay (ELISA) and immunodiffusion to determine when antibodies to *P. cuniculi* and *P. ovis* antigens developed in rabbits infested for the first time with *P. cuniculi* and *P. ovis*, respectively. He found that ear canker developed before a detectable serum antibody response to antigens of these two mites. It was also reported that serum antibodies to either *P. cuniculi* or *P. ovis* antigens were detected earlier during the infestation by ELISA than by immunodiffusion and that serum antibody reactivity to *P. cuniculi* was higher in those rabbits parasitized for a longer period than in those infested for a shorter period (Fisher, 1983a). Fisher (1983b) used ELISA to examine naive and previously infested calves for serum antibody activity to *P. ovis* to ascertain when specific serum antibody activity began to develop during the infestation. He found that serum antibody reactivity to *P. ovis* antigens in previously non-parasitized calves usually began to develop at the same time or just after the mites and lesions had been detected; he also detected serum antibody activity to *P. ovis* antigens four weeks after calves were treated in a toxaphene dip to reduce the mite infestation or when *P. ovis* populations began to decline after the calves were allowed to groom themselves.

denHollander and Allen (1986) demonstrated by direct immunofluorescence assay that rabbits with short-term experimental infestations and rabbits naturally infested for long periods of time with *P. cuniculi* had antibodies that were reactive with tissues of *Psoroptes* as well as those of the American dog tick, *Dermacentor variabilis*. The observation that sera from BALB/c mice subjected to multiple tick infestations did not cross-react with mite antigens led denHollander and Allen (1986) to speculate that tick saliva did not contain tick antigens which were shared with mite antigens whereas saliva of *P. ovis* did include these cross-reactive antigens.

Stromberg *et al.* (1986) studied the association of the severity of clinical dermatitis with *P. ovis* population density on Hereford calves and elucidated histopathologic changes and clinical characteristics that could reveal mechanisms of systemic disease in cattle with advanced mange. The extent of the mange was dependent on the degree of infestation and calves which had 50 to 70% of the skin affected with dermatitis developed a mild anaemia and lymphopenia, primarily a neutropenia and variable eosinophilia. They found significant correlations with increases in the extent of dermatitis for at least 15 clinical variables. Lymphocyte proliferation assays yielded a 100% increase in responsiveness to concanavalin A (Con A) and phytohaemagglutinin (PHA), but changes in response to pokeweed mitogen (PWM) did not occur. They believed that the lack of evidence of systematic lesions did not support the hypothesis that *P. ovis* secreted toxic products.

Stromberg and Fisher (1986) infested naive and previously exposed Hereford calves with *P. ovis* and examined these hosts over a seven-week period. The lesions which developed in both groups were qualitatively similar, but they occurred early in calves previously exposed and progressed slowly, whereas in naive calves lesions appeared later but progressed rapidly. Histopathologically, the mange appeared as a chronic exudative superficial perivascular dermatitis. *P. ovis* populations on naive calves increased exponentially, had a high fecundity, and exhibited high densities, but mite populations on previously exposed hosts had low growth rates, low fecundity and population densities were 100 to 1000 times less than those observed on naive calves. There seemed to be a hypersensitivity response to mite antigens and this response might have imparted humoral immunity which could have contributed to acquired resistance by reducing *P. ovis* densities, partly, at least, as a result of reduced fecundity (Stromberg and Fisher, 1986). It may be that as mites feed on blood of resistant cattle, immunoglobulins ingested might precipitate some component of the digestive fluids and reduce nutrients for egg production (Stromberg and Fisher, 1986).

In a study of feeding behaviour, Rafferty and Gray (1987) noted that the mouthparts of *P. cuniculi* from rabbits and *P. ovis* from sheep had identical morphology and were adapted for surface feeding and not piercing the epidermis. They found haemoglobin in both species when these mites infested rabbits but not when feeding on sheep. They concluded that the mites ingested erythrocyte fragments rather than whole cells and that the haemoglobin was obtained from small haemorrhages on the surface of inflamed rabbit skin. Based on immunofluorescent studies, cross-reactions between rabbit anti-mite serum and sheep anti-*P. ovis* serum indicated that *P. cuniculi* and *P. ovis* shared the same antigenic determinants. Rafferty and Gray (1987) concluded that *P. cuniculi* and *P. ovis* were the same species because of identical morphology, ingestion of haemoglobin when feeding on rabbits, their ability to cross-mate and produce viable offspring and the ability for both to live and reproduce on sheep and cattle.

The reduced fecundity found by Stromberg and Fisher (1986) was substantiated by Guillot and Stromberg (1987) who observed that naive Hereford calves developed mange with *P. ovis* densities three times higher after seven weeks' exposure than those calves previously exposed to these mites, and egg production by female mites on naive calves was significantly greater than that recorded for *P. ovis* on previously infested hosts. As mange progressed on naive calves, Guillot and Stromberg (1987) detected a decrease in the proportion of ovigerous females which they concluded was probably due to an immune response.

ELISA was used by Fisher *et al.* (1986) to monitor the development and decline of serum antibody activity, which followed the increase and subsequent decline of *P. ovis* populations in Hereford steer calves maintained under the stressful conditions of Texas feedlots in endemic (Bushland) and non-endemic (Kerrville) psoroptic mange areas. Cattle in the endemic area developed more lesions as well as a higher incidence of mange and more cattle had specific serum antibody reactivity at the height of infestations in late January and early February than those experimental herds in the non-endemic area. The antibody activity declined in both areas as the usual summer decline of *P. ovis* populations occurred. When mites on cattle in the non-endemic area were no longer detected, antibody activity could not be verified; in the endemic area calves, low numbers of *P. ovis* were still on some cattle in August and October and serum antibody was reactive with *P. ovis* antigen. These authors deemed ELISA a useful tool to detect specific serum antibody activity in calves with only light mite infestations.

Pruett *et al.* (1986) studied stanchioned Hereford heifers infested with *P. ovis* and with the aid of ELISA found a correlation of the development of circulating antibodies specific for *P. ovis* antigens and the onset of dermatitis with an increase in total numbers of mites. The reaction was typical of a type I immediate-type hypersensitivity reaction and probably accounted for an increase in mite numbers and dermatitis. Their studies based on the mitogens Con A, PHA and PWM revealed that along with the developing infestation, there was a suppression of T-cell function that was probably due to the stress of stanchioning. These authors also provided a working hypothesis that attempted to correlate a reaginic type antibody as a component of the developing humoral response and suppression of T-lymphocyte function with both an increasing dermatitis and mite numbers.

Pruett *et al.* (1987) studied and described the immunosuppressive effect of the synthetic gluococorticoid, dexamethasone, on Hereford steer T-lymphocyte function in response to the mitogens PHA and Con A and the T-dependent antigen, keyhole limpet haemocyanin (KLH). They considered the pharmacokinetics of dexamethasone and optimal and suboptimal doses of Con A, PHA and KLH. Responses of haematological variables including leukocyte and red blood cell counts, packed cell volumes and haemoglobin concentrations to dexamethasone exhibited only slight variations. Suboptimal doses of Con A and PHA resulted in suppression of T-lymphocyte function, and antibody production to KLH was suppressed in dexamethasone-injected calves. The results of the study by Pruett *et al.* (1987) suggested that the immunosuppressive effect of dexamethasone in the bovine depended on the dose administered, its pharmacological level *in vivo*, the ability of the host to adapt physiologically, the dose of mitogen administered and when the evaluation is conducted after the drug is administered.

Stromberg and Guillot (1987) concluded that Hereford heifer calves with chronic dermatitis caused by *P. ovis* infestation developed myeloid hyperplasia and that mange-related neutropenia was not due to bone marrow suppression by either the mites or some unidentified factor in the scab. Decreases they observed in granulocyte numbers in the circulating pool were due to rapid emigration of these cells into the marginal pool or lungs, intestines or scab on the skin surface. They further stated that the emigration rate was influenced by concentration of inflammatory mediators and chemotactic stimuli, and perhaps, also by the amount of antigen present, number of mast cells and the nature of the reaginic antibody responses in infested calves.

Wassall et al. (1987) reported on the development of an ELISA for a simple and reliable serodiagnosis that could be used for detecting sheep infested with P. ovis relatively quickly. Losson et al. (1988) studied the dermal reactivity of cattle injected with P. ovis antigen. After a first infection none of the cows exhibited clinical psoroptic mange but there was a leukocytosis, due primarily to eosinophoilia, and some lymphocytosis. Phytohaemagglutinin, Con A and PWM used in lymphocyte transformation assays demonstrated that infestation with P. ovis induced a marked cell-mediated immune response and skin tests of cattle infested with this mite suggested that an immediate and delayed hypersensitivity developed after challenge infection (Losson et al., 1988).

Stromberg and Guillot (1989) infested Hereford heifer calves with P. ovis and after seven weeks the cattle were treated with ivermectin. Mast cell degranulation was observed in the skin of infested calves and active and degenerate neutrophils were found in the scab of infested animals. Infested calves developed microscopic dermal ulcers with mites present, but not after ivermectin treatment. Increased numbers of circulating neutrophils were detected one week after treatment, whereas numbers of dermal neutrophils and plasma cells decreased. These workers presumed that neutrophil emigration decreased and allowed the bone marrow to match the normal demand for circulating granulocytes.

Wikel (1989) used immunoblotting to identify extract components of whole P. cuniculi reactive with antibodies stimulated by experimental infestation of rabbits. He observed that immunoglobulins reacted with extract polypeptides with molecular masses ranging from 31.0 to 155.0 kDa and 15 of those that reacted with antibodies of at least four of seven infested rabbits had molecular masses ranging from 42.0 to 150.0 kDa.

Bubenik (1989) examined P. ovis infestations in captive male white-tailed deer, Odocoileus virginanus, that were suffering from hypoandrogenism. He speculated that the severe psoroptic mange observed in some of those hypoandrogenic deer resulted from either a reduced immunological response in the skin or a change in the ecological nature of the skin that favoured an increase in numbers of P. ovis.

Pruett et al. (1989) examined the effects of dexamethasone on Hereford calves to determine if the immunosuppressive treatment would normalize susceptibility to infestation by P. ovis. They utilized lymphocyte proliferation assays and compared P. ovis-infested hosts that had either been treated or not treated with the T-cell suppressent. T-lymphocyte function and anti-*P. ovis* antibody production were suppressed by dexamethasone and its presence also delayed the onset of scab lesions. Calves were not found to be equally susceptible to *P. ovis* because of dexamethasone treatment, rather, the onset of lesion development and an increase in the number of mites paralleled the suppression of antibody to *P. ovis*.

Psoroptic mange infests both bighorn sheep (*Ovis candensis nelsoni, O. c. canadensis, O. c. mexicana*) and mule deer (*Odocoileus hemionus*) in the western United States and morphometric analyses can detect differences between mites collected from allopatric populations of rabbits, bighorn sheep and cattle and place them into distinct groups (Boyce *et al.*, 1990). However, mites infesting sympatric populations of bighorn sheep and mule deer cannot be placed into distinct groups based on morphology nor can differences be detected in mites from the body and ears of bighorn sheep and rabbits (Boyce *et al.*, 1990). As a result of these findings, Boyce *et al.* (1990) concluded that other methods such as immunological and molecular characterization of these mites were needed to determine taxonomic relationships among species of *Psoroptes*.

Based on immunoassays for detecting small numbers of *P. ovis* infestations on cattle and sheep (Fisher, 1983b; Fisher *et al.*, 1986; Wassall *et al.*, 1987), Boyce *et al.* (1991a) used ELISA, Western blots and antigen from rabbit-reared *P. cuniculi* to study antibody responses of *Psoroptes* sp.-infested bighorn sheep (three subspecies) in the western United States. Serum antibodies from 20 sheep in California, Nevada, New Mexico and Idaho consistently reacted with *P. cuniculi* antigens that had molecular masses ranging from 12 to 34 kDa whereas sera taken from 35 non-infested bighorns from the same region did not react or, at most, exhibited a weak reaction with a few antigens in the 34 to 164 kDa range. The antibody response to antigens in the 12–34 kDa range was a rather sensitive indicator of *Psoroptes* sp. infestation as emphasized by four asymptomatic sheep which were parasitized by fewer than ten mites each (Boyce *et al.*, 1991a). These workers concluded that bighorn sheep were similar to cattle since antibody responses to species of *Psoroptes* could be detected only during an infestation, or a recent infestation, and that *P. cuniculi* antigens could be used for detecting *Psoroptes* sp.-specific antibody reactivity in bighorn sheep.

Boyce and Brown (1991) used antigen–antibody interactions to investigate antigenic relationships between defined populations as well as known species of mites collected from four species of hosts: *P. cuniculi* from a New Zealand white rabbit and *P. ovis* from a bighorn sheep, mule deer and cow. Immunoblotting with defined antigens and antisera of the extracts from mites from each of the four hosts revealed extensive and nearly complete antigenic cross-reactivity with heterologous antisera and essentially no antigenic differences were detected between extracts of mites on the bighorn sheep and those on the sympatric mule deer. Minor antigenic differences between cow mite and rabbit mite extracts were revealed by immunoblot and further antigenic differences were found between these mite antigens and those antigens from the bighorn sheep and mule deer mites. The study by Boyce and Brown (1991) supports the idea that populations and putative species of *Psoroptes* are not reproductively or ecologically isolated.

Boyce et al. (1991b) reported on the development of a kinetic ELISA that used *P. cuniculi* extract and sera from uninfested and *Psoroptes* sp.-infested bighorn sheep (three species) from western North American (nine geographic areas). These workers deemed kinetic ELISA as useful for prospective and retrospective serologic studies of psoroptic mite infestation in bighorn sheep because the assay provided for the quantitative measurement of antibody response to *Psoroptes* sp. Their study confirmed and extended observations and practical applications of their earlier (Boyce et al., 1991a) qualitative study. They maintained that the use of kinetic ELISA would enhance the evaluation of the extent of psoropitc mite infestation in bighorn sheep in western North America and, in turn, could provide input for management and control strategies.

Uhlir (1991) studied naive rabbits (Chinchilla strain) experimentally infested with *P. cuniculi* (Group A) and naturally heavily infested rabbits with an unknown history of infestation (Group B) to determine specific serum antibody activity (ELISA) and responsiveness of peripheral blood lymphocytes to Con A, PHA and *P. cuniculi* antigen. Ivermectin was administered during the 13th week to both groups with a reinfestation the 17th week. Group A rabbits developed a resistance to infestation as evidenced by a smaller extension of lesions after a secondary infestation when compared to the primary infestation. An antigen-specific lymphocyte response was observed only in the Group A rabbits. Uhlir (1991) found that Group B rabbits neither developed resistance to the mites nor was there an antigen-stimulated response which he speculated could be due to these rabbits being predisposed to a stress factor, as a result of infestation or being exposed to stress for a long period. Uhlir (1991) observed decreased lymphocyte responsiveness to Con A and PHA in both groups of rabbits, especially those of Group B which he attributed to stress. He stated that the partial inhibition of lymphocyte responsiveness to Con A and PHA observed could have been associated with high levels of antibody.

Uhlir and Volf (1992) investigated the influence of ivermectin on the specific immune response in rabbits (Chinchilla) infested with *P. cuniculi*. The experimental design included different combinations of immunization and ivermectin treatment of naive and lightly and heavily infested rabbits. A modified ELISA was used to detect specific serum antibody reactivity against *P. cuniculi* whole body extract and Con A and PHA were used in a lymphocyte proliferation assay to evaluate the *in vivo* lymphocyte responsiveness to mite antigens. Antigens were subjected to SDS–PAGE and components were detected by immunoblotting. Uhlir and Volf (1992) observed that rabbits developed a different profile of specific antibodies prior to and after ivermectin treatment as revealed by immunoblotting. They suggested that the cause for this effect was probably the result of a massive release of antigens upon death of the mites and not due to immunoenhancement by ivermectin. They concluded that ivermectin treatment did not affect specific humoral and cellular responsiveness in rabbits artificially immunized with whole body *P. cuniculi* extract.

Uhlir (1992) compared uninfested rabbits (Chinchilla strain) with rabbits artificially immunized on days 0, 10, 17 and 24 with whole body extract of *P. cuniculi*; both groups were challenged with live mites on day 31. Immunized rabbits developed partial immunity to *P. cuniculi* and lesions due to mite activity were not as extensive as in non-immunized rabbits. Both antigen-specific serum antibody activity (by ELISA) and the antigen-specific lymphocyte responsiveness were higher in immunized hosts. The loss of cellular responsiveness to Con A, which occurred when mite-caused lesions began to appear, was attributed to factors associated with the lesions. The pattern of specific antibody as revealed by immunoblotting was different in immunized rabbits when compared to controls after mite infestation. Serum from immunized rabbits reacted with up to six protein components ranging from 10 to 48 kDa 7 days before and 14 days after mite infestation. Sera from uninfested rabbits reacted with up to five proteins ranging from 18 to 59 kDa 14 days after infestation, but, just prior to infestation only one non-specific band was observed at 30 kDa. Uhlir (1992) also found one component of whole body extract at 48 kDa in the sera of both immunized and non-immunized rabbits 14 days after mite infestation. He speculated that this protein component was in the mite haemolymph and that it represented the main antigen involved in the immune-mediated resistance of rabbits to *P. cuniculi*.

Uhlir (1993) used several approaches to describe and partially characterize the immunological reations of protein components in *P. cuniculi* whole body extract. An immunoblot analysis of mite extract with sera from naive rabbits (Chinchilla strain), and rabbits either mildly or heavily infested (infestation history unknown) with *P. cuniculi* revealed two non-specific bands of 28 and 30 kDa in non-infested rabbits. Sera from both mildly and heavily infested rabbits reacted at 48 and 59 kDa, but sera from mildly infested hosts also reacted at 16.5, 18.5 and 25 kDa. Thus, Uhlir's results show that a larger number of protein components were recognized by sera from mildly infested rabbits and he speculated that stress in heavily infested hosts caused by long exposure to mites was involved and that the higher antibody serum activity he observed in mildly infested rabbits corresponded with the wider breadth of specific serum antibodies. Uhlir noted the differences in the results of his study compared to those by Wikel (1989) who detected 12 to 26 *P. cuniculi* extract components ranging from 31 to 155 kDa. He attributed the differences, in part – in the spectrum of specific antibodies – to different rates of immunoresponsiveness in the two rabbit strains (Chinchilla versus New Zealand White) or the differences, if any, in grooming behaviour because during grooming more mite proteins could enter the host through the skin and stimulate the production of specific antibodies.

An immunogenic antigen isolated by immunoaffinity chromatography and identified by crossed immunoelectrophoresis (CIE) was a glycoprotein of the complex or high mannose type with a relative molecular mass of 48 kDa (Uhlir, 1993). A second protein component, detected by immunoaffinity chromatography but not CIE, had a relative molecular mass of 18.5 kDA and it probably lacked saccharide residues. Uhlir (1993) also used blotting with six plant lectins and demonstrated that eight protein components of whole body extract of *P. cuniculi* were glycoproteins with different glycan structures.

## OTODECTIC MANGE MITE

*Otodectes cynotis*, the ear canker mite of carnivores is an obligatory, non-burrowing mite that passes all stages in the ear canal near the eardrum or less frequently on the body of the host (Sweatman, 1958b). They are especially a problem with dogs and cats and other carnivores including foxes and ferrets. The mites apparently feed on tissue fluids and in the process form a scab over the lesions they produce. Transmission is by direct contact, especially during nursing (Sweatman, 1958b; Pence, 1984).

Weisbroth *et al.* (1974b) studied random-source cats with and without active clinical signs of infestation by *O. cynotis* as well as naive barrier-reared cats. In skin tests, aqueous whole *O. cynotis* extracts were injected intradermally as were *Psoroptes cuniculi* antigen and a sterile, non-pyrogenic saline solution. Skin tests were observed at 15 minutes and at 2 to 4, 24 and 48 hours post-injection. Sera were evaluated by agar-gel immunodiffusion precipitin tests. About 87% of 40 random-source infested and non-infested cats had 15 minute (immediate) wheal and flare skin reactivity to *O. cynotis* antigens. Similar results were obtained for *P. cuniculi* antigen. About 94% of 19 infested cats had a two to four hour (Arthus-type) reaction but only 54% of 21 uninfested individuals had this reaction. Similar results were observed with *P. cuniculi* antigen. No evidence was found for delayed hypersensitivity. Serum antibodies to *O. cynotis* were detected by Ouchterlony analysis in most infested cats and in some mite-free cats. Weisbroth *et al.* (1974b) noted that evidence of immediate and Arthus-type hypersensitivity in clinically normal cats was an indication of prior infestation with *O. cynotis*. Barrier-reared cats neither reacted to intradermal injections of antigens nor did they have serum precipitins.

Barrier-reared mite-free cats were used by Powell *et al.* (1980) in a study to determine the sequence of humoral and cellular immune reactivity to *O. cynotis* and *P. cuniculi* antigens. Four pathogen-free cats were infested with about 30 mites in one ear and the infestation was allowed to develop and observed for more than nine months. By 14 days post-infestation, the cats exhibited immediate wheal and flare skin reactions to intradermal injections of antigen of both mite species. After 35 to 40 days the cats showed 15 minute reactions to both antigens and two-hour skin reactions only to *O. cynotis* antigen, but, after day 60 post-infestation, two-hour skin reactions to *P. cuniculi* antigen were observed. Serum-precipitating antibodies were not detected by agar-gel immunodiffusion until day 45 and only with *O. cynotis* antigen. Based on size of swellings it was noted that skin reactions to *O. cynotis* were more intense than those observed for *P. cuniculi*. Delayed hypersensitivity reactions were not observed at 24 and 72 hours. Powell *et al.* (1980) also determined that *O. cynotis* fed on host blood and lymph and not epithelial components. They speculated that this feeding mode provided the means by which the cats became exposed to and immunized against *O. cynotis* antigens.

Moriello (1991) noted that physical examinations of most cats with intense head and ear pruritus often reveal only a few *O. cynotis* or none and that the pruritus responds to prophylactic treatment for ear mites. Moriello (1991) speculated that intermittent exposure to mites, as is probably the case, might predispose cats with long histories of chronic otitis and *O. cynotis* infestations to hypersensitivity reactions.

## DEMODECTIC MANGE MITE

Hair follicle mites of the genus *Demodex* are species specific obligate parasites of mammals worldwide. All active stages usually feed on the cutaneous tissues by penetrating cells with their needle-like chelicerae and taking in cell contents. Demodectic mange of dogs (demodicosis, demodicidosis, red mange, follicular mange, acarus mange) is caused by *Demodex canis* and it is generally believed that most dogs harbour this acarine.

Scott *et al.* (1974) reviewed information pertaining to the biology of the mite, as well as pathogenesis and treatment of the disease. The mite inhabits the hair follicles and, on occasion, the sebaceous glands in the skin. It has been reported in most areas of the world and clinical signs are most common in short-haired dogs less than one year old. Two types of lesions are found in demodectic mange (Scott *et al.*, 1974; Miller, 1984). The squamous or localized type involves a localized alopecia and dry, scaly dermatitis with a small amount of induration. The generalized or pustular type occurs as a primary condition or as a sequel to the localized type; it is characterized by alopecia, erythema, oedema, seborrhoea, and pruritus over large areas of the body and there is usually a secondary bacterial infection. It is of interest that some individual dogs develop clinical disease but others do not even though their care is the same (Miller, 1984). A plethora of predisposing factors have been proposed that might enhance the increase in mite populations on a dog (Scott et al., 1974). Major findings concerned with immunoresponsiveness of canines to *D. canis* are summarized below.

Owen (1972) injected a total of eight puppies from three litters with antilymphocyte serum (ALS) prepared in pigs and horses and seven puppies from the same litters were not injected. All those dogs that were injected with ALS developed generalized demodectic mange whereas those not injected, remained healthy. Owen (1972) concluded that immunosuppression produced by ALS led to conditions that enabled the mite to produce a severe clinical disease.

Scott *et al.* (1974) reported on attempts to assess cellular immunity with *in vitro* lymphocyte transformation studies and skin tests for delayed hypersensitivity using the mitogens PHA, PWM and Con A. The results of intradermal injection of PHA and Con A suggested T-cell dysfunction and cell-mediated immunodeficiency and *in vitro* lymphocyte transformation experiments with PHA and PWM also suggested a T-cell dysfunction. Scott *et al.* (1974) speculated that the presence of mites or their products, or both, interfered with cell-mediated immune responsiveness.

Hirsh *et al.* (1975) also demonstrated that the response to PHA of peripheral lymphocytes from normal dogs or dogs with generalized demodectic mange cultured in the presence of normal dog serum (no lymphocyte suppression) was statistically different from the response in the presence of serum from dogs with demodectic mange (lymphocyte suppression). T-cell suppression was apparently triggered by some component in sera of demodectic dogs. Lymphoctye suppression was not observed when the dogs were in remission (Hirsh *et al.*, 1975).

Corbett *et al.* (1975) also studied lymphocyte responsiveness to PHA and found in dogs with generalized demodicosis that cell proliferation was markedly suppressed and sera of dogs with the localized type also suppressed lymphocytes. Further, these workers noted the presence of IgG attached to the surface of a proportion of lymphocytes from dogs with demodectic mange.

Scott *et al.* (1976), in a follow-up study of their 1974 work, found that most of their experimental dogs with generalized demodicosis had reduced lymphocyte responsiveness to the mitogens PHA, PWM and Con A whereas dogs with the localized type exhibited normal lymphocyte blastogenesis. Their attempts to treat infected dogs with levamisole, an immunostimulant effective in restoring T-cell function, did not kill the mites but lymphocyte responsiveness was at last partially restored. Scott *et al.* (1976) concluded that generalized demodicosis was probably based on a genetically controlled T-cell defect which caused cell-mediated immunodeficiency as a result of an increased number of feeding *D. canis*.

Healey and Gaafar (1977) exposed beagle pups to *D. canis* and then injected them with rabbit anti-canine lymphocyte serum (ALS); in a second experiment littermates were infested with either mites alone or ALS alone or kept as controls. Localized demodicosis developed in all dogs that received both mites and ALS and in two dogs that received only mites; no lesions were observed in any other puppies. The number of peripheral lymphocytes declined in those dogs injected with ALS and, three weeks after the treatment ended, the numbers of cells returned to normal. The localized lesions on the pups injected with ALS and infested with mites developed into generalized demodicosis. This study by Healey and Gaafar (1977) also established that ALS reduced the number of T lymphocytes with the result that a favourable environment was established that supported an increase in *D. canis* numbers. The suppression of the ability of the dogs to mount an adequate immune response led to the production of the generalized form of the disease.

Krawiec and Gaafar (1980) used normal canine lymphocytes and normal serum for globulin fractionation from clinically normal purebred beagles and sera from dogs affected with generalized demodectic mange in lymphocyte blastogenesis studies with Con A. Two littermate puppies whose parents had demodectic mange lesions were used for studies on the development of immunosuppression. In another series of experiments they used two dogs with generalized demodectic mange to study the effects of lymphocyte dialysate therapy. Krawiec and Gaafar (1980) found that the serum beta-proteins of dogs with generalized demodicosis depressed the Con A responses but serum fractions of demodectic dogs supplemented with fetal bovine serum stimulated blastogenesis. Serum from puppies whose parents had generalized demodicosis increased the Con A reactivity initially, but blastogenesis was reduced when these dogs developed generalized demodicosis. The lymphocytes of three-week-old puppies did not respond normally to mitogen, but blastogenesis occurred normally at six months of age. Their results supported the idea at that time of the existence of an initial primary lymphocyte non-responsiveness in dogs susceptible to demodectic mange and, that when the lymphocytes could respond normally, the suppressive effect of the serum was operational which in turn permitted an increase in mite populations on the host. Krawiec and Gaafar also (1980) determined that lymphocyte dialysate could be an effective immunotherapy because the severity of demodicosis was reduced in two experimental dogs.

Barta *et al.* (1983) studied the *in vitro* lymphocyte response of three demodectic dogs and nine dogs with demodectic mange as well as pyoderma to the mitogens Con A, PHA and PWM; they also tested for the presence of serum lymphocyte immunoregulatory factors (SLIF) suppressing blastogenesis. Lymphocytes of those dogs with only demodectic mange responded normally to mitogens and there was no indication of suppression of blastogenesis by SLIF. Those dogs with both demodicosis and pyoderma had varying responses: four dogs had high levels of the blastogenesis-suppressing SLIF for Con-A-sensitive cells, two dogs for PHA-sensitive cells and only one of three dogs for PWM-sensitive cells. Also, lymphocyte dysfunction was found in three dogs with demodicosis and bacterial infection and these either died or were deemed untreatable and destroyed. Results of the study by Barta et al. (1983) indicated that the serum-mediated immunosuppression in demodectic mange was the result of pyoderma and was absent in dogs with only demodicosis.

In a study of immunosuppression by D. canis, Barriga et al. (1992) used three groups of three beagles: clinically normal dogs; a second group with localized demodectic mange; and a third group had generalized demodicosis. At one to three weeks from the appearance of clinical signs (experiment 1), reactivity of peripheral lymphocytes to Con A and PHA was investigated and at four to six weeks (experiment 2) reactivities of peripheral lymphocytes of the same dogs to Con A and PHA as well as E. coli lipopolysaacharide (LPS) were studied. In a third experiment, they compared the reactivity of peripheral lymphocytes of a mixed breed dog exhibiting uncomplicated general demodicosis with a mixed breed dog with pyogenic dermatitis to Con A, PHA, LPS and purified protein derivatives (PPD) of Mycobacterium.

The results obtained by Barriga et al. (1992) demonstrated that the lymphocytes of dogs with localized demodectic mange reacted normally to mitogens within three weeks (experiment 1) of the appearance of clinical signs of mange. Also, during this period, lymphocytes of dogs with generalized demodectic mange were completely comparable to lymphocytes of healthy dogs. Barriga et al. (1992) stated that this was a demonstration that immunosuppression was not a prerequisite for dogs to develop clinical demodicosis. In experiment 2 at six weeks, the lymphocytes of dogs with localized demodicosis were only moderately depressed to Con A stimulation and slightly depressed to PHA stimulation; response to LPS was normal. Dogs with generalized demodicosis exhibited severe depressions of the Con A, PHA and LPS reactivities and blastogenesis was the same as cells that had not been stimulated. Barriga et al. (1992) maintained this was an indication that immunosuppression proceeds with the clinical signs of the infection and that in dogs with the generalized type of mange it is more severe. In addition, they noted that the mites or the reactivity of the host to the mites and not a prior condition caused the immunosuppression. These workers also noted that their results suggested that the occurrence was related to the number of mites or the strength of the response mounted by the host. Barriga et al. (1992) also found that both generalized demodicosis and pyoderma decreased the lymphocyte response to PHA, LPS and PPD but that bacterial infection seemed to heighten the effect on the reactivities of LPS and PPD. Generalized demodectic mange, but not pyoderma, apparently inhibited Con A reactivity. Barriga et al. (1992) concluded that localized demodectic mange had only a moderate inhibiting effect on suppressive and cytotoxic T cells whereas the generalized form of mange strongly inhibited these two types of lymphocytes as well as having a mild effect on B lymphocytes in an early stage of differentiation.

# CHIGGERS

Chiggers are mites whose larvae parasitize all major groups of vertebrates except fish. Other stages in the life cycle are free-living. Of the more than 3000 known species only a few are of known medical or veterinary importance. Chigger-borne rickettsiosis, also called scrub typhus, tsutsugamushi fever or mite-borne typhus fever is known to be transmitted by seven species in the genus *Leptotrombidium*, and at least six additional species in the genus have been incriminated as vectors (Traub and Wisseman, 1974; Wang and Wang, 1988). The disease, or the occurrence of the pathogen, *Rickettsia tsutsugamushi*, in chiggers or hosts, has been reported from east and southeast Asia, Pakistan, islands of the eastern Pacific, northern Australia and New Zealand. Chiggers are also known for the cutaneous reactions that can result when they bite humans or other animals. The most severe cutaneous reactions in mammals occur with species that ordinarily infest reptiles or birds (Audy, 1951) and only about 15 species are reported as frequently biting humans. In North America, five species of *Eutrombicula* are known to cause severe cutaneous reactions at the site of attachment on humans (W.J. Wrenn, unpublished data; Audy, 1951). Those species that normally infest mammals (including scrub typhus vectors) seem to cause only slight skin responsiveness in their hosts when feeding.

Relatively little work has dealt with host immune responsiveness as a result of chigger feeding activity and only a very few papers have appeared. A major reason for this paucity relates to the fact that live, unengorged chigger larvae are not readily available for experimentation. Maintenance of laboratory cultures is labour intensive and larvae are not easily manipulated. And finally, the taxonomy of trombiculids is based on the larval stage and species identifications are, at best, difficult and detailed studies have shown that in many instances different species have been identified as the same taxon; for example, about 20 different species of *Eutrombicula* in the Western Hemisphere have been called *E. alfreddugesi* (Brennan and Reed, 1974; Loomis and Wrenn, 1984; Wrenn and Loomis, 1984).

The first immunological study of chiggers was conducted by Morrow (1940). Live larvae of *Eutrombicula cinnabaris* (=*Trombicula irritans*) were ground in and mixed with distilled water so that each millilitre represented antigen from 100 chiggers and, after centrifugation, the supernatant was autoclaved or the infusion of ground chiggers was sterilized by passage through a Seitz filter. Intradermal injections of human subjects with both autoclaved and filtered antigens resulted in a sharp stinging sensation during injection, and development of an intermediate wheal with a surrounding area of erythema. Only the filtered antigen, however, produced an itching and the resulting wheal and erythematous areas were larger.

Morrow (1940) also used an extract of larvae in a Prausnitz–Kustner test with sera from three subjects each with a different reactivity and sensitivity to chigger bites as follows: very sensitive, moderately sensitive and relatively insensitive. Sera from the three donors were injected intradermally in nine areas on the back of an individual not sensitive to chigger bites; after 24 hours, three different dilutions of antigen were injected into those same areas. A control used the same dilutions of antigen injected into the skin. This passive transfer revealed an association between the response and the estimated sensitivity of the donor as shown by the fact that the wheal and erythema varied with the different sera. Antigen injections into the skin of these donors also produced the same order of differences in the sizes of the resulting wheal and erythematous areas. Morrow (1940) suggested that cutaneous reactivity was due to previous sensitization to chigger bites and, that heat affected the antigen rendering it less irritating and not as reactive.

Audy (1951) listed 12 species of trombiculids found in the world whose bites were known to cause an intense specific delayed reaction in humans. He noted that there was much variation in sensitivity of individuals to these chiggers and that there were records of some acquired immunity to their bites. Chigger infestation of humans is accidental and no species is known which depends upon humans or other large primates for a host. Those species that do bite humans and cause a reaction generally parasitize reptiles and birds but they do have a wide host range and seem to attack any available terrestrial vertebrate (Audy, 1951; Loomis and Wrenn, 1984).

Jones (1950) noted that during feeding by the larva of *Neotrombicula autumnalis*, a chigger which commonly infests rabbits and birds in Britain, that saliva was injected into the Malpighian layer of experimentally infested field mice where it caused the cells to disintegrate thus enabling the mite to ingest tissue fluid as well as cell contents. Cutaneous reactivity was the result of the irritant effect of the saliva (Jones, 1950).

Wright *et al.* (1988) studied host immune responsiveness to *Eutrombicula cinnabaris*, the common pest chigger in the eastern two thirds of the United States. Adult male BALB/c mice were given four infestations of laboratory reared larvae. Each exposure was separated by a 14-day mite-free period. Interpretation of the results of the study led to the following conclusions.

Attachment studies revealed that the average time for larval engorgement was nine to ten days for first exposure but four to five days for the fourth exposure. Host reactions to chigger feeding changed histologically with repeated exposures. Macroscopic changes at attachment sites generally were not manifest in an initial exposure whereas all third and fourth exposure mice exhibited strong responses including erythema, epidermal thickening and serous exudation. At the attachment sites, third and fourth exposure hosts had infiltrates of lymphocytes, eosinophils, basophils and neutrophils, which were characteristic of murine cell-mediated immune responses. Adoptive transfer of responsiveness elicited a microscopic response to larval challenge similar to that for fourth infestation hosts.

Wright et al. (1988) concluded that antibody was probably involved in reactivity as a result of adoptive transfer because both B and T lymphocytes were transferred. Passive transfer studies suggested that antibody was involved in acquired reactivity and induration reactions at 24–48 hours after skin testing of reactive animals suggested involvement of delayed hypersensitivity in the response to chigger infestation. Wright et al. (1988) also determined that the results of skin test and active cutaneous anaphylaxis studies indicated the absence of homocytotropic antibody. The results of skin test studies, which revealed induration at 24 and 48 hours after antigen injection to fourth exposure animals, were consistent with a cell-mediated component of responsiveness and macroscopic changes within 24 to 48 hours of attachment to reactive hosts and supported a delayed hypersensitivity response to chigger feeding. Erythema at five hours on fourth exposure hosts suggested the presence of circulating, extract binding, immunoglobulin. The study by Wright et al. (1988) provided evidence that the cutaneous response to chigger feeding has an immunological basis.

## FUR MITE OF MOUSE

The fur mite of mouse, *Myobia musculi*, is an obligate parasite inhabiting the fur on the laboratory and the wild house mouse throughout the world. Yunker (1973) and Weisbroth et al. (1974a) presented information concerning the veterinary importance of this mite. The levels of pathogenicity to different strains of mice is variable and light infestations are not obvious. Heavy infestations cause dermatitis, alopecia, pruritus, self-inflicted trauma, secondary amyloidosis and an increase in mitotic activity of affected skin. Active stages feed on extracellular tissue fluids and transmission is by direct contact.

Friedman and Weisbroth (1975) infested mice of four strains or strain crosses with *M. musculi* to determine if there was a genetic basis for host response between strains. Mite numbers were monitored periodically by microscopic examination for about 130 days. They did detect significant differences in levels of mite populations between strains, but variability within strains was evident. They observed a general pattern of an increase in numbers of mites on day 40 post-infestation to peak numbers between the 60th and 80th days and then a decline to a stable level. They noted the possibility that differential host responses may have limited mite numbers, but they did not provide any direct evidence that host immune responsiveness had an affect on *M. musculi* in their experiments.

Dawson et al. (1986) studied the susceptibility of 16 inbred strains of mice to mite-associated ulcerative dermatitis due to *M. musculi* infestation. They found significant differences in the frequency of this condition in certain C57BL mice and associated these with differences in genetic background and in the H-2 type. They suggested that the occurrence of ulcerative skin lesions in inbred mice parasitized by *M. musculi* was controlled by at least two genes and that susceptibility to lesions was affected by a non-H-2-linked gene or gene combination which was shared by all of their C57BL background strains. Dawson *et al.* (1986) also suggested that H-2-linked genes may influence many of the stages in the pathogenesis of murine allergic diseases such as mite-saliva-specific IgE production, antigen-specific T-helper cell proliferation and function or processes involved with antigen presentation by macrophages.

## MYOCOPTIC MANGE MITE

The myocoptic mange mite, *Myocoptes musculinus*, the most common ectoparasite of laboratory mice, is found in the fur of laboratory and wild mice over the world. All stages occur on pelage of the host and the mite is transmitted by direct contact. They apparently feed on epidermal tissue, but not on tissue fluids (Yunker, 1973). Little work has been done on host immune responsiveness to this rather common species.

Laltoo *et al.* (1979) used SWR mice infested with *M. musculinus* to determine if these hosts would develop an antibody response to whole extracts of this mite. Active cutaneous anaphylaxis revealed the development of homocytotropic antibody, IgE, after five weeks in mice infested with mites at birth and the titre continued to rise over three months. Laltoo *et al.* (1979) speculated that the delayed appearance of IgE anti-mite antibody in mouse sera was related to the time necessary for a large increase in the number of mites infesting the mice. They also observed a similar response in mite-infested adult mice. Further observations revealed a degranulation of most cells in the connective tissue of skin upon contact with mite antigen.

Laltoo and Kind (1979) challenged oxazolone-sensitized SWR mice infested with *Myocoptes musculinus* and mite-free SWR mice with oxazolone on the skin of the neck and shoulder. When compared to naive mice, contact sensitivity to oxazolone was significantly reduced in mite-infested mice. These workers also observed that sera from mite-infested mice suppressed the influx of cells into the skin test sites of non-parasitized mice challenged with oxazolone and that there was a decrease of movement of cells in parasitized hosts into areas where mites were present (skin of neck and shoulder) and the ears, which were not infested. Mite-infested mice also exhibited an increase in vascular permeability. Laltoo and Kind (1979) suggested that histamine from sensitized mast cells activated a non-specific suppressive factor from lymphocytes that suppressed the cellular influx in contact sensitivity.

# NORTHERN FOWL MITE

The northern fowl mite, *Ornithonyssus sylviarum* (Canestrini and Fanzago), as an obligate parasite, feeds on the blood of a diverse assemblage of wild and domestic birds and is an important pest of poultry. The species is most common in the north temperate latitudes but it is also known from domestic fowl in Australia and New Zealand (Desch, 1984; van Bronswijk, 1984). Heavy infestations of chickens and other domestic fowl can result in anaemia, a decrease in both weight gain and egg production and, in some cases, death (DeVaney et al., 1977; DeVaney, 1978a; Pence, 1984). Based on a national survey, DeVaney (1978b) established that the northern fowl mite was the most important parasite of the chicken industry. The entire life cycle is completed on the host and only the protonymphs and adults are known to feed on blood (Yunker, 1973; Burg et al., 1989) and heavy infestations (50,000 or more mites) have been estimated to remove 6% of the total blood volume of a 1.5 kg hen (DeLoach and DeVaney, 1981).

Loomis et al. (1970) observed in heavy infestations of White Leghorn pullets by *O. sylviarum* that most birds were parasitized, but others remained free of mites. Hall et al. (1978) noted that Payne (1930) also had observed this in commercial layer flocks. Loomis et al. (1970) further observed that the mites quickly adapted to more than half of their experimental pullets in two to seven weeks after infestation, and that there was a general decrease in the number of heavily infested birds with a rise when the birds were 44 to 50 weeks old. The number of non-infested birds generally increased from the beginning of the experiment until the age of 44 to 50 weeks when there was a slight decrease. These relationships suggested that some birds developed immunity to northern fowl mite infestation whereas, in other pullets, the immune response was slower or did not occur (Loomis et al., 1970).

Matthysse et al. (1974) reported that infested White Leghorn chickens slowly developed *O. sylviarum*-specific circulating antibodies and skin-sensitizing antibodies. Agar-gel double diffusion revealed two precipitin bands in sera between the ninth and tenth weeks of infestation. Mite extract was used in skin tests of infested birds and wheals were evident within one hour after the antigen was administered and these persisted for up to nine hours. These authors did not find a correlation between the mite population and immune response, but they did observe a negative correlation between the presence of precipitating antibodies and positive skin-test reactivity.

DeVaney et al. (1977) compared the effect of *O. sylviarum* on several variables for infested and non-infested White Leghorn roosters. These variables included body weight, several dealing with reproductive potential and total red blood cell counts and packed cell volume. They found that the greatest decrease in red blood cell counts and packed cell volume occurred from the fourth to the eighth week when the mite populations were increasing.

Hall et al. (1978) found that 16-week-old Cornell Randombred Leghorn roosters from a line artificially selected for high initial antibody response to sheep red blood cells were more resistant to increases of *O. sylviarum* infestations than birds that were selected for low antibody response. High antibody line roosters had 30 times fewer mites than did those birds in the low antibody line. Increases in mite numbers were not affected by oral administration of varying doses of corticosterone, but their observation of a reduced lymphocyte mass in test birds led Hall et al. (1978) to conclude that antibody competency alone may not have resulted in differences in mite infestations. In an effort to determine the feasibility of developing genetically mite-resistant birds, Eklund et al. (1980) used eight- to ten-month-old White Leghorn hens to investigate genetic variability to northern fowl mite resistance. They found a significant difference in average mite infestations between different lines and also that body weight was positively correlated with infestation levels but egg production and infestation level were not associated and each seemed to vary independently.

DeVaney and Ziprin (1980a) used agar-gel immunodiffusion tests to study the reactivity of sera of naive White Leghorns, mite-infested roosters and two levels of mite-infested hens to whole *O. sylviarum* extract. They concluded that both hens and roosters mounted an immune response that could be correlated with an increase in the numbers of mites infesting the host. These workers also noted that even with declining mite members, the immune response was retained by the birds for at least 12 weeks. Heavier precipitation bands developed with rooster sera, and the sera of some birds that were rather heavily infested produced more pronounced precipitation bands than birds of the same sex with fewer numbers of mites.

In a second study dealing with the relationship of infestation level and the immune response, DeVaney and Ziprin (1980b) maintained three experimental groups of White Leghorn hens infested with *O. sylviarum* at low levels, high levels and mite-free. After 24 weeks, hens were treated to kill the mites and the birds were reinfested 9 weeks later with the same levels of mites as previously and mite numbers were monitored for 12 weeks. Those hens that initially had the highest infestation levels, had the fewest numbers of mites 12 weeks after reinfestation and the numbers did not increase as rapidly as they did on the other two groups. Those hens with low infestations maintained a lower level of mites after reinfestation than the mite-free hens but greater than that of the heavily infested group. Immunodiffusion tests revealed that sera of about 33% of the heavily infested hens retained serum antibodies for at least seven months after mite removal; the sera of those hens infested at low levels lost their reactivity within two months. DeVaney and Ziprin (1980b) concluded that their White Leghorn hens acquired a degree of immunity to *O. sylviarum* and that this immunity resulted in lower numbers of mites on hens after reinfestation and, further, that this immunity was apparently dependent on previous mite infestation levels.

Studies of chickens by Chang *et al.* (1957, 1958, 1959) that dealt with antibody production by the bursa of Fabricius and the spleen against bacteria prompted DeVaney *et al.* (1984) to investigate the effects of removal of these organs from White Leghorn roosters on the population levels of the northern fowl mite. They did not detect differences in mite numbers when birds that had been bursectomized at one day of age, splenectomized at an age of 21 days, or both, were infested with mites at 22 to 23 weeks of age and monitored for 24 weeks.

DeVaney and Augustine (1988) used non-infested White Leghorn hens and birds with differing infestation levels to investigate the association of the increase and decline of northern fowl mite numbers with the development of specific antibodies to mite-extract moieties. Western blot analysis revealed that sera of hens with low, moderate and high infestations reacted with mite extract polypeptides with molecular masses ranging from about 20 to 200 kDa. They hypothesized that the reactivity represented binding of rabbit anti-chicken IgG to antigens in chicken blood that were present in the mite extract rather than to mite antigens. In support of their hypothesis, they noted that bands of mite extract components were reactive with pre-infestation sera, and similar moieties ranging from 20 to 200 kDa were reactive in the chicken blood antigen with pre-infestation sera and 42 days post-infestation sera. The previous exposure of hens to infestation was ruled out because the intensity of the reactions did not seem to increase over the 84 days of the study and there was no history of infestation prior to experimentation.

DeVaney and Augustine (1988) also found that only sera from heavily infested hens (>1001 mites) taken during a 2 to 12 week post-infestation period reacted with mite-extract components at 8 to 10 kDa. They noted that this band probably represented a specific antibody response to *O. sylviarum* infestation and that the responsiveness was approximately proportional, both in intensity and when it appeared, to the mite burdens supported by the hens. Furthermore, DeVaney and Augustine (1988) suggested that an antigen from the northern fowl mite could be identified and used in a vaccine for control of this mite on chickens.

Burg *et al.* (1988) used two-dimensional electrophoresis and rocket electrophoresis to compare the effects of crude mite extract injections on *O. sylviarum* population numbers and on precipitating antibody concentration response in White Rock and Fayoumi hens. Bovine serum albumin (BSA) was used as the positive control antigen for the antibody concentration response aspect of their study. The mean anti-BSA antibody concentrations of the two breeds were similar. The results of a rocket electrophoresis study of pooled serum samples analysed by regression revealed significant differences between the slopes of pools of the two breeds for mite extract and BSA injections suggested to Burg *et al.* (1988) that this was an indication of differences in antibody–antigen reactions. As stated by Burg *et al.* (1989), rocket electrophoresis can be used to correlate precipitating antibody concentrations with antigen volumes, thus providing a method to quantify these concentrations. Migration of antigen into agarose containing antibody produces a rocket-shaped precipitate at the site of optimum antigen and antibody concentrations. The area under the precipitate is proportional to antigen concentration and inversely proportional to precipitating antibody concentration. Additionally, Burg et al. (1988) did not find that injections of mite extract had an effect on developing mite populations on White Rock hens, but that mite extract-injected Fayoumi hens supported statistically fewer mite numbers than hens treated with BSA. They noted that the lower mite populations on Fayoumi hens might have been due to these birds being genetically more resistant than White Rock hens and that this difference was expressed in mite-specific precipitating antibodies.

Murano et al. (1989) adapted a microscope slide modification of the Ouchterlony double-gel diffusion method to monitor circulating antibodies in White Leghorn hens and roosters infested with northern fowl mite. At 12 weeks post-infestation, all birds were dipped in an aqueous 0.5% solution of carbaryl to kill the mites. These workers noted that antigen in positive reference sera reacted at dilutions of 1:1 to 1:8, but not above a 1:16 dilution; in their experiments, antigen was diluted to 1:4. After four weeks of mite infestation, precipitating antibodies were detected in sera, but not in sera of naive birds. They observed that as long as the mites were feeding on the birds, the antibodies could be detected but disappeared in most chickens a few weeks after the hosts were dipped in carbaryl. Murano et al. (1989) observed, as had earlier workers (Mattysse et al., 1974; DeVaney, 1978a; DeVaney and Ziprin, 1980a,b), that mite numbers on infested chickens increased exponentially to a peak, then gradually declined. Murano et al. (1989) confirmed that there was a close association between mite population numbers and antibody development which had been suggested by earlier workers including Mattysse et al. (1974), DeVaney and Ziprin (1980a,b), and DeVaney and Augustine (1988). Murano et al. (1989) also did not detect any differences in the antibody development and disappearance between hens and roosters as did DeVaney and Ziprin (1980a).

Burg et al. (1989) used rocket electrophoresis to quantify precipitating antibody concentrations and to correlate these with changes in density of *O. sylviarum* on White Rock hens. Mite population densities were estimated weekly for 28 weeks and sera were recovered at two-week intervals for 28 weeks. Mite-specific antibodies resulted in measurable precipitates in about 74% of their serum samples. They found a highly significant negative correlation between mite population density and precipitating antibody concentration which indicated that as mite numbers increased there was a decline in precipitating antibody concentration. This decrease in antibody concentration suggested to Burg et al. (1989) that partial immunological tolerance or mite-induced immunosuppression might have been involved in the humoral response of the hens to northern fowl mite feeding activity and that this response was initiated by the second week post-infestation.

Wikel et al. (1989) performed a series of experiments to identify northern fowl mite extract components that were reactive with antibodies of parasitized White Leghorn hens. Each of 12 birds was infested with 50 mites and 12 other hens were each infested with 2000 mites. Mite numbers were estimated weekly for 12 weeks and venous blood was sampled from birds on day two and weeks 1, 2, 3, 4, 5, 6, 8, 10 and 12 post-infestation. Sera from four hens that exhibited representative mite infestations were analysed for development of antibodies over the 12-week study. Mite extract was fractionated by SDS–PAGE and protein immunoblotting was used to identify serum-antibody-reactive polypeptides. Wikel et al. (1989) also used lectin blotting to identify mite extract glycoconjugates probed with different carbohydrate specificities (Wikel, 1988).

The four hens revealed that, in general, mite infestations reached maximum numbers (heavy to extra heavy) then declined by week 12 to light to moderate infestations. Based on immunoblot reactivity, these infested hens developed antibodies to several extract moieties ranging from 31.5 to 160 kDa over the course of the study. A few extract components complexed with immunoglobulins obtained from three hens on the second day of infestation and one hen at week 1. These included extract polypeptides of 36, 77 and 117 kDa which complexed with immunoglobulins of all four hens; three birds had antibodies reactive with extract components of 80.5 and 103 kDa; and two birds exhibited immunoglobulins that complexed a 38 kDa component.

A variety of extract polypeptides were reactive with antibodies in sera obtained from hens during the 12 weeks of the study. Different extract components bound by host antibodies appeared and were observed in some hens and not others and at different times during the infestation. Fifteen polypeptides of *O. sylviarum* extract with molecular masses ranging from 40 to 160 kDa bound antibodies that most birds developed.

Lectin blotting of mite extract revealed that most of the antigens identified by immunoblotting were glycoconjugates. The lectins utilized in the study by Wikel et al. (1989) and their carbohydrate specificities included: concanavalin A binding D-mannose; soybean agglutinin binding 2-acetamido-2-deoxy-D-galactose; *Ulex europaeus* agglutinin binding L-fructose; and wheat germ agglutinin binding 2-acetamido-2-deoxy-D-glucose.

Wikel et al. (1989) did not detect a northern fowl mite extract polypeptide in the 8 to 10 kDa band as did DeVaney and Augustine (1988). They speculated that differences in mite-extract preparation and electrophoresis conditions could have accounted for lack of this band.

Minnifield et al. (1993) reported on research leading to the development of a vaccine against northern fowl mite. They purified antigen proteins from mites by affinity chromatography, immunized male Sex-Sals chickens with the antigens (at 2, 3, 4, 5 and 7 weeks of age) and infested the birds with mites at 9, 13, and 20 weeks of age. Western blots were used to determine antibody activity in chicken sera at varying times post-infestation. These workers identified mite extract proteins with molecular masses ranging from 45 to 160 kDa and several of these moities were the same as, or similar to, those identified by Wikel *et al.* (1989) as components reactive to serum antibodies from mite-infested White Leghorn hens. Also, as Wikel *et al.* (1989), they did not detect the presence of an 8 to 10 kDa protein in *O. sylviarum* extract. Western blot analysis of sera of chickens that were immunized or infested or both, revealed antibody reactivity with a 100 kDa protein. The response appeared to be stronger if birds were immunized with the antigen and infested with mites. Both experimentally immunized and infested birds exhibited rather small decreases in mite infestations. Minnifield *et al.* (1993) concluded that even though their study did not identify a specific northeren fowl mite protein that could be used as an antigen, the 100 kDa component was a likely candidate for further study as to its use for a vaccine.

Minnifield *et al.* (1993) also conducted *in vitro* mite feeding studies over a 72 week period to assess survival of blood fed *O. sylviarum* after ingestion of chicken blood. The experimental design utilized parafilm sacs containing either whole blood from non-injected controls or serum containing Evans blue dye from saline-injected or antigen-injected birds. Feeding success of mites fed on blood from antigen-injected was not different from those fed on saline-injected blood; significant survival differences were detected between mites fed on blood from antigen- and saline-injected birds and those fed on the control group; no differences in mite success were found between the three experimental groups. Based on these results, they also concluded that the chickens did develop antibodies to northern fowl mite antigens, but that this did not decrease the mite infestation levels or *in vitro* feeding success.

## REFERENCES

Audy, J.R. (1951) Trombiculid mites and scrub itch. *Australian Journal of Science* 14, 94–96.

Barriga, O.O., Al-Khalidi, N.W., Martin, S. and Wyman, M. (1992) Evidence of immunosuppresion by *Demodex canis*. *Veterinary Immunology and Immunopathology* 32, 37–46.

Barta, O., Waltman, C., Oyekan, P.P., McGrath, R.K. and Hribernik, T.N. (1983) Lymphocyte transformation suppression caused by pyoderma – failure to demonstrate it in uncomplicated demodectic mange. *Comparative Immunology, Microbiology and Infectious Diseases* 6, 9–17.

Boyce, W.M. and Brown, R.N. (1991) Antigenic characterization of *Psoroptes* spp. (Acari: Psoroptidae) mites from different hosts. *Journal of Parasitology* 77, 675–679.

Boyce, W.M., Elliot, L., Clark, R. and Jessup, D. (1990) Morphometric analysis of *Psoroptes* spp. mites from bighorn sheep, mule deer, cattle, and rabbits. *Journal of Parasitology* 76, 823–828.

Boyce, W.M., Jessup, D.A. and Clark, R.K. (1991a) Serodiagnostic antibody responses to *Psoroptes* sp. infestations in bighorn sheep. *Journal of Wildlife Diseases* 27, 10–15.

Boyce, W.M., Mazet, J.A.K , Mellies, J., Gardner, I., Clark, R.K. and Jessup, D.A. (1991b) Kinetic ELISA for detection of antibodies to *Psoroptes* sp. (Acari: Psoroptidae) in bighorn sheep (*Ovis canadensis*). *Journal of Parasitology* 77, 692–696.

Brennan, J.M. and Reed, J.T. (1974) The genus *Eutrombicula* in Venezuela (Acarina: Trombiculidae). *Journal of Parasitology* 60, 699–711.

van Bronswijk, J.E.M.H. (1984) Temperature and boreal arachnid diseases of the Old World. In: Nutting, W.B. (ed.), *Mammalian Diseases and Arachnids*, Volume 2. CRC Press, Inc., Boca Raton, Florida, pp. 59–81.

Bubenik, G.A. (1989) Can androgen deficiency promote an outbreak of psoroptic mange mites in male deer? *Journal of Wildlife Diseases* 25, 639–642.

Burg, J.G., Collison, C.H. and Mastro, A.M. (1988) Comparative analysis of precipitating antibodies in White Rock and Fayoumi hens injected with bovine serum albumin or crude mite extract with resulting effects on northern fowl mite, *Ornithonyssus sylviarum* (Acari: Macronyssidae) population densities. *Poultry Science* 67, 1015–1019.

Burg, J.G., Collison, C.H. and Mastro, A.M. (1989) Precipitating antibody concentrations and visual estimates correlated to absolute northern fowl mite (Acari: Macronyssidae) densities from white rock hens. *Journal of Agricultural Entomology* 6, 153–157.

Chang, T.S., Rheins, M.S. and Winter, A.R. (1957) The significance of the bursa of Fabricius in antibody production in chickens: 1. Age of Chickens. *Poultry Science* 36, 735–738.

Chang, T.S., Rheins, M.S. and Winter, A.R. (1958) The significance of the bursa of Fabricius of chickens in antibody production: 2. Spleen relationship. *Poultry Science* 37, 1091–1093.

Chang, T.S., Rheins, M.S. and Winter, A.R. (1959) The signficance of the bursa of Fabricius of chickens in antibody production: 3. Resistance to *Salmonella typhimurium* infection. *Poultry Science* 38, 174–176.

Corbett, R., Banks, K., Hinrichs, D. and Bell, T. (1975) Cellular immune responsiveness in dogs with demodectic mange. *Transplantation Proceedings* 7, 557–559.

Culbertson, J.T. (1935) Antibody production by the rabbit against an ectoparasite. *Proceedings of the Society for Experimental Biology and Medicine* 32, 1239–1240.

Dawson, D.V., Whitmore, S.P. and Bresnahan, J.F. (1986) Genetic control of susceptibility to mite-associated ulcerative dermatitis. *Laboratory Animal Science* 36, 262–267.

DeLoach, J.R. and DeVaney, J.A. (1981) Northern fowl mite, *Ornithonyssus sylviarum* (Acari: Macronyssidae), ingests large quantities of blood from White Leghorn hens. *Journal of Medical Entomology* 18, 374–377.

DeLoach, J.R. and Wright, F.C. (1981) Ingestion of rabbit erythrocytes containing $^{51}$Cr-labeled hemoglobin by *Psoroptes* spp. (Acari: Psoroptidae) that originated on cattle, mountain sheep, or rabbits. *Journal of Medical Entomology* 18, 345–348.

Desch, C.E., Jr (1984) Biology of biting mites (Mesostigmata). In: Nutting, W.B. (ed.), *Mammalian Diseases and Arachnids*, Volume 1. CRC Press, Inc., Boca Raton, Florida, pp. 83–109.

DeVaney, J.A. (1978a) Effects of the northern fowl mite, *Ornithonyssus sylviarum* (Canestrini and Fanzago), on fertility and hatchability of eggs from artificially inseminated White Leghorn hens. *Poultry Science* 57, 1189–1191.

DeVaney, J.A. (1978b) A survey of poultry ectoparasite problems and their research in the United States. *Poultry Science* 57, 1217–1220.

DeVaney, J.A. and Augustine, P.C. (1988) Correlation of estimated and actual northern fowl mite populations with the evolution of specific antibody to a low molecular weight polypeptide in the sera of infested hens. *Poultry Science* 67, 549–556.

DeVaney, J.A. and Ziprin, R.L. (1980a) Detection and correlation of immune responses in White Leghorn chickens to northern fowl mite, *Ornithonyssus sylviarum* (Canestrini and Fanzago), populations. *Poultry Science* 59, 34–37.

DeVaney, J.A. and Ziprin, R.L. (1980b) Acquired immune response of White Leghorn hens to populations of northern fowl mite, *Ornithonyssus sylviarum* (Canestrini and Fanzago). *Poultry Science* 59, 1742–1744.

DeVaney, J.A., Elissalde, M.H., Steel, E.G., Hogan, B.F. and var Petersen, H. Del (1977) Effect of the northern fowl mite, *Ornithonyssus sylviarum* (Canestrini and Fanzago) on White Leghorn roosters. *Poultry Science* 56, 1585–1590.

DeVaney, J.A., Martin, B.W. and Harvey, R.B. (1984) Effects of bursectomy and splenectomy of White Leghorn roosters on subsequent northern fowl mite populations. *Poultry Science* 63, 1276–1278.

DeVos, J., Glaze, R.L. and Bunch, T.D. (1980) Scabies (*Psoroptic ovis*) in Nelson Desert Bighorn of northwestern Arizona. *Desert Bighorn Council 1980 Transactions*, pp. 44–46.

Eklund, J., Loomis, E. and Abplanalp, H. (1980) Genetic resistance of White Leghorn chickens to infestation by the northern fowl mite, *Ornithonyssus sylviarum*. *Archiv für Geflügelkunde*. 44, 195–199.

Fisher, W.F. (1972) Precipitating antibodies in sheep infested with *Psoroptes ovis* (Acarina:Psoroptidae), the sheep scab mite. *Journal of Parasitology* 58, 1218–1219.

Fisher, W.F. (1983a) Detection of serum antibodies to psoroptic mite antigens in rabbits infested with *Psoroptes cuniculi* or *P. ovis* (Acari: Psoroptidae) by enzyme-linked immunosorbent assay and immunodiffusion. *Journal of Medical Entomology* 20, 257–262.

Fisher, W.F. (1983b) Development of serum antibody activity as determined by enzyme-linked immunosorbent assay to *Psoroptes ovis* (Acarina: Psoroptidae) antigens in cattle infested with *P. ovis*. *Veterinary Parasitology* 13, 363–373.

Fisher, W.F. and Wilson, G.I. (1977) Precipitating antibodies in cattle infested by *Psoroptes ovis* (Acarina: Psoroptidae). *Journal of Medical Entomology* 14, 146–151.

Fisher, W.F., Guillot, F.S. and Cole, N.A. (1986) Development and decline of serum antibody activity to *Psoroptes ovis* antigens in infested cattle in an endemic and non-endemic scabies area of Texas. *Experimental and Applied Acarology* 2, 239–248.

Fox, I., Bayona, I.G., Umpierre, C.C. and Morris, J.M. (1967) Circulating precipitating antibodies in the rabbit from mite infection as shown by agar-gel tests. *Journal of Parasitology* 53, 402–405.

Friedman, S. and Weisbroth, S.H. (1975) The parasitic ecology of the rodent mite *Myobia musculi*. II. Genetic factors. *Laboratory Animal Science* 25, 440–445.

Guillot, F.S. (1981a) Population increase of *Psoroptes ovis* (Acari:Psoroptidae) on stanchioned cattle during summer. *Journal of Medical Entomology* 18, 44–47.

Guillot, F.S. (1981b) Susceptibility of Hereford cattle to sheep scab mites after recovery from psoroptic scabies. *Journal of Economic Entomology* 74, 653–657.

Guillot, F.S. and Stromberg, P.C. (1987) Reproductive success of *Psoroptes ovis* (Acari: Psoroptidae) on Hereford calves with a previous infestation of psoroptic mites. *Journal of Medical Entomology* 24, 416–419.

Guillot, F.S. and Wright, F.C. (1981) Evaluation of possible factors affecting degree of ear canker and numbers of psoroptic mites in rabbits. *Southwestern Entomologist* 6, 245–352.

Hall, R.D., Gross, W.B., Turner, E.C., Jr. and Siegel, P.B. (1978) Initial observations on the effect of corticosterone and inbred antibody competency in chickens on population development of the northern fowl mite. *Poultry Science* 57, 1728–1732.

Healey, M.C. and Gaafar, S.M. (1977) Immunodeficiency in canine demodectic mange. 1. Experimental production of lesions using antilymphocyte serum. *Veterinary Parasitology* 3, 121–131.

Hirsh, D.C., Baker, B.B., Wiger, N., Yuskulski, S.G. and Osburn, B.I. (1975) Suppression of *in vitro* lymphocyte transformation by serum from dogs with generalized demodicosis. *American Journal of Veterinary Research* 36, 1591–1595.

denHollander, N. and Allen, J.R. (1986) Cross-reactive antigens between a tick, *Dermacentor variabilis* (Acari:Ixodidae), and a mite, *Psoroptes cuniculi* (Acari: Psoroptidae). *Journal of Medical Entomology* 23, 44–50.

Jones, B.M. (1950) The penetration of host tissue by the harvest mite, *Trombicula autumnalis* Shaw. *Parasitology* 40, 247–260.

Kirkwood, A.C. (1986) History, biology and control of sheep scab. *Parasitology Today* 2, 302–307.

Krawiec, D.R. and Gaafar, S.M. (1980) Studies on the immunology of canine demodicosis. *Journal of the American Animal Hospital Association* 16, 669–676.

Laltoo, H. and Kind, L.S. (1979) Reduction of contact sensitivity reactions to oxazolone in mite-infested mice. *Infection and Immunity* 26, 30–35.

Laltoo, H., Voost, T.V. and Kind, L.S. (1979) IgE antibody response to mite antigens in mite infested mice. *Immunological Communications* 8, 1–9.

Loomis, E.C., Bramhall, E.L., Allen, J.A., Ernst, R.A. and Dunning, L.L. (1970) Effects of the northern fowl mite on White Leghorn chickens. *Journal of Economic Entomology* 63, 1885–1889.

Loomis, R.B. and Wrenn, W.J. (1984) 5.3 Systematics of the pest chigger genus *Eutrombicula* (Acari: Trombiculidae). In: Griffiths, D.A. and Bowman, C.E. (eds), *Acarology IV*, Volume I. Ellis Horwood, Chichester, England, pp. 152–159.

Losson, B., Detry-Pouplard, M. and Pouplard, L. (1988) Haematological and immunological response of unrestrained cattle to *Psoroptes ovis*, the sheep scab mite. *Research in Veterinary Science* 44, 197–201.

Matthysse, J.G., Jones, G.J. and Purnasiri, A. (1974) Development of northern fowl mite populations on chickens, effects on the host, and immunology. *Search Agriculture* 4, 1–39.

Miller, W.H. (1984) Diseases of domestic animals. In: Nutting, W.B. (ed.), *Mammalian Diseases and Arachnids*, Volume 2. CRC Press, Inc., Boca Raton, Florida, pp. 115–126.

Minnifield, N.M., Carroll, J., Young, K. and Hayes, D.K. (1993) Antibody development against northern fowl mites (Acari: Macronyssidae) in chickens. *Journal of Medical Entomology* 30, 360–367.

Moriello, K.A. (1991) Parasitic hypersensitivity. *Seminars in Veterinary Medicine and Surgery (Small Animal)* 6, 286–289.

Morrow, A.S. (1940) Allergic reactions to an antigen from the chigger (*Trombiculum irritans*). *Proceedings of the Society for Experimental Biology and Medicine* 43, 303–305.

Murano, T., Namiki, K., Uchino, T., Shimizu, S. and Fujisaki, K. (1989) Research note: Development of precipitating antibody in chickens experimentally infested with northern fowl mite, *Ornithonyssus sylviarum* (Acari: Macronyssidae). *Poultry Science* 68, 842–845.

Owen, L.N. (1972) Demodectic mange in dogs immunosuppressed with antilymphocyte serum. *Transplantation* 13, 616–617.

Payne, L.F. (1930) Feather mites and their control. *Bulletin of the Alabama Polytechnical Institute* 25, 61–63.

Pence, D.B. (1984) Diseases of laboratory animals. In: Nutting, W.B. (ed.), *Mammalian Diseases and Arachnids*, Volume 2. CRC Press, Inc., Boca Raton, Florida, pp. 129–187.

Powell, M.B., Weisbroth, S.H., Roth, L. and Wilhelmsen, C. (1980) Reaginic hypersensitivity in *Otodectes cynotis* infestation of cats and mode of mite feeding. *American Journal of Veterinary Research* 41, 877–882.

Pruett, J.H., Guillot, F.S. and Fisher, W.F. (1986) Humoral and cellular immunoresponsiveness of stanchioned cattle infested with *Psoroptes ovis*. *Veterinary Parasitology* 22, 121–133.

Pruett, J.H., Fisher, W.F. and DeLoach, J.R. (1987) Effects of dexamethasone on selected parameters of the bovine immune system. *Veterinary Research Communications* 11, 305–323.

Pruett, J.H., Fisher, W.F. and DeLoach, J.R. (1989) Dexamethasone-induced bovine T-lymphocyte suppression and the effect upon susceptibility to sheep scab mite (Acari: Psoroptidae) infestation. *Journal of Economic Entomology* 82, 175–179.

Rafferty, D.E. and Gray, J.S. (1987) The feeding behavior of *Psoroptes* spp. mites on rabbits and sheep. *Journal of Parasitology* 73, 901–906.

Scott, D.W., Farrow, B.R.H. and Schultz, R.D. (1974) Studies on the therapeutic and immunologic aspects of generalized demodectic mange in the dog. *Journal of the American Animal Hospital Association* 10, 233–244.

Scott, D.W., Schultz, R.D. and Baker, E. (1976) Further studies on the therapeutic and immunologic aspects of generalized demodectic mange in the dog. *Journal of the American Animal Hospital Association* 12, 203–213.

Stromberg, P.C. and Fisher, W.F. (1986) Dermatopathology and immunity in experimental *Psoroptes ovis* (Acari: Psoroptidae) infestation of naive and previously exposed Hereford cattle. *American Journal of Veterinary Research* 47, 1551–1560.

Stromberg, P.C. and Guillot, F.S. (1987) Bone marrow response in cattle with chronic dermatitis caused by *Psoroptes ovis*. *Veterinary Pathology* 24, 365–370.

Stromberg, P.C. and Guillot, F.S. (1989) Pathogenesis of psoroptic scabies in Hereford heifer calves. *American Journal of Veterinary Research* 50, 594–601.

Stromberg, P.C., Fisher, W.F., Guillot, F.S., Pruett, J.H., Price, R.E. and Green, R.A.

(1986) Systemic pathologic responses in experimental *Psoroptes ovis* infestation of Hereford calves. *American Journal of Veterinary Research* 47, 1326–1331.

Sweatman, G.K. (1958a) On the life history and validity of the species in *Psoroptes*, a genus of mange mites. *Canadian Journal of Zoology* 36, 905–929.

Sweatman, G.K. (1958b) Biology of *Otodectes cynotis*, the ear canker mite of carnivores. *Canadian Journal of Zoology* 36, 849–862.

Traub, R. and Wisseman, C.L., Jr (1974) The ecology of chigger-borne rickettsiosis (Scrub Typhus). *Journal of Medical Entomology* 11, 237–303.

Uhlir, J. (1991) Humoral and cellular immune response of rabbits to *Psoroptes cuniculi*, the rabbit scab mite. *Veterinary Parasitology* 40, 325–334.

Uhlir, J. (1992) Immunization of rabbits with antigens from *Psoroptes cuniculi*, the rabbit scab mite. *Folia Parasitologica* 39, 375–382.

Uhlir, J. (1993) Isolation and partial characterisation of an immunogen from the mite, *Psoroptes cuniculi*. *Veterinary Parasitology* 45, 307–317.

Uhlir, J. and Volf, P. (1992) Ivermectin: its effect on the immune system of rabbits and rats infested with ectoparasites. *Veterinary Immunology and Immunopathology* 34, 325–336.

Wang, L.-l. and Wang, D.-q. (1988) Studies on the karyotypes of five chigger mites (Acari: Trombiculidae and Leeuwenhoekidee). *Acta Entomologica Sinica* 31, 171–175.

Wassall, D.A., Kirkwood, A.C., Bates, P.G. and Sinclair, I.J. (1987) Enzyme-linked immunosorbent assay for the detection of antibodies to the sheep scab mite *Psoroptes ovis*. *Research in Veterinary Science* 43, 34–35.

Weisbroth, S.H., Wang, R., Scher, S., Spohr, B. and Luft, B. (1972) Immunopathology of psoroptic otitis in laboratory rabbits. A model for the study of allergic mechanisms in acariasis. *Federation Proceedings* 31, 614.

Weisbroth, S.H., Friedman, S. and Powell, M. (1974a) The parasitic ecology of the rodent mite *Myobia musculi*. I. Grooming factors. *Laboratory Animal Science* 24, 510–516.

Weisbroth, S.H., Powell, M.B., Roth, L. and Scher, S. (1974b) Immunopathology of naturally occurring otodectic otoacariasis in the domestic cat. *Journal of the American Veterinary Medical Association* 165, 1088–1093.

Wikel, S.K. (1982) Immune responses to arthropods and their products. *Annual Review of Entomology* 27, 21–48.

Wikel, S.K. (1988) Immunological control of hematophagous arthropod vectors: Utilization of novel antigens. *Veterinary Parasitology* 29, 235–264.

Wikel, S.K. (1989) Mite antigens recognized by antibodies from rabbits infested with *Psoroptes cuniculi*. *Medical Science Research* 17, 455–456.

Wikel, S.K., DeVaney, J.A. and Augustine, P.C. (1989) Host immune response to northern fowl mite: Immunoblot and lectin blot identification of mite antigens. *Avian Diseases* 33, 668–675.

Wrenn, W.J. and Loomis, R.B. (1984) 5.4 Host selectivity in the genus *Eutrombicula* (Acari: Trombiculidae). In: Griffiths, D.A. and Bowman, C.E. (eds), *Acarology IV*, Volume I. Ellis Horwood, Chichester, UK, pp. 160–165.

Wright, F.C. and DeLoach, J.R. (1980) Ingestion of erythrocytes containing $^{51}$Cr-labeled hemoglobin by *Psoroptes cuniculi* (Acari: Psoroptidae). *Journal of Medical Entomology* 17, 186–187.

Wright, F.C. and DeLoach, J.R. (1981) Feeding of *Psoroptes ovis* (Acari: Psoroptidae) on cattle. *Journal of Medical Entomology* 18, 349–350.

Wright, F.C., Riner, J.C. and Guillot, F.S. (1983) Cross-mating studies with *Psoroptes ovis* (Hering) and *Psoroptes cuniculi* Delafond (Acarina: Psoroptidae). *Journal of Parasitology* 69, 696–700.

Wright, S.M., Wikel, S.K. and Wrenn, W.J. (1988) Host immune responsiveness to the chigger, *Eutrombicula cinnabaris*. *Annals of Tropical Medicine and Parasitology* 82, 283–293.

Yunker, C.E. (1973) Mites. In: Flynn, R.J. (ed.), *Parasites of Laboratory Animals*. Iowa State University Press, Ames, pp. 425–492.

# 12

# Immunological-based Control of Blood-feeding Arthropods

## Stephen K. Wikel, Douglas K. Bergman and Rangappa N. Ramachandra

*Department of Entomology, Oklahoma State University, Stillwater, Oklahoma 74078, USA*

## INTRODUCTION

Haematophagous arthropods and the disease-causing agents they transmit are significant human and veterinary public health concerns. Effective control of arthropod-borne pathogens relies largely upon vector suppression (Laird, 1985). Development of resistance to insecticides and acaricides poses a threat to continued effective control of vectors (Nolan, 1990; Hemingway, 1992). Rapid development of arthropod resistance to new control compounds might discourage efforts to discover new insecticides and acaricides in light of the high costs of research, development and registration. Novel vector suppression strategies are needed. A particularly promising approach is immunological based control, anti-arthropod vaccines (Wikel 1981, 1982, 1988, 1996a, b; Wikel *et al.*, 1992; Kay and Kemp, 1994; Opdebeeck, 1994). Advantages of anti-arthropod vaccines include: target species specificity, environmental safety, absence of human health risks, no meat/milk residues, ease of administration and cost. Immunological control of arthropods has moved from a possibility to a reality during the past decade.

The most significant advances in anti-arthropod vaccine research involve tick–host associations (Wikel *et al.*, 1992; Kay and Kemp, 1994; Opdebeeck, 1994). The elegant research of Willadsen and colleagues resulted in successful development of a recombinant anti-*Boophilus microplus* vaccine based upon 'concealed' immunogens of digestive tract origin (Willadsen and Kemp, 1988; Tellam *et al.*, 1992; Riding *et al.*, 1994).

© CAB INTERNATIONAL 1996. From Wikel, S.K. (ed.) *The Immunology of Host–Ectoparasitic Arthropod Relationships.* CAB INTERNATIONAL, Wallingford.

Anti-insect vaccine research lags behind that focused on immunological control of ticks (Kay and Kemp, 1994; Jacobs-Lorena and Lemos, 1995). Both rapidly engorging insects and long-term feeding ticks are susceptible to immunological control strategies. A triatomine bug may feed for a few minutes to an hour (Lavoipierre *et al.*, 1959; Friend and Smith, 1971), while mosquito engorgement is completed in less than ten minutes (Ribeiro, 1987). These durations of exposure to appropriately directed host immune effector elements should be capable of damaging an insect. Anti-insect vaccine research is in its infancy. Limited successes achieved to date by vaccination with crude extracts of tissues should not be interpreted to mean that effective anti-insect vaccines cannot be developed. Efforts need to be directed towards the use of defined antigens from digestive tract and other tissues (Kay and Kemp, 1994; Jacobs-Lorena and Lemos, 1995). Differential expression of antigens during engorgement should be determined. Bloodmeal digestion in the gut lumen of insects might complicate immunogen selection as well as resulting in proteolytic breakdown of host antibodies and complement (Kay and Kemp, 1994). Maximal *Aedes aegypti* protease activity occurred at 24 hours after blood-feeding, which might provide a window of opportunity prior to digestion for host antibody to complex with mosquito immunogen (Briegel and Lea, 1975).

An understanding of host immune responses induced by infestation is essential for development of any anti-arthropod vaccine. Candidate immunogens must be effectively processed and presented to specifically reactive B and T lymphocytes and elicit effective immune responses in genetically diverse populations. Immunogens that stimulate strong antibody responses during infestation are not necessarily good candidate immunogens for a vaccine. Antibodies arising during a response where infestation is reduced might be effective probes for vaccine immunogens. Do not assume that protection rests only with antibodies. T-cell-mediated responses are not only key regulatory elements, but could prove to be important effectors of protection. T lymphocytes should be as important probes for the identification of immunogens.

Characterization of the mechanisms by which an arthropod evades or suppresses host immunity can be of great value. Molecules that induce host immunosuppression are possibly essential for arthropod survival. Arthropod suppression of specific host defences indicates the importance of those components of the immune response. Full expression of suppressed host immune regulatory and/or effector pathways would likely be deleterious to arthropod feeding and impair pathogen transmission. Arthropod molecules that suppress host immunity should be considered as vaccine targets. An anti-immunosuppressant vaccine might enhance resistance to blood-feeding and pathogen transmission by neutralizing the effects of arthropod-introduced immunosuppressive molecules, allowing full expression of host immunity (Titus and Ribeiro, 1990; Bergman *et al.*, 1995; Wikel, 1996a, b).

Advances in immunology, cell biology, and biotechnology facilitate anti-arthropod vaccine development. Recombinant immunogens are not the only basis for an effective vaccine. Novel cell culture strategies can prove to be a valuable source of protection-inducing immunogens. An understanding of cytokine networks allows for development of novel adjuvants and delivery systems that target specific cell populations. Vaccines can become a valuable component in the array of methods for control of blood-feeding arthropods and vector-borne diseases.

## ANTI-ARTHROPOD VACCINES

Immunization-based control of arthropods can be traced to the visionary tick research of Trager (1939a, b). Guinea pigs developed immunity against infestation with *Dermacentor variabilis* larvae after intracutaneous inoculation of a *D. variabilis* larval extract (Trager, 1939a). In addition, protection-inducing antigens were reported to be present in 'cephalic glands' and digestive tract of partially engorged *D. variabilis*, and both fed and unfed salivary glands (Trager, 1939b). These two reports provided the foundation for subsequent studies, virtually all of which, to date, utilized whole arthropod or tissue extracts as sources of protection-inducing immunogens. Only recently have specific 'concealed' tick gut immunogens been identified (Tellam *et al.*, 1992; Opdebeeck, 1994).

### Insects

Vaccine-based control of arthropods is challenging due to their ectoparasitic nature, intermittent feeding and ability to utilize a variety of different hosts (Opdebeeck, 1994). Insects possess several characteristics that must be considered in anti-insect vaccine development: the generally brief period of blood-feeding (lice are a notable exception), extracellular digestion and the presence of a peritrophic membrane. Results of attempts to vaccinate against insects have been variable (Jacobs-Lorena and Lemos, 1995). However, these results are not sufficient cause to abandon efforts to develop anti-insect vaccines. In fact, they should provide a stimulus for further investigations due to the indications that anti-insect immunity can be induced.

#### *Mosquitoes*

Initial attempts to artificially induce immunity to mosquitoes involved administration of *Anopheles quadrimaculatus* homogenate to rabbits (Dubin *et al.*, 1948). Antigen I consisted of whole female mosquito homogenate. Antigen II was an extract of antigen I prepared by repeated freezing, thawing and passage through a Seitz filter. Both antigens were administered to rabbits by subcutaneous inoculation. Mosquitoes were exposed to immunized or control rabbits for five minutes. Mosquitoes were killed upon removal from the host and microscopically examined for engorgement. Immunization with mosquito homogenate did not adversely affect engorgement. Unfortunately, post-engorgement viability and fecundity were not evaluated in this study. The influence of anti-mosquito antibodies would likely not be evident within a few minutes of feeding.

Rabbits were immunized with *Anopheles stephensi* tissue homogenates (Alger and Cabrera, 1972). Antigen preparations consisted of: (i) $1063 \times g$ supernatant of whole ground mosquitoes; (ii) $1063 \times g$ pellet of whole ground mosquitoes; and (iii) homogenized mosquito midgut. All homogenates were prepared from seven- to ten-day-old female *An. stephensi* fed only on dextrose. Mortality rate of mosquitoes which obtained a bloodmeal from rabbits immunized with *An. stephensi* midgut was significantly higher on feedings one, three, four, five and seven than that of adjuvant-immunized controls or those mosquitoes fed on rabbits immunized with whole mosquito homogenates. Variability in mosquito feeding was observed when rabbits were challenged after different booster inoculations of antigen. On some occasions, viability was greater after feeding upon mosquito-antigen-immunized rabbits than controls. Although results were variable, the feasibility of anti-mosquito immunization was indicated (Table 12.1).

Rabbits and guinea pigs were immunized with an homogenate prepared from sugar fed *A. aegypti* prior to challenge infestations with *A. aegypti* and *Culex tarsalis* (Sutherland and Ewen, 1974). All animals immunized with mosquito homogenate stopped drinking, eating, urinating and defecating within 48 hours of the first injection and remained in that condition for six to ten days. Rabbits recovered, one guinea pig died prior to mosquito challenge, and the second guinea pig died post-challenge. Control animals were not affected. The influence of these severe reactions on subsequent tests was not ascertained. Fecundity of *A. aegypti* was reduced approximately 24% after feeding on immunized guinea pigs and 31% after engorging on immunized rabbits, when compared with controls. Mosquito mortality was not affected by a bloodmeal obtained from an immunized host. Re-sensitization of two rabbits and challenge with *A. aegypti* resulted in reduced egg production. The anti-mosquito response was not evident ten days later. Challenge mosquito fecundity was greater than that of mosquitoes that fed upon controls, when the same sensitized rabbits were infested ten days later with *A. aegypti*. Absence of reactivity was attributed to a decline in antibody titre below an effective threshold. Fecundity and viability of *Cx. tarsalis* was not affected by antibodies to *A. aegypti*.

Mosquito immunogens induced production of anti-mosquito antibodies that reduced fecundity and viability of progeny of *A. aegypti*, which obtained a bloodmeal from vaccinated rabbits (Ramasamy *et al.*, 1988). Mosquitoes were blood fed for 24 hours prior to preparation of antigen extracts consisting

**Table 12.1.** Immunological-based control of mosquitoes.

| Species | Results | Reference |
|---|---|---|
| Anopheles quadrimaculatus | Mosquitoes exposed to immunized or control rabbits for five minutes. Mosquitoes killed upon removal. Did not adversely affect engorgement | Dubin et al., 1948 |
| Anopheles stephensi | Mortality of mosquitoes significantly higher after feeding on rabbits immunized with mosquito midgut. Anti-mosquito affect variable | Alger and Cabrera, 1972 |
| Aedes aegypti | Fecundity reduced for Aedes aegypti fed on hosts immunized with whole mosquito homogenates. Culex tarsalis challenge not affected | Sutherland and Ewen, 1974 |
| Aedes aegypti | Mosquitoes fed on rabbits immunized with extracts of either head/thorax, midgut or remainder of abdomen had variable levels of reduced fecundity. No significant difference in mortality | Ramasamy et al., 1988 |
| Aedes aegypti | Mosquitoes fed on mice immunized with whole mosquito or midgut antigens had slight increase in mortality, but no change in fecundity | Hatfield, 1988 |
| Anopheles tessellatus | Mosquitoes fed on rabbits immunized with either head/thorax, midgut or abdomen had no significant reduction in mortality, but reduced fecundity was reported. Culex quinquefasciatus fed on rabbits immunized with Anopheles tessellatus head/thorax or midgut displayed increased mortality | Ramasamy et al., 1992 |

of: (i) head/thorax; (ii) midgut; and (iii) the remainder of the abdomen. Eight immunizations were given at three to four week intervals. Serum antimosquito antibodies were detected by immunodiffusion ten days after the fifth immunization. Very likely, the use of a more sensitive assay would have resulted in earlier detection. Mosquitoes readily obtained a bloodmeal from immunized rabbits. No significant difference in mortality was reported for mosquitoes fed on immunized and control rabbits. However, fecundity was significantly reduced for one rabbit immunized with head/thorax, two rabbits immunized with midgut and one rabbit immunized with abdomen. Mosquitoes feeding on other rabbits within those groups were not adversely affected. Immunofluorescence was used to detect rabbit antibodies in mosquito tissues. Mosquitoes which fed on immunized rabbits had host antibodies, or antibody fragments, associated with oocytes.

*Aedes aegypti* fed upon mice immunized with whole mosquito or midgut antigens had a slight increase in mortality, when compared with mosquitoes obtaining a bloodmeal from controls (Hatfield, 1988). No changes in fecundity were observed. Higher anti-mosquito enzyme-linked immunosorbent assay (ELISA) titre was directly correlated with more rapid mortality of feeding mosquitoes. Depending upon the study, anti-mosquito antibodies affected viability and fecundity.

Antigen preparations consisting of *Anopheles tessellatus*: (i) head/thorax; (ii) midgut; and (iii) rest of the abdomen were homogenized in phosphate buffered saline (Ramasamy *et al.*, 1992). Antigen homogenates were administered to rabbits in four intramuscular injections given at three to four week intervals, resulting in development of high antibody titres. *Anopheles tesselatus* engorged readily on both immunized and control rabbits. No significant mortality differences were detected at 48 hours post-engorgement. Increased mortality of *Cx. quinquefasciatus* was observed for mosquitoes fed on antibodies to *An. tessellatus* head/thorax or midgut. Egg production by *An. tessellatus* was significantly lower after exposure to anti-mosquito antibodies. Fecundity reductions were 15% (head/thorax), 23% (abdomen), and 20% (midgut) after feeding on blood containing antibodies to mosquito antigens. A third experiment resulted in variable changes in fecundity with some groups displaying increased egg production. Antibodies to *An. tessellatus* did not alter egg production by *Cx. quinquefasciatus*. Mouse antibodies to *Cx. quinquefasciatus* did not reduce fecundity of that species. Mortality was observed against a different species rather than the one used to sensitize rabbits. Observed variations in susceptibility to anti-mosquito antibodies are possibly related to differences in feeding patterns of the mosquito species studied.

Antibodies to mosquito digestive tract might influence pathogen development within the mosquito (Table 12.2). Antibodies to midgut homogenate of *An. stephensi* were tested for their ability to alter the development of *Plasmodium berghei* when both the anti-midgut immunoglobulins and parasites were taken up simultaneously in the bloodmeal (Lal *et al.*, 1994). Infection rate and number of *P. berghei* oocysts were significantly reduced for those mosquitoes that fed upon anti-mosquito-midgut antibodies. A salivary gland infection rate of 42.9% was observed for control mosquitoes between 17 and 21 days post-infection, while no sporozoites were detected through day 29 in salivary glands of mosquitoes that consumed anti-midgut antibodies. Earlier studies demonstrated that anti-mosquito antibodies reduced infection of *Anopheles farauti* with the oocyst stage of *P. berghei* (Ramasamy and Ramasamy, 1990). In addition, anti-mosquito midgut antibodies reduced susceptibility of *A. aegypti* to arbovirus infection (Ramasamy *et al.*, 1990). Specific immunogens and mechanisms by which anti-mosquito midgut antibodies disrupt pathogen development have not been identified.

A vaccine directed against only the vector (stand-alone anti-vector vaccine), or an anti-vector component in a vaccine against one or several

**Table 12.2.** Influence of antibodies to mosquito digestive tract on pathogen development.

| Species | Results | Reference |
| --- | --- | --- |
| *Anopheles stephensi* | Reduced number of *Plasmodium berghei* oocysts | Lal *et al.*, 1994 |
| *Anopheles farauti* | Reduced number of *Plasmodium berghei* oocysts | Ramasamy and Ramasamy, 1990 |
| *Aedes aegypti* | Reduced susceptibility to arbovirus infection | Ramasamy *et al.*, 1990 |

pathogens transmitted by a given arthropod, represents novel methods for controlling vector-borne diseases. This strategy should be vigorously investigated to identify specific immunogens, mechanisms responsible for disruption of vector-borne pathogen development, and the basis for altered arthropod feeding.

### Biting flies

Immunological based control is a feasible strategy for control of biting flies. Rabbits were immunized with selected tissues of the stable fly, *Stomoxys calcitrans*, in an intriguing study of immunological based control (Schlein and Lewis, 1976). The following tissues were isolated, washed and homogenized prior to administration of two subcutaneous injections given seven days apart: (i) cuticle and adhering hypodermal cells; (ii) thoracic muscles; (iii) abdominal tissues; and (iv) wing buds. *Stomoxys calcitrans* mortality was higher in all immunization groups than controls. Most significant mortality was associated with immunization with thoracic muscle. In addition, flies obtaining bloodmeals from immunized rabbits displayed paralysis of legs or wings and difficulty in probing. Difficulty in probing was expressed as bending of the mouthparts as flies attempted to pierce rabbit skin. This effect was especially observed in flies that fed on rabbits immunized with wing bud homogenate.

*Glossina morsitans* were allowed to feed on rabbits immunized with *S. calcitrans* tissues (Schlein and Lewis, 1976). *Glossina morsitans* mortality was significantly higher than controls for flies that obtained bloodmeals from rabbits immunized with cuticle/adhering hypodermal cells and wing buds. The wing bud homogenate consisted largely of cuticle and hypodermal cells. Antibodies consumed in the bloodmeal appeared to effectively disrupt fly physiological processes. Shared epitopes exist between the two biting fly species examined. Serum immunoglobulins pass through the gut of *Sarcophaga falculata* and selected mosquito species into the haemolymph, where they can affect tissues (Schlein *et al.*, 1976; Vaughan and Azad, 1988).

## Myiasis

Development of vaccines to control myiasis has advanced significantly in recent years, and the reader is referred to the treatment of this topic by R.M. Sandeman in this book (Chapter 8). The most promising prospects for vaccine-based control are the warble flies, *Hypoderma bovis* and *Hypoderma lineatum*, and the sheep blowfly, *Lucilia cuprina* (Sandeman, 1992).

Cattle acquired resistance to *Hypoderma* spp. infestation (Gingrich, 1982). Resistance did not correlate with anti-fly antibodies. Cell-mediated immunity appeared to be correlated with the protective response. Vaccination against *Hypoderma* spp. increased *in vivo* killing of larvae in the bovine host (Khan *et al.*, 1960). Hypodermin A purified from first instar larvae induced the greatest level of long-lived protection (Pruett *et al.*, 1987). Availability of recombinant hypodermins (Moire *et al.*, 1994) opened the way for determination of the utility of these molecules as vaccine immunogens.

Significant advances have been made towards development of a vaccine for control of *L. cuprina*. Immunogens were identified in the peritrophic membrane which induced antibodies capable of both *in vivo* and *in vitro* inhibition of larval growth (East *et al.*, 1993). Larval killing was not significant *in vivo*. However, *in vitro* larval killing was enhanced by increasing the concentration of antibody. Efficacy of the vaccine was related to immunoglobulin complexing with peritrophic membrane epitopes and blocking passage of nutrients to the midgut cells of the larva (Eisemann and Binnington, 1994). In another study, a monoclonal antibody reactive with a *L. cuprina* midgut epitope significantly impaired *in vitro* growth of larvae (Fry *et al.*, 1994).

Immunoglobulins produced by B lymphocytes obtained from regional lymph nodes of sheep infested with *L. cuprina* were used to identify candidate immunogens by immunoblotting (Meeusen and Brandon, 1994). Four immunogens were identified and used for vaccination. Sheep immunized with these antigens had fewer infections than controls. Protection against *L. cuprina* should not be considered to be solely dependent upon antibody production. The role of effector T lymphocytes in vaccination-induced protection needs to be investigated. The selection of appropriate immunogens and adjuvants might stimulate both antibody and cell-mediated responses that provide an acceptable level of protection against infestation and lesion development.

## Sucking lice

Cox/Swiss and C3H/HeSN mice acquired resistance to *Polyplax serrata* within 50 days of infestation (Ratzlaff and Wikel, 1990). This systemic, anamnestic resistance had an immunological basis. Second infestation louse burdens were reduced by 78% and 98% for C3H/HeSn and Cox/Swiss mice, respectively.

Louse-specific immune responses included: louse-antigen-specific *in vitro* proliferation of draining lymph node cells, and immediate and delayed skin reactions to louse antigens by infested animals. Mice developed variable levels of immunity to infestation with a mouse louse.

If acquired resistance develops to infestation, could mice be immunized against lice? Murine resistance to *P. serrata* infestation was induced by vaccination (Ratzlaff and Wikel, 1990). Lice were placed into 0.15 M phosphate buffered saline (pH 7.2) and sonicated prior to centrifugation at 12,000 × g. Mice were administered 100 µg of louse immunogen daily for six days by subcutaneous injection without adjuvant into the base of the tail. Control mice were divided into two groups and administered either bovine serum albumin or phosphate buffered saline. Louse resistance was induced by administration of *P. serrata* immunogens in two separate immunization experiments. Louse burden/host weight ratio was reduced significantly to 62% for louse-immunogen-vaccinated mice. Slight reduction in louse burden induced by vaccination with bovine serum albumin was not significant. Non-immunized mice expressing acquired resistance during a second infestation reduced their louse burden by 94%.

Resistance to infestation can be induced by immunization for a natural louse–host association. This type of study needs to be extended to louse–host associations of medical and veterinary importance. Defined candidate immunogens need to be identified. The long-term association of sucking lice with their hosts increases the likelihood that immunization could be an effective control strategy.

## *Fleas*

Fleas are important disease vectors (Hoogstraal, 1980) and their bites elicit cutaneous hypersensitivity (Feingold *et al.*, 1968). Immunological control of fleas and desensitization of flea bite hypersensitivity would be of great value. Cherney *et al.* (1939) reported that susceptible persons were protected against flea bites by a vaccine prepared from whole fleas. Immunogens prepared from the gut of unfed *Ctenocephalides felis felis* were used to vaccinate cats (Opdebeeck and Slacek, 1993). A 100,000 × g pellet was administered subcutaneously with Quil A adjuvant. Depression, oedema, inflammation and fever were associated with use of Quil A. The change was made to Ribi adjuvant, which did not induce a reaction upon administration. Vaccinated cats developed anti-flea antibodies. The number of fleas recovered from vaccinated and control cats did not differ over the six challenge infestations given to each group. Production of oocytes did not differ for either group. Although the results of this study were negative, efforts to develop an anti-flea vaccine should not be abandoned. Attention might need to be focused upon candidate immunogens derived from the engorged flea.

Despite the often contradictory results of studies to date (Jacobs-Lorena and Lemos, 1995), vaccination-based control of haematophagous insects remains a distinct possibility that must be more thoroughly investigated. A critical step will be the fractionation of appropriate insect tissues to identify potential protection-inducing immunogens derived from unfed or fed insects. The use of whole tissue extracts might be inducing responses to immunodominant molecules with limited efficacy as a vaccine immunogen, while inhibiting responses to other less immunogenic, potentially protection-inducing molecules.

## Ticks

Anti-tick vaccine research evolved from the initial induction of guinea pig resistance to *D. variabilis* larvae by immunization with whole larval extract (Trager, 1939a). Also, protection-inducing immunogens were present in a variety of tissues: 'cephalic glands', digestive tract of partially engorged *D. variabilis*, and salivary glands from either fed or unfed ticks (Trager, 1939b). The logical starting place for subsequent investigations into artificial induction of anti-tick immunity was the use of extracts of whole ticks or specific tissues, particularly salivary glands (Wikel, 1981, 1982, 1996a, b; Wikel *et al.*, 1992). Early efforts were largely focused on inducing responses that mimicked those observed during expression of acquired resistance to infestation. Variable levels of resistance were often induced among a group of animals by immunization with the same antigenic extract. A second problem was vaccination-induced sensitization for development of cutaneous hypersensitivity responses upon challenge infestation. Both of these situations are unacceptable in a commercial vaccine.

Development of vaccines that utilize 'novel' or 'concealed' immunogens has proven to be a potentially highly effective strategy for induction of anti-tick immunity (Wikel, 1988; Tellam *et al.*, 1992; Riding *et al.*, 1994). The most dramatic success to date is development of a recombinant vaccine against *B. microplus* (Tellam *et al.*, 1992). Advances in the fields of immunology and biotechnology provide the foundation for rapid progress in this area of investigation, which has such vast potential for improving the quality of life in many regions of the world.

### Whole tick extracts

Guinea pig resistance to *D. variabilis* was induced by intracutaneous inoculation of a supernatant of whole larvae homogenized in saline (Trager, 1939a). Immunity was systemic. Induced resistance resulted in engorgement levels of 0 to 44%. Saline-vaccinated controls allowed 25 to 42% engorgement. Resistance to infestation was stimulated by immunization with 'cephalic glands' and digestive tract from partially engorged ticks, as well as salivary glands from fed or unfed ticks (Trager, 1939b). All fractions contained antigenic material, but the best responses were elicited with salivary gland preparations. Again, variable levels of resistance were observed. Immunization with unfed salivary gland antigens resulted in challenge infestation engorgement between 23 and 97%, while engorgement values post-immunization with fed salivary gland antigens were 25 to 62%. Saline-control-immunized host allowed 91% of challenge ticks to engorge. Induction of variable levels of resistance has been a continuing problem associated with immunization regimens using whole tick extracts.

Variable levels of guinea pig resistance to larvae of *Ixodes holocyclus* were induced by immunization with larval extract of that species (Bagnall, unpublished). Larval challenge rejection was 38 to 68%. Administration of an extract prepared from unfed, adult *D. variabilis* did not induce immunity to a subsequent challenge with adults of the same ixodid species (Ackerman *et al.*, 1980). Immunizations with 2 mg homogenate per injection were given at weekly intervals for three weeks. No significant differences from controls were observed for mean infestation period and engorgement weight of ticks collected from immunized hosts.

Hereford and Jersey calves were immunized with 18 or 3 mg $kg^{-1}$, respectively, with homogenate of whole adult *A. americanum* (McGowan *et al.*, 1981). These very high doses of immunogen induced resistance that was expressed by reduced female *A. americanum* engorgement weights, when compared with controls. Cattle were given a repeat infestation, which resulted in even lower mean engorgement weights. Infestation heightened expression of the vaccination-induced resistance. The possibility of synergism between vaccination-induced and acquired resistance was not investigated. A 'natural booster' effect from tick challenge would be a highly desirable characteristic of an anti-tick vaccine.

Resistance to *Rhipicephalus appendiculatus* infestation was induced in rabbits by immunization with antigen–antibody complexes derived from immunodiffusion gels prepared with whole female tick antigen and antibodies induced by immunization with unfed tick homogenate (Mongi *et al.*, 1986a). The antibodies served as the detection system for tick homogenate immunogens. Induced resistance was characterized by delayed attachment, prolonged feeding time and delayed drop-off from the host. Final engorgement weight was not altered. Several candidate *R. appendiculatus* immunogens were subsequently identified (Mongi *et al.*, 1986b).

Nymphal homogenates of *Amblyomma hebraeum* and *Amblyomma marmoreum* prepared by grinding, sonication, centrifugation and passage through a 0.45 μm filter were used to immunize rabbits (Tembo and Rechav, 1992). Immunized animals developed a significant level of resistance to infestation with the sensitizing species, when compared to naive and Quil A adjuvant controls. Cross-resistance to the heterologous species was not enhanced by immunization.

## Tick salivary glands

A logical approach to the induction of resistance similar to acquired resistance to infestation was vaccination with salivary-gland-derived antigens. Numerous investigators pursued this line of investigation. Magnitude of resistance induced was generally variable and hypersensitivity reactions often occurred at tick attachment sites on vaccinated animals. Many experiments involved laboratory animal species rather than natural hosts. Levels of resistance induced were often not compatible with those needed for a commercially acceptable vaccine. Despite difficulties, very valuable information was obtained from these studies.

Two one-day-old calves were immunized with salivary gland antigen of *B. microplus*, inducing development of antibodies and resistance to infestation (Brossard, 1976). Salivary gland antigen prepared from four day fed female *D. andersoni* stimulated an immune response that significantly reduced the number and engorgement weight of challenge larvae of the same species (Wikel, 1981). Significant resistance to larval infestation was induced by immunization with 1 µg of salivary gland antigen administered on days 0 and 12 by intradermal inoculation with Freund's complete adjuvant. A similar vaccination regimen with either 10 µg or 1 µg of salivary gland extract administered in Freund's incomplete adjuvant induced solid resistance to larval infestation. Percutaneous application of antigen extract did not stimulate resistance to larval challenge.

*Bos indicus* × *Bos taurus* calves were immunized subcutaneously with salivary gland antigens prepared from *H. anatolicum anatolicum* and challenged with adults ticks of the same species (Banerjee et al., 1990). Three salivary gland antigen formulations were prepared as follows: (i) a 10,000 × g supernatant of homogenized and sonicated glands was used as whole salivary gland antigen; (ii) stored salivary glands thawed, homogenized, centrifuged at 10,000 × g and the supernatant used as antigen; and (iii) the resuspended pellet of the 10,000 × g centrifugation described in item (ii) was used as antigen. Calves immunized with the first antigen preparation developed a response that significantly increased the engorgement period, reduced engorgement weight, increased the pre-oviposition period, reduced the egg mass and decreased the number of eggs laid, when compared with controls. Immunization with antigen (ii) induced immunity expressed by significantly reduced engorgement weight, increased pre-oviposition period and reduced egg mass. Immunization with antigen (iii) induced responses that did not differ from those of controls. Salivary gland and whole tick extracts were prepared from *H. anatolicum anatolicum* and administered to rabbits by two intramuscular injections on days 0 and 14 with a subcutaneous injection on day 21 (Manohar and Banerjee, 1992). Resistance was induced by immunization with salivary gland extract, whole adult female tick extract, and 30,000 × g supernatant of whole tick extract. Vaccination-induced resistance resulted in decreased engorgement and egg mass weights, as well as increased engorgement and pre-oviposition periods.

A novel vaccination approach utilizing salivary gland antigens is the development of an 'anti-immunosuppressant' vaccine. Tick salivary glands contain immunosuppressants that modulate host T-lymphocyte proliferation and cytokine elaboration by macrophages and T lymphocytes (Wikel et al., 1994; Wikel, 1996a). Readers interested in this topic are directed to Chapter 5 in this volume. An anti-immunosuppressant vaccine would be based on a chemical/physical modification of the tick-derived immunosuppressant, or identification of epitopes unique to the immunosuppressant, so that immunization of the host would induce an immune response that would neutralize the native immunosuppressant introduced by the tick. Vaccination must not in itself induce immunosuppression.

### Immunogens other than whole tick and salivary gland extracts

Tick molecules not introduced during the feeding process are extremely appealing candidates for incorporation into anti-tick vaccines. A vaccine immunogen must be accessible to host immune effector elements in the bloodmeal. An obvious source of immunogens is tick digestive tract. However, other tick tissues should not be discounted as vaccine immunogens, since host antibodies can transverse the tick gut and be detected in haemolymph (Ackerman et al., 1981; Ben-Yakir, 1989; Wang and Nuttall, 1994). Possible target immunogens could include epitopes of tick hormones, hormone receptors, nervous tissue, reproductive tissue, muscle and haemolymph proteins.

Attention has been focused on the use of immunogens associated with the tick digestive tract. These molecules would not be introduced into the host in the same way as saliva during engorgement. Hosts could be under less initial pressure to direct immune responses away from 'novel' or 'concealed' immunogens that are not, or rarely, encountered during infestation. Gut epitopes are accessible to host immune effector elements in the bloodmeal. An important question is whether or not potential protection-inducing epitopes are differentially expressed by gut cells during the feeding cycle. Attention is most often focused on development of a vaccine consisting of a recombinant protein.

Trager (1939b) was the first to utilize digestive tract of partially fed ticks as an immunogen (see Table 12.3). The usefulness of the digestive tract and internal organs as immunogens was further established by Allen and Humphreys (1979). Antigen I was prepared from midgut and reproductive organs of female *D. andersoni* that had engorged for five days, while antigen II consisted of all internal organs from similar ticks. Guinea pigs were immunized by subcutaneous injection of antigen and Freund's complete adjuvant into footpads on days 0 and 14. Guinea pigs were given a challenge

**Table 12.3.** Immunization with tick digestive tract immunogens.

| Tick species | Immunogen preparation | Reference |
|---|---|---|
| *Dermacentor variabilis* | Digestive tract | Trager, 1939b |
| *Dermacentor andersoni* | Midgut/reproductive organs<br>All internal organs | Allen and Humphreys, 1979 |
| *Dermacentor variabilis* | Midgut | Ackerman et al., 1980 |
| *Amblyomma americanum* | Midgut brush border membranes | Wikel, 1988 |
| *Rhipicephalus apendiculatus* | Solubilized midgut proteins | Essuman et al., 1991, 1992 |
| *Boophilus microplus* | Digestive tract, defined immunogens | Refer to Opdebeeck, 1994 |
| *Boophilus microplus* | Gut, recombinant Bm86, Bm89 | Willadsen et al., 1988<br>Tellam et al., 1992<br>Riding et al., 1994 |

infestation with four pairs of adult *D. andersoni* on day 28. Ticks engorged on four of six controls and produced ova, which yielded large numbers of larvae. Ticks derived from antigen-I-immunized hosts produced relatively few ova, and none of the ova hatched to produce viable larvae. Ticks infesting antigen-II-immunized guinea pigs neither engorged nor produced ova. Immunization with internal organs produced a consistently effective level of anti-tick immunity. Vaccination of cattle with midgut/reproductive tract immunogen did not induce as dramatic an anti-tick response. Each calf was given 67 mg of antigen by intramuscular injection on days 0 and 16 with Freund's complete adjuvant and again on day 25 without adjuvant. No significant difference was observed in the numbers of ticks recovered from immunized and control hosts. However, immunized hosts yielded significantly smaller live ticks, fewer ova and fewer larvae than controls.

An extract prepared from midgut tissue of female and male *D. variabilis* fed for 48 to 72 hours and a 10,000 × g supernatant or pellet of midgut extract were used as immunogens (Ackerman et al., 1980). Ticks fed on rats immunized with midgut antigen preparations displayed delayed attachment, reduced engorgement weights, lengthened pre-oviposition period, altered egg production and reduced numbers of ova that hatched.

*Amblyomma americanum* digestive tract brush border fragments were used to induce resistance to infestation with adults of the same species (Wikel, 1988). Female engorgement weights were reduced up to 69.8%. Mortality of ticks feeding on two treatment groups immunized with gut brush border fragment immunized guinea pigs was between 37.5 and 71.5%. Although brush border immunogen was prepared from female *A. americanum*, anti-tick immunity affected both female and male ticks. Biotinylated lectins were used to probe SDS–PAGE electrophoretograms of fractionated tick gut brush border that were electroeluted to nitrocellulose (Wikel, 1988). Virtually all proteins/polypeptides were glycosylated.

Midgut-soluble membrane-bound immunogens of female *R. appendiculatus* were partially purified by gel filtration chromatography and used to vaccinate cattle (Essuman *et al.*, 1991). Immunized cattle were significantly resistant to all life cycle stages of *R. appendiculatus*. The number of ticks which fed was not significantly reduced. However, engorgement weights and egg viability of ticks feeding on immunized cattle were significantly lower. Ticks that engorged on immunized animals produced a significantly higher number of eggs.

A comparison was made of rabbit immunity to *R. appendiculatus* induced by infestation or infestation followed by immunization with solubilized midgut proteins (Essuman *et al.*, 1992). Immunization resulted in significantly larger reduction in engorgement, egg batch and percentage of eggs hatched, when compared to infestation alone. The greatest reduction in engorgement weight was evident when immunization occurred after two experimental infestations. All experimental infestations occurred prior to immunization. An effective commercial vaccine should be one that could be safely administered to animals previously exposed to ticks. In addition, an important question that needs to be answered regarding any vaccine is whether or not subsequent infestation in the field boosts the level of immunity induced by vaccination.

Opdebeeck and her colleagues performed a series of well-designed studies to develop an anti-*B.-microplus* vaccine utilizing tick midgut immunogens (Opdebeeck, 1994). Hereford cattle were immunized with membrane and soluble components extracted from *B. microplus* digestive tracts isolated 14 to 17 days after infestation (Opdebeeck *et al.*, 1988a). The $100,000 \times g$ midgut pellet and supernatant were used as candidate immunogens. Soluble supernatant immunogen was administered with Freund's incomplete adjuvant, while membrane pellet antigen was given with Quil A. Immunogen–adjuvant preparations administered on days 0, 14 and 42 induced significant antibody responses. Vaccine-induced resistance was measured by the mean weight of eggs produced by ticks collected from individual hosts. Vaccination with membrane pellet provided 91% protection and administration of combined membrane pellet/supernatant resulted in 82% protection. Immunization with soluble antigen stimulated a response that reduced tick egg weights only 33%.

Immunization with *B. microplus* gut or combined gut–synganglion immunogens induced bovine resistance that resulted in cattle dropping 87% and 80% fewer ticks respectively, than controls (Opdebeeck *et al.*, 1988b). Egg production was reduced by 95% and 91% for the two treatment groups.

Immunization with synganglion alone was ineffective. Although declining in effectiveness, vaccination-induced protection was detectable for at least seven months.

Gut cell membranes of B. microplus were solubilized with a variety of detergents as well as high and low ionic strength buffers (Wong and Opdebeeck, 1989). All solubilized membrane extracts administered as immunogens, with the exception of the high ionic strength buffer preparation, induced significant protection against tick challenge. Tick gut cell membranes extracted with either non-ionic detergents or low ionic strength buffer were the best source of protection inducing immunogens. Low ionic strength buffer solubilized only 19% of gut membrane proteins. After booster immunization, the weight of ticks collected was reduced 69% for gut-membrane-immunized and 95% for the low ionic strength buffer-extract-immunized bovines. An important aspect of this study was correlation of protective immune responses with the bovine major histocompatibility class I antigens MB5, MB14 and MB20 (Wong and Opdebeeck, 1989). A subsequent study revealed that protection against tick infestation was not significantly affected by the presence or absence of major histocompatibility class I antigen MB5 (Wong et al., 1990).

Sheep and cattle were immunized with different concentrations of B. microplus midgut membrane antigens and antibody responses were evaluated by ELISA (Jackson and Opdebeeck, 1989). Total amounts of immunogen administered to sheep ranged from 0.05 to 5000 µg. Immunization with 1 µg of midgut membrane antigen induced a sheep antibody response comparable to administration of 500 µg of the same antigen extract. Hereford steers were equally protected against challenge infestation after immunization with 2 × 500 µg or 3 × either 50 or 500 µg of midgut membrane antigen. Protection measured by reduction of egg mass weight was 89%, 80% and 95%, respectively, for these three treatment groups.

A panel consisting of 18 monoclonal antibodies was raised against gut membrane antigen of B. microplus (Lee and Opdebeeck, 1991). Eleven of these monoclonals precipitated Triton X-100 solubilized tick gut membrane molecules. Monoclonal QU13-precipitated tick gut components induced bovine resistance to infestation. QU13-precipitated material was separated by SDS–PAGE and the resulting electrophoretogram was silver stained, revealing bands of <30, 57, 62, 74, 80 and 200 kDa. The QU13-reactive epitope was repeated on several molecules. Treatment of the gut membrane extract with 10 mM periodate resulted in the failure of 17 of the 18 monoclonals to react with antigen on ELISA, indicating carbohydrate specificity of antibodies.

Protection-inducing epitopes were found to be associated with larval membranes of B. microplus (Wong and Opdebeeck, 1990). Based on a comparison with egg production by ticks feeding on control Hereford steers, larval-membrane-vaccinated animals were protected 78% against challenge infestation. Larval membranes were solubilized with Triton X-100 and sera from gut-membrane-vaccinated steers were used to immunoaffinity purify potentially protection-inducing immunogens. Cattle vaccinated with immunoaffinity-purified immunogens from detergent-solubilized larval membranes were protected in two separate experiments by 80% and 89%, respectively.

Immunization-induced protection against infestation was positively correlated with levels of anti-tick antibodies (Jackson and Opdebeeck, 1990). Post-immunization production of IgG, particularly IgG1, correlated with development of protection. In addition, complement fixation correlated directly with induction of anti-tick immunity. IgG2 and IgM isotypes did not correlate with vaccination-induced protection.

Hereford steers vaccinated with gut membrane antigen and Quil A developed significant antibody responses to soluble salivary gland extract, salivary gland membranes, soluble larval extract and larval membrane antigens of B. microplus (Opdebeeck and Daly, 1990). Lymphocytes obtained from the same vaccinated steers proliferated in vitro in the presence of B. microplus gut and salivary gland antigens (Opdebeeck and Daly, 1990). Lymphocytes collected from unimmunized cattle, infested with 40,000 larvae, developed in vitro lymphocyte blastogenesis responses induced by soluble salivary gland extract.

Immunization with antigens solubilized from membranes of larval B. microplus induced resistance to infestation (Wong and Opdebeeck, 1990). Egg production of ticks which fed upon immunized Hereford steers was reduced by 78%, when compared with controls. Larval membranes were solubilized with Triton X-100 and antigens were isolated by immunoaffinity chromatography, using antibodies obtained from steers immunized with midgut antigens of partially engorged female B. microplus. Cattle vaccinated with immunoaffinity-isolated immunogens were protected in two separate experiments 80% and 89%, respectively. Membranes of B. microplus eggs induced formation of high levels of anti-egg antibodies, but they did not protect Hereford cattle against challenge infestation (Kimaro and Opdebeeck, 1994).

The applied goal of these investigations is development of a controlled release delivery device consisting of a formulated pellet implant that is biodegradable and biocompatible (Opdebeeck, 1994). Protection-inducing immunogen would be released over a prolonged period with minimal handling of the animal. A biodegradable implant consisting of cholesterol and cholesterol–phospholipid was developed to provide an initial pulse of immunogen followed by trickle release (Opdebeeck and Tucker, 1993). This formulation was designed to enhance and sustain protective immunity.

Research of Opdebeeck and her co-workers is particularly noteworthy, since consideration was given early on in development to both the antibody and cell-mediated components of immune reactivity and the genetic background of the target population.

The most dramatic success to date in anti-tick vaccine research has been the development of a recombinant anti-*B.-microplus* vaccine by Willadsen and colleagues (Tellam *et al.*, 1992). This research firmly established the feasibility of developing an anti-arthropod vaccine. Immunization with extracts prepared from adult female *B. microplus* induced partial immunity in both *B. taurus* and *B. taurus* × *B. indicus* (Johnston *et al.*, 1986). Resistance induced by immunization with this extract varied from good to poor. Immunization-induced resistance remained evident after 14 weeks of daily challenges with 1000 larvae. Tick burdens on vaccinated animals were reduced 70%, when compared with controls. Levels of tick-extract-reactive antibodies of individual vaccinated bovines did not correlate with anti-tick resistance. The primary site of damage to ticks collected from cattle vaccinated with *B. microplus* extract was the digestive tract (Agbede and Kemp, 1986). Host leukocytes, which entered the haemolymph, damaged muscle and Malpighian tubules, but not salivary glands.

A minor membrane component of *B. microplus* was identified, which induced protection (Willadsen *et al.*, 1988). Approximately, 1.2 kg of partially engorged ticks yielded 100 μg of immunogen. As previously stated, the term 'concealed' antigen refers to target immunogens on tissues not normally introduced into the host during feeding (Willadsen and Kemp, 1988). However, these tissues are exposed to elements of the host immune system in the bloodmeal.

The protection-inducing antigen, Bm86, was purified and characterized as a glycoprotein with a molecular mass of 89 kDa and isoelectric point of 5.1 to 5.6 (Willadsen *et al.*, 1989). This molecule was immunolocalized to the digest cell surface. Endocytotic activity of digest cells was rapidly inhibited by interaction with this antibody. Bm86 was shown by immunohistochemical staining to be conserved among several strains of *B. microplus* (Penichet *et al.*, 1994).

Recombinant Bm86 has been expressed in bacterial, fungal and insect cell expression systems (Tellam *et al.*, 1992). A eukaryotic expression system was determined to be optimal due to the need for post-translational processing and glycosylation. Host protection induced by Bm86 varied with the expression system used (Tellam *et al.*, 1992). Vaccination with native Bm86 induced host immunity that reduced challenge tick feeding success by 61 to 70% and reduced reproductive capacity by 91 to 93%. *Escherichia-coli*-expressed Bm86 was less effective than the native molecule, reducing feeding by 24 to 27% and reproductive capacity by 77 to 89%. Baculovirus-expressed Bm86 induced immunity that reduced the tick burden by 13 to 34% and reproductive capacity by 63 to 68%.

A second 'concealed' antigen has been identified that might be a suitable candidate for inclusion in a 'cocktail' of recombinant immunogens for induction of more intense protection against *B. microplus* (Riding *et al.*, 1994). Bm91 is a glycoprotein with a molecular mass of 86 kDa and an isolectric point of 4.8 to 5.2, located predominantly in the salivary glands and gut.

Partial amino acid sequence data revealed homology with mammalian angiotensin-converting enzyme. Natural infestation with *B. microplus* does not result in generation of antibodies to this immunogen.

'Concealed' or 'novel' immunogens other than those derived from digestive tract can be used to induce protection against infestation. Over 50% of sera collected from mice infested with *I. ricinus* nymphs and rabbits exposed to *I. ricinus* adults bound in immunoblotting to a 25 kDa protein/polypeptide in the integument and in an extract of whole *I. ricinus* (Rutti and Brossard, 1989). The same antibodies that reacted with the 25 kDa *I. ricinus* integument molecule complexed with a 20 kDa molecule in the integument of female *R. appendiculatus*. Vaccination of cattle with the 20 kDa integumental molecule induced significant resistance to challenge infestation with adult *R. appendiculatus* (Rutti and Brossard, 1992).

Primary tissue culture cells prepared from developing larvae of *A. americanum* were used to successfully induce guinea pig resistance to infestation with adults of the same ixodid species (Wikel, 1985). Mean engorgement weight of females derived from immunized hosts was reduced by 74.8% compared with ticks infesting controls. Mortality among female *A. americanum* engorging on immunized animals was 54.6%. Vaccination with tick tissue culture cells induced cross-resistance to a heterologous species. Guinea pigs immunized with *A. americanum* primary tissue culture cells were resistant to a challenge infestation with adult *D. andersoni*.

Tick digestive tract brush border was used to effectively induce resistance to *A. americanum* infestation (Wikel, 1988). The concept of combining tick digestive tract immunogens and tissue culture was recently used to develop an effective anti-tick vaccine. Tick digestive tract cells were successfully immortalized to create a source of protection-inducing immunogens that elicited high levels of anti-tick immunity (Figs 12.1 and 12.2) (S.K. Wikel, R.N. Ramachandra, D.K. Bergman, T.J. Miller and D.A. Brake, unpublished observation).

## OVERVIEW

Immunological control of arthropods can be achieved for both rapid and long-term blood feeders. Clearly, research related to ticks has advanced well beyond studies related to insects. Work should be focused upon the identification of specific immunogens. The findings to date indicate that extracts of tissues induce variable levels of immunity. The effect of anti-arthropod immunity on pathogen transmission must be clearly defined. An anti-arthropod vaccine might stimulate responses that result in impaired or inhibited feeding, reduced or blocked reproduction and/or the disruption of physiological processes needed for pathogen transmission.

Anti-arthropod vaccine development must be based on a thorough

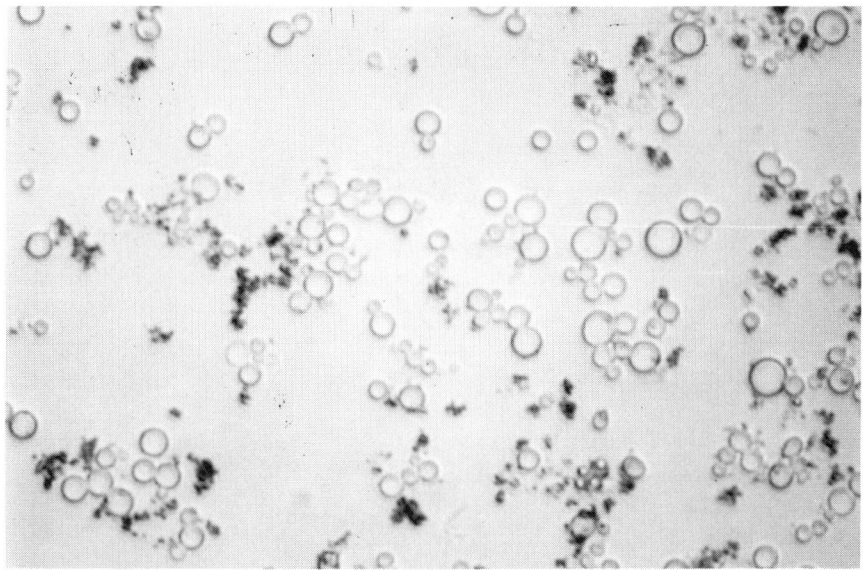

**Fig. 12.1.** Immortalized digestive tract cells of *Dermacentor andersoni* grown *in vitro*. These cells have been successfully used to induce host immunity to challenge infestations with adult ticks of the same and heterologous species.

understanding of the host immunological responses induced by feeding and arthropod countermeasures to those responses. Immunogen selection must take into consideration the effective processing and presentation of immunogen by hosts with diverse genetic backgrounds. T-lymphocyte stimulatory epitopes must be incorporated. An adjuvant can be selected to recruit the appropriate immunocompetent cell populations for induction of the desired immune response.

The tools are available and the fundamental knowledge base exists to provide the foundation for the discovery and development processes needed to create effective anti-arthropod vaccines.

## ACKNOWLEDGEMENTS

Research of S.K.W. is supported by the US Department of Agriculture, Oklahoma Center for Advancement of Science and Technology, Pfizer Animal Health, and Oklahoma Agricultural Experiment Station Project OKL02174.

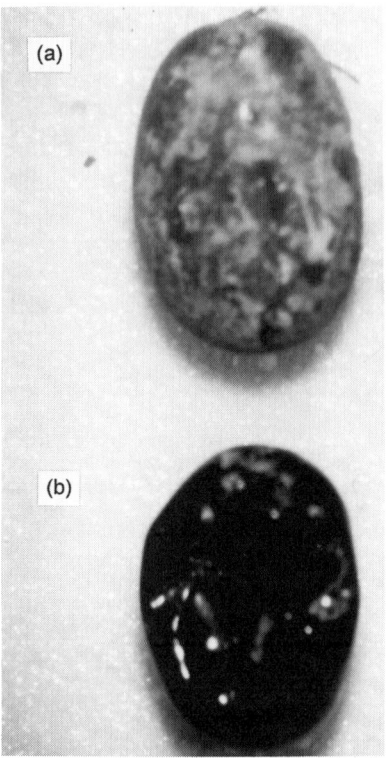

**Fig. 12.2.** Female *Dermacentor andersoni* removed from a control animal (a) and from a host vaccinated with immortalized *D. andersoni* digestive tract cells (b). Tick removed from control animal has grey colour associated with normal engorgement. Tick removed from the host vaccinated with immortalized tick digestive tract cells has a black appearance, sunken cuticle, and this tick is dead.

# REFERENCES

Ackerman, S., Floyd, M. and Sonenshine, D.E. (1980) Artificial immunity to *Dermacentor variabilis* (Acari:Ixodidae): vaccination using tick antigens. *Journal of Medical Entomology* 17, 391–397.

Ackerman, S., Clare, F.B., McGill, T.W. and Sonenshine, D.E. (1981) Passage of host serum components, including antibody, across the digestive tract of *Dermacentor variabilis*. *Journal of Parasitology* 67, 737–740.

Agbede, R.I.S. and Kemp, D.H. (1986) Immunization of cattle against *Boophilus microplus* using extracts derived from adult female ticks: histopathology of tick feeding on vaccinated cattle. *International Journal for Parasitology* 16, 35–41.

Alger, N.E. and Cabrera, E.J. (1972) An increase in death rate of *Anopheles stephensi* fed on rabbits immunized with mosquito antigen. *Journal of Economic Entomology* 65, 165–168.

Allen, J.R. and Humphreys, S.J. (1979) Immunisation of guinea pigs and cattle against ticks. *Nature* 280, 491–493.

Banerjee, D.P., Momin, R.R. and Samantaray, S. (1990) Immunization of cattle (*Bos indicus* × *Bos taurus*) against *Hyalomma anatolicum anatolicum* using antigens derived from tick salivary gland extracts. *International Journal for Parasitology* 20, 969–972.

Ben-Yakir, D. (1989) Quantitative studies of host immunoglobulin G in the hemolymph of ticks (Acari). *Journal of Medical Entomology* 26, 243–246.

Bergman, D.K., Ramachandra, R.N. and Wikel, S.K. (1995) *Dermacentor andersoni*: salivary gland proteins suppressing T-lymphocyte responses to concanavalin A *in vitro*. *Experimental Parasitology* 81, 262–271.

Briegel, H. and Lea, A.O. (1975) Relationship between protein and proteolytic activity in the midgut of mosquitoes. *Journal of Insect Physiology* 21, 1597–1604.

Brossard, M. (1976) Relations immunologiques entre Bovins et Tiques, plus particulierement entre Bovins et *Boophilus microplus*. *Acta Tropica* 33, 15–36.

Cherney, L.S., Wheeler, C.M. and Reed, A.C. (1939) Flea-antigen in prevention of flea bites. *American Journal of Tropical Medicine* 19, 327–332.

Dubin, I.N., Reese, J.D. and Seamans, L.A. (1948) Attempt to produce protection against mosquitoes by active immunization. *Journal of Immunology* 58, 293–297.

East, I.J., Fitzgerald, C.J., Pearson, R.D., Donaldson, R.A., Vuocolo, T., Cadogan, L.C., Tellam, R.C. and Eisemann, C.H. (1993) *Lucila cuprina*: inhibition of larval growth induced by immunisation of host sheep with extracts of larval peritrophic membrane. *International Journal for Parasitology* 23, 221–229.

Eisemann, C.H. and Binnington, K.C. (1994) The peritrophic membrane: its formation, structure, chemical composition and permeability in relation to vaccination against ectoparasitic arthropods. *International Journal for Parasitology* 24, 15–26.

Essuman, S., Dipeolu, O.O. and Odhiambo, T.R. (1991) Immunization of cattle with a semi-purified fraction of solubilized membrane-bound antigens extracted from the midgut of the tick *Rhipicephalus appendiculatus*. *Experimental and Applied Acarology* 13, 65–73.

Essuman, S., Hassanali, A., Nyindo, M. and Ole-Sitayo, E.N. (1992) Augmentation of host's naturally acquired immunity by solubilized membrane-bound midgut proteins of the tick *Rhipicephalus appendiculatus*. *Journal of Parasitology* 78, 466–470.

Feingold, B.F., Benjamini, E. and Michaeli, D. (1968) The allergic responses to insect bites. *Annual Review of Entomology* 13, 137–158.

Friend, W.G. and Smith, J.J.B. (1971) Feeding in *Rhodnius prolixus*: mouthpart activity and salivation, and their correlation with changes in electrical resistance. *Journal of Insect Physiology* 17, 233–243.

Fry, K.J., Seaton, D.S. and Sandeman, R.M. (1994) The production and characterization of monoclonal antibodies to *Lucilia cuprina* larval antigens. *International Journal for Parasitology* 24, 379–387.

Gingrich, R.E. (1982) Acquired resistance to *Hypoderma lineatum*: comparative immune response of resistant and susceptible cattle. *Veterinary Parasitology* 9, 233–242.

Hatfield, P.R. (1988) Anti-mosquito antibodies and their effects on feeding, fecundity

and mortality of *Aedes aegypti*. *Medical and Veterinary Entomology* 2, 331–338.
Hemingway, J. (1992). Genetics of insecticide resistance in mosquito vectors of disease. *Parasitology Today* 8, 296–298.
Hoogstraal, H. (1980) The roles of fleas and ticks in the epidemiology of human diseases. In: Traub, R. and Starcke, H. (eds), *Proceedings of the International Conference on Fleas*. A.A. Balkema, Rotterdam, pp. 241–244.
Jackson, L.A. and Opdebeeck, J.P. (1989) The effect of antigen concentration and vaccine regimen on the immunity induced by membrane antigens from the midgut of *Boophilus microplus*. *Immunology* 68, 272–276.
Jackson, L.A. and Opdebeeck, J.P. (1990) Humoral immune responses of Hereford cattle vaccinated with midgut antigens of the cattle tick, *Boophilus microplus*. *Parasite Immunology* 12, 141–151.
Jacobs-Lorena, M. and Lemos, F.J.A. (1995) Immunological strategies for control of insect disease vectors: a critical assessment. *Parasitology Today* 11, 144–147.
Johnston, L.A.Y., Kemp, D.H. and Pearson, R.D. (1986) Immunization of cattle against *Boophilus microplus* using extracts derived from adult female ticks: effects of induced immunity on tick populations. *International Journal for Parasitology* 16, 27–34.
Kay, B.H. and Kemp, D.H. (1994) Vaccines against arthropods. *American Journal of Tropical medicine and Hygiene* 50, 87–96.
Khan, M.A., Connell, R. and Darcel, C. leQ. (1960) Immunization and parenteral chemotherapy for the control of cattle grubs *Hypoderma lineatum* (De Vill) and *H. bovis* (L.) in cattle. *Canadian Journal of Comparative Medicine* 24, 177–180.
Kimaro, E.E. and Opdebeeck, J.P. (1994) Tick infestations on cattle vaccinated with extracts from the eggs and guts of *Boophilus microplus*. *Veterinary Parasitology* 52, 61–70.
Laird, M. (1985) New answers to malaria problems through vector control? *Experientia* 41, 446–456.
Lal, A.A., Schriefer, M.E., Sacci, J.B., Goldman, I.F., Louis-Wileman, V., Collins, W.E. and Azad, A.F. (1994) Inhibition of malaria parasite development in mosquitoes by anti-mosquito-midgut antibodies. *Infection and Immunity* 62, 316–318.
Lavoipierre, M.M.J., Dickerson, G. and Gordon, R.M. (1959) Studies on the methods of feeding of blood sucking arthropods. I. The manner in which triatomine bugs obtain their blood meal as observed in the tissues of the living rodent, with some remarks on the effects of the bite on human volunteers. *Annals of Tropical Medicine and Parasitology* 53, 235–250.
Lee, R.P. and Opdebeeck, J.P. (1991) Isolation of protective antigens from the gut of *Boophilus microplus* using monoclonal antibodies. *Immunology* 72, 121–126.
Manohar, G.S. and Banerjee, D.P. (1992) Effects of immunization of rabbits on establishment, survival, and reproductive biology of the tick *Hyalomma anatolicum anatolicum*. *Journal of Parasitology* 78, 77–81.
McGowan, M.J., Barker, R.W., Homer, J.T., McNew, R.W. and Holscher, K.H. (1981) Success of tick feeding on calves immunized with *Amblyomma americanum* (Acari: Ixodidae) extract. *Journal of Medical Entomology* 18, 328–332.
Meeusen, E.N.T. and Brandon, M.R. (1994) Antibody secreting cells as specific probes for antigen identification. *Journal of Immunological Methods* 172, 71–76.
Moire, N., Bigot, Y., Periquet, G. and Boulard, C. (1994) Sequencing and gene expression of Hypodermins A, B and C in larval stages of *Hypoderma lineatum*.

*Molecular and Biochemical Parasitology* 66, 233–240.
Mongi, A.O., Shapiro, S.Z., Doyle, J.J. and Cunningham, M.P. (1986a) Immunization of rabbits with *Rhipicephalus appendiculatus* antigen–antibody complexes. *Insect Science Applications* 7, 471–477.
Mongi, A.O., Shapiro, S.Z., Doyle, J.J. and Cunningham, M.P. (1986b) Characterization of antigens from extracts of fed ticks using sera from rabbits immunized with extracted tick antigen and by successive tick infestation. *Insect Science Applications* 7, 479–487.
Nolan, J. (1990) Acaricide resistance in single and multi-host ticks and strategies for control. *Parassitologia* 32, 145–153.
Opdebeeck, J.P. (1994) Vaccines against blood-sucking arthropods. *Veterinary Parasitology* 54, 205–222.
Opdebeeck, J.P. and Daly, K.E. (1990) Immune responses of infested and vaccinated Hereford cattle to antigens of the cattle tick, *Boophilus microplus*. *Veterinary Immunology and Immunopathology* 25, 99–108.
Opdebeeck, J.P. and Slacek, B. (1993) An attempt to protect cats against infestation with *Ctenocephalides felis felis* using gut membrane antigens as a vaccine. *International Journal for Parasitology* 23, 1063–1067.
Opdebeeck, J.P. and Tucker, I.G. (1993) A cholesterol implant used as a delivery system to immunize mice with bovine serum albumin. *Journal of Controlled Release* 23, 271–279.
Opdebeeck, J.P., Wong, J.Y.M., Jackson, L.A. and Dobson, C. (1988a) Hereford cattle immunized and protected against *Boophilus microplus* with soluble and membrane-associated antigens from the midgut of ticks. *Parasite Immunology* 10, 405–410.
Opdebeeck, J.P., Wong, J.Y.M., Jackson, L.A. and Dobson, C. (1988b) Vaccines to protect Hereford cattle against the cattle tick, *Boophilus microplus*. *Immunology* 63, 363–367.
Penichet, M., Rodriguez, M., Castellano, O., Mandado, S., Rojas, Y., Rubiera, R., Sanchez, P., Lleonart, R. and De La Fuente, J. (1994) Detection of Bm86 antigen in different strains of *Boophilus microplus* and effectiveness of immunization with recombinant Bm86. *Parasite Immunology* 16, 493–500.
Pruett, J.H., Barrett, C.C. and Fisher, W.F. (1987) Kinetic development of serum antibody to purified *Hypoderma lineatum* proteins in vaccinated and non-vaccinated cattle. *Southwestern Entomologist* 12, 79–88.
Ramasamy, M.S. and Ramasamy, R. (1990) Effect of antimosquito antibodies on the infectivity of the rodent malaria parasite *Plasmodium berghei* to *Anapholes farauti*. *Medical and Veterinary Entomology* 4, 161–166.
Ramasamy, M.S., Ramasamy, R., Kay, B.H. and Kidson, C. (1988) Anti-mosquito antibodies decrease the reproductive capacity of *Aedes aegypti*. *Medical and Veterinary Entomology* 2, 87–93.
Ramasamy, M.S., Sands, M., Kay, B.H., Fanning, I.D., Lawrence, G.W. and Ramasamy, R. (1990) Anti-mosquito antibodies reduce the susceptibility of *Aedes aegypti* to arbovirus infection. *Medical and Veterinary Entomology* 4, 49–55.
Ramasamy, M.S., Srikrishnaraj, K.A., Wijekoone, S., Jesuthasan, L.S.B. and Ramasamy, R. (1992) Host immunity to mosquitoes: effect of antimosquito antibodies on *Anopheles tessellatus* and *Culex quinquefasciatus* (Diptera: Culicidae). *Journal of Medical Entomology* 29, 934–938.

Ratzlaff, R.E. and Wikel, S.K. (1990) Murine immune responses and immunization against *Polyplax serrata* (Anoplura: Polyplacidae). *Journal of Medical Entomology* 27, 1002–1007.

Ribeiro, J.M.C. (1987) Role of saliva in blood-feeding by arthropods. *Annual Review of Entomology* 32, 463–478.

Riding, G.A., Jarmey, J., McKenna, R.V., Pearson, R., Cobon, G.S. and Willadsen, P. (1994) A protective 'concealed' antigen from *Boophilus microplus*. Purification, localization and possible function. *Journal of Immunology* 153, 5158–5166.

Rutti, B. and Brossard, M. (1989) Repetitive detection by immunoblotting of an integumental 25-kDa antigen in *Ixodes ricinus* and a corresponding 20-kDa antigen in *Rhipicephalus appendiculatus* with sera from pluriinfested mice and rabbits. *Parasitology Research* 75, 325–229.

Rutti, B. and Brossard, M. (1992) Vaccination of cattle against *Rhipicephalus appendiculatus* with detergent solubilized tick tissue proteins and purified 20 kDa protein. *Annals Parasitologie Humane et Comparative* 67, 50–54.

Sandeman, R.M. (1992) Biotechnology and the control of myiasis diseases. In: Yong, W.K. (ed.), *Animal Parasite Control Utilizing Biotechnology*. CRC Press, Boca Raton, Florida, pp. 275–301.

Schlein, Y. and Lewis, C.T. (1976) Lesions in haematophagous flies after feeding on rabbits immunized with fly tissues. *Physiological Entomology* 1, 55–59.

Schlein, Y., Spira, D.T. and Jacobson, R.L. (1976) The passage of serum immunoglobulins through the gut of *Sarcophaga falculata*, Pand. *Annals of Tropical Medicine and Parasitology* 70, 227 230.

Sutherland, G.B. and Ewen, A.B. (1974) Fecundity decrease in mosquitoes ingesting blood from specifically sensitized mammals. *Journal of Insect Physiology* 20, 655–660.

Tellam, R.L., Smith, D., Kemp, D.H. and Willadsen, P. (1992) Vaccination against ticks. In: Yong, W.K. (ed.), *Animal Parasite Control Utilizing Biotechnology*. CRC Press, Boca Raton, Florida, pp. 303–331.

Tembo, S.D. and Rechav, Y. (1992) Immunization of rabbits against nymphs of *Amblyomma hebraeum* and *A. marmoreum* (Acari: Ixodidae). *Journal of Medical Entomology* 29, 757–760.

Titus, R.G. and Ribeiro, J.M.C. (1990) The role of vector saliva in transmission of arthropod-borne disease. *Parasitology Today* 6, 157–160.

Trager, W. (1939a) Acquired immunity to ticks. *Journal of Parasitology* 25, 57–81.

Trager, W. (1939b) Further observations on acquired immunity to the tick *Dermacentor variabilis* Say. *Journal of Parasitology* 25, 137–139.

Vaughan, J.A. and Azad, A.F. (1988) Passage of host immunoglobulin G from blood meal into hemolymph of selected mosquito species (Diptera: Culicidae). *Journal of Medical Entomology* 25, 472–474.

Wang, H. and Nuttall, P.A. (1994) Excretion of host immunoglobulin in tick saliva and detection of IgG-binding proteins in tick haemolymph and salivary glands. *Parasitology* 109, 525–530.

Wikel, S.K. (1981) The induction of host resistance to tick infestation with a salivary gland antigen. *American Journal of Tropical Medicine and Hygiene* 30, 284–288.

Wikel, S.K. (1982) Immune responses to arthropods and their products. *Annual Review of Entomology* 27, 21–48.

Wikel, S.K. (1985) Resistance to ixodid tick infestation induced by administration of

tick-tissue culture cells. *Annals of Tropical Medicine and Parasitology* 79, 513–518.
Wikel, S.K. (1988) Immunological control of hematophagous arthropod vectors: utilization of novel antigens. *Veterinary Parasitology* 29, 235–264.
Wikel, S.K. (1996a) Host immunity to ticks. *Annual Review of Entomology* 41, 1–22.
Wikel, S.K. (1996b) Immunologic control of vectors. In: Marquardt, W.C and Beaty, B.J. (eds), *Biology of Disease Vectors: A Molecular, Physiological and Quantitative Approach*. University Press of Colorado, Boulder, pp. 575–594.
Wikel, S.K., Ramachandra, R.N. and Bergman, D.K. (1992) Immunological strategies for suppression of vector arthropods: novel approaches in vector control. *Bulletin of the Society for Vector Ecology* 17, 10–19.
Wikel, S.K., Ramachandra, R.N. and Bergman, D.K. (1994) Tick-induced modulation of the host immune response. *International Journal for Parasitology* 24, 59–66.
Willadsen, P. and Kemp, D.H. (1988) Vaccination with 'concealed' antigens for tick control. *Parasitology Today* 4, 196–198.
Willadsen, P., McKenna, R.V. and Riding, G.A. (1988) Isolation from the cattle tick, *Boophilus microplus* of antigenic material capable of eliciting a protective immunological response in the bovine host. *International Journal for Parasitology* 18, 183–189.
Willadsen, P., Riding, G.A., McKenna, R.V., Kemp, D.H., Tellam, R.L., Nielsen, J.N., Lahnstein, J., Cobon, G.S. and Gough, J.M. (1989) Immunologic control of a parasitic arthropod. Identification of a protective antigen from *Boophilus microplus*. *Journal of Immunology* 143, 1346–1351.
Wong, J.Y.M. and Opdebeeck, J.P. (1989) Protective efficacy of antigens solubilized from gut membranes of the cattle tick, *Boophilus microplus*. *Immunology* 66, 149–155.
Wong, J.Y.M. and Opdebeeck, J.P. (1990) Larval membrane antigens protect Hereford cattle against infestation with *Boophilus microplus*. *Parasite Immunology* 12, 75–83.
Wong, J.Y.M., Dufty, J.H. and Opdebeeck, J.P. (1990) The expression of bovine lymphocyte antigen and response of Hereford cattle to vaccination against *Boophilus microplus*. *International Journal for Parasitology* 20, 677–679.

# 13

# A Synthesis of Current Concepts Regarding the Immunology of the Host—Arthropod Interface

## Stephen K. Wikel

*Department of Entomology, Oklahoma State University, Stillwater, Oklahoma 74078, USA*

As I approached the writing of this chapter, I was faced with a dilemma. I did not want to simply reiterate the findings of the authors of individual chapters. That would be inappropriate given the fine jobs that they did describing the research that has been performed in their respective areas of interest. The second consideration was to develop a general model of host immunity to ectoparasitic arthropods. Patterns of host immune reactivity were described for various host–ectoparasitic arthropod associations. Due to the magnitude of variation among the immune interactions of different host–ectoparasitic arthropod associations, a highly meaningful, universally applicable, model could not be developed. The course that I have decided to take is to describe areas of research, based upon the reports in this book, that appear to have promise at present. These ideas are intended to stimulate discussion and investigation. Many of you likely can propose any number of equally promising, or more exciting, areas of study. These proposed areas of investigation are merely the thoughts of an immunologist who has devoted his career to this fascinating area of study.

Fortunately, the molecular and cellular immunology of a wider variety of host–ectoparasitic arthropod relationships are being examined. This trend needs to be continued, particularly for short-term blood feeders such as mosquitoes and biting flies. Although results to date have been variable, efforts should continue to develop vaccines for the control of short-term blood feeders. Attention should be given to identifying and using specific immunogens rather than crude extracts of whole arthropods or selected tissues.

---

© CAB INTERNATIONAL 1996. From Wikel, S.K. (ed.) *The Immunology of Host–Ectoparasitic Arthropod Relationships.* CAB INTERNATIONAL, Wallingford.

Fractionation might be essential for initial identification of appropriate candidate immunogens. Extracts often contain immunodominant molecules that have little, or nothing, to do with host resistance. A molecule to which little response occurs might be essential for successful feeding or infestation by the arthropod.

Approaches to immunogen identification need to be examined in a broader fashion. The majority of studies have utilized anti-arthropod antibodies generated during the course of infestation to identify immunogens by immunoblotting. This approach has been successful and will likely continue to be so. However, consideration should be given to using T lymphocytes derived from infested animals for immunogen identification. Screening could be performed by immunogen driven *in vitro* proliferation of T lymphocytes and/or the elaboration of the cytokines, such as IL-2, IL-4, IL-10 and IFN-γ. Lymphocytes collected at different stages during the course of infestation could be used in these analyses. Consideration should be given to the development of ectoparasitic arthropod saliva-specific T lymphocyte clones. What are the contributions of $T_H1$ and $T_H2$ lymphocyte subpopulations to the immune response to arthropod infestation?

Arthropod immunogens could be screened against macrophages from infested and normal hosts to assess their ability to stimulate the release of cytokines, such as IL-1, IL-6, IL-12 and TNF. Identification of the host cytokine repertoire induced during infestation can provide valuable clues regarding the nature of immune responses induced by ectoparasitic arthropods.

Which macrophages and T lymphocytes should be studied? Most often peripheral blood mononuclear cells are used for these analyses. Consideration should be given to cells derived from lymph nodes draining the infestation site. In addition, B lymphocytes collected from draining lymph nodes could be analysed for production of anti-arthropod antibodies. Both systemic and 'local' cell-mediated and humoral immune responses are worthy of attention.

Valuable insights can be gained by examining the findings and current lines of investigation in areas of research that might seem peripheral to our own primary focus of study. Particular attention should be directed towards the vast amount of research on the immunology of the skin in regard to normal cutaneous immune interactions and changes that occur in pathological conditions. Arthropods infesting the cutaneous interface encounter a variety of cells of potentially vast importance: Langerhans cells, keratinocytes, dendritic epidermal T cells, epidermotropic T lymphocytes, macrophages, mast cells, granulocytes and endothelial cells. How do ectoparasitic arthropods influence the following: antigen processing and cytokine elaboration by Langerhans cells; keratinocyte expression of class II major histocompatibility antigens and their ability to present immunogen; keratinocyte production of cytokines; expression of adhesion molecules by endothelial cells; cellular trafficking; the role of mast cells; and stimulation of dendritic epidermal T lymphocytes bearing their limited array of antigen specificities?

How does the host immune response to infestation directly, or indirectly, affect the ectoparasitic arthropod? What effects do ingested cellular elements and mediators have on the arthropod? Considerable insight will be gained by elucidating the roles of prostaglandins, leukotrienes, thromboxanes, major basic protein and other bioactive molecules in the response to infestation.

An intriguing area of study that promises to provide substantial insights into the development of host–ectoparasitic arthropod relationships is that of arthropod modulation of host immune competence. Arthropods and their hosts co-evolve and an important element in those interactions is host immunity and arthropod countermeasures to those defences. Suppressed host immune pathways would potentially be damaging to the arthropod if fully expressed. Otherwise, why develop a countermeasure? Reduced host immune competence likely arose to facilitate blood-feeding and as a secondary, and extremely important, consequence the establishment of vector-borne pathogens in the host.

Host-acquired immunity to pathogen-free ticks results in reduced transmission of two virulent bacteria species when infected ticks feed upon those hosts. Does the host immune response to feeding by pathogen-free ticks result in a response that neutralizes some factor(s) essential for successful pathogen transmission and establishment?

Can neutralization of arthropod produced immunosuppressant(s) enhance host resistance to both infestation and vector-borne pathogens? A given arthropod might be capable of transmitting a variety of potential pathogens: viruses, rickettsiae, bacteria, protozoa and/or helminths. Does successful infection of a host species by any vector-borne pathogen depend, at least in part, on a common pathway of arthropod immunosuppression of the host? Will neutralization of arthropod-derived molecules that reduce host immune competence provide protection against a broad array of different vector-borne pathogens? These are fascinating questions from both a purely biological and a public health perspective. I believe that a 'vector-blocking' vaccine that neutralizes arthropod immunosuppression of the host can be a practical reality. Target the vector rather than every possible type of pathogen that could be transmitted by that arthropod. Examination of arthropod immunomodulation of hosts has the potential of being an extremely productive area of research.

Our understanding of the immunology of the host–ectoparasitic arthropod interface (immunoentomology) is increasing and has the potential to expand greatly in the next few years. Hopefully, this book will attract new investigators and stimulate new avenues of research that will increase our basic knowledge and provide new approaches to the control of arthropods of medical and veterinary importance and the diseases they transmit.

# Index

Note: Page numbers in *italic* refer to figures and/or tables

acanthosis 156
Acari
   feeding 51–54
   saliva 89
acute-phase proteins 3
ADP in platelet aggregation 85, 86, 87, 88–89
*Aedes*
   digestive enzymes 140
   feeding 44
   *A. aegypti*
      antibodies to digestive tract 295, 296
      bloodmeal size *137*
      coagulation inhibition 98
      digestive enzymes 140
      feeding 42–43, 44
      haemolysis 139
      longevity 45
      maximal protease activity 291
      modulation of host immune response 114–115
      peritrophic matrix 134–135
      regulation of salivary secretion 70
      saliva 87, *88*, 90, 92, 181
      salivary glands 66, 67
      vaccine against 182, 293–295
   *A. albopictus* 43
   *A. communis* 179
   *A. dorsalis* 45
   *A. triseriatus* 92
allergens 236
allergic hypersensitivity *see* immediate hypersensitivity
allergy dermatitis to flea bite 156
*Amblyomma*
   *A. americanum* 2
      acquired resistance to 206, 209
      dynamic relationship with host 68
      regulation of salivary secretion 71
      saliva 87, 90, 91, 211–212
      vaccine against 300, 303–304, 308
   *A. hebraeum*
      regulation of salivary secretion 73
      saliva 213
      vaccine against 300
   *A. marmoreum* 300
   *A. variegatum* 218
aminopeptidase 139
anaphylaxis 236
*Anopheles*
   digestive enzymes 140
   feeding 44
   *A. albimanus* 88, 93
   *A. arabiensis* 88
   *A. farauti* 295, 296

*Anopheles contd*
    *A. freeborni*
        feeding 44
        saliva 88
    *A. gambiae*
        digestive enzymes 140
        saliva 88, 93
    *A. maculipennis* 42
    *A. melas* 88
    *A. merus* 88
    *A. quadriannulatus* 88
    *A. quadrimaculatus*
        bloodmeal size 137
        feeding 43
        vaccine against 292–293, 294
    *A. stephensi*
        antibodies to digestive tract 295, 296
        peritrophic matrix 134
        saliva 88
        salivary gland morphology 64
        vaccine against 144, 292–293, 294
    *A. tessellatus* 182, 294, 295
Anoplura 150
    clinical response to 167
    evolution 163–164
    feeding 33–34, 35
    host responses to 164–166
    population dynamics 166–167
    vaccines against 297–298
antibodies 5
antibody response
    primary 5–6
    secondary/anamnestic 5
anticoagulation factors
    midgut 139
    saliva 97–99, 181
antigen 5
antigen–presenting cells 4, 5
    in acquired resistance 207–208
    keratinocytes as 10
    Langerhans cells as 8
antigenic determinant 5
antihistamines 179
antithrombins in saliva 97–99
Antryp 1 140
Antryp 2 140

apyrase 86–87, *88–89*, 160–161, 181, 217
arachidonic acid 16, 68
Argasidae
    acquired resistance to 204–223
    feeding 51, 53–54
    vaccines against 299–308
arthropods
    feeding 31–54
    haematophagous 30–31, *32–33*
    mouthpart derivation 31
    salivary gland physiology 62–78
ASC probes 192
ATP 8
*Atyphloceras multidentatus* 37
avoidance behaviour 178


B7–1 expression by Langerhans cells 8
B lymphocytes 4, 317
    in acquired resistance 208
    antigen-specific 5
    memory 6
basophils 4, 14–17
    in acquired resistance 206, 207
bat flies 39, 40, 47, 48–49, 50–51
bedbugs  *see Cimex*
Birbeck granules 8
biting midges  *see* Ceratopogonidae
blackflies  *see* Simuliidae
bloodmeal
    constituents *138*, 139
    dependence on 141
    digestion 139–140
    haemolysis 138
    size of 136–138
blowflies, regulation of salivary
        secretion 69–70
Bm86 307
Bm91 307–308
*Boophilus microplus*
    acquired resistance to 209
    modulation of host immune
        response 120–121, 218
    saliva 90, 93, 211
    vaccine against 144, 290, 299, 301, *303*, 304–308
*Borrelia burgdorferi* 108, 124

botflies 176
*Bovicola bovis*
    host response to 166
    population dynamics 166
Brachycera, feeding 38, 39, 45–51
bradykinin 107–108, 217

C3
    host serum levels in *S. scabiei*
        infection 245–246
    host skin lesion deposits in *S. scabiei*
        infection 247–248
    receptors 8
C3a 4, 111, 116, 208, 217
C3b 4, 116, 217
C4 208, 245–246
C5a 4, 108, 111, 208
C5b 116, 217
calcitonin gene-related peptide 9
*Calliphora* 70
    *C. erythrocephala* 134, 135
carboxypeptidase A 139
carboxypeptidase B 139, 217
cardia 134
cardo 31
catechol oxidase/peroxidase in saliva 93
CBH response 5, 14, 110, 206
Ceratopogonidae
    feeding mechanisms 38, 40–45
    feeding strategies 67
cetirizine 179
chelicerae 31
chiggers 259, 274–276
*Chiroptonyssus robustipes* 51, 53
chitin 135
chitinases 142–144
chondroitin sulphate E 15
chromogenic substrate assay 98–99
*Chrysops* 87
    *C. callidus* 46
chymotrypsins 188
*Cimex*
    haemolysis 139
    *C. hemipterus* 36
    *C. lectularius*
        bloodmeal size *137*
        coagulation inhibition 99

        feeding 36
        host imuune response to 159
        saliva 91, 94
    *C. pilosellus* 159
Cimicidae 150
    evolution 158
    feeding 34, 36
    host immune responses to 159
coagulation 86, 94–97
    inhibition 95, 97–99
    interaction with immune
        response 107–109
*Cochliomyia hominivorax* 176, 192
cockroach 69
colostrum, antibody transfer
    in 250–251
complement 3–4
    in acquired resistance 208–209
    alternative pathway 111
        inhibition by *I. dammini* 116–117,
            217–218
    classical pathway 5, 111
    host serum levels in *S. scabiei*
        infection 245–246
    host skin lesion deposits in *S. scabiei*
        infection 247–248 concanavalin
        A 118–119
crop 131–132, *133*
*Ctenocephalides felis*
    feeding mechanism 37
    feeding strategy 151
    host clinical response to 156
    host desensitization 156–158
    host immune response to 153–154
    saliva 153
    vaccine against 298
*Ctenophthalmus*, bloodmeal size 137
*Culex*
    saliva 87
    *C. pipiens*
        feeding 43
        host preference 141
    *C. quinquefasciatus*
        bloodmeal size *137*
        vaccine against 182, 295
Culicidae
    feeding 38
    saliva *88*

*Culicoideae*, saliva 88
*Culicoides*
    feeding 40–41, 43, 45
    hypersensitivity responses
        to 178–179
    *C. sanguisuga* 40, 41–42
    *C. varipennis*
        coagulation inhibition 98
        saliva 88
cutaneous basophil hypersensitivity
    response 5, 14, 110, 206
cyclosporin A 215
cytokines 4, 5, 7
    in acquired resistance 208
    in eosinopoiesis 17
    expression
        by eosinophils 19–20
        by keratinocytes 11–12
        by Langerhans cells 9
        by neutrophils 21
    mast cell derived 15, 16–17
    modulation of interactions 219

deer flies *see* Tabanidae
delayed type hypersensitivity response 5, 8, 206, 234
    to *S. scabiei* 235, 240–244, 252–254
*Demodex canis* 115, 270–273
dendritic cells 4
*Dermacentor*
    *D. andersoni*
        acquired resistance to 206, 207, 210, 214–215
        coagulation inhibition 95, 97–98
        engorgement *210*
        feeding 54
        modulation of host immune response 117, 118, 119–120, 121, 122–123, 217, 218, 219
        saliva 213, 219
        vaccine against 301, 302–303, 308, *309–310*
    *D. reticulatus* 124
    *D. variabilis*
        acquired resistance to 206–207, 209, 292
        vaccine against 299–300, 303
Dermanissidae, feeding 51–52
*Dermanyssus gallinae*, feeding 52
dermatitis
    allergic 156
    feline miliary 156
*Dermatobia hominis* 192
*Dermatophagoides farinae* 233–234
*Dermatophagoides pteronyssinus* 233–234
dermis 2
desensitization 156–158, 179, 193, 234, 236
dexamethasone 264, 265–256
diacylglycerol 70
digestion 139–140
    haemolysis in 139
digestive proteases 139, 143–144
*Dipetalogaster maxima* 163
Diptera 193–194
    feeding 35, 38–51
    haematophagous 175
        hypersensitivity to 177 180
        immune responses in resistant hosts 180–181
        modulation of host responses to 181–182
        vaccines against 176, 182–184, 193
    myiasis-causing 175–176
        host immune responses to 184–189
        vaccines against 176–177, 189–193, 297
    saliva 88, 181–182
disagregin 90
disintegrins 90
diverticula 131–132, *133*
dopamine 69, 71, *72, 73*
*Drosophila melanogaster* 66

E-cadherin 8
E-selectin 7, 18, 20
*Echidnophaga gallinacea* 137
ECMLs 136
ELAM-1 7, 18, 20
eosinophil cationic protein 18–19

eosinophil-derived neurotoxin 18, 19
eosinophil peroxidase 19
eosinophil protein X 18, 19
eosinophils 4, 5, 17–20
    in acquired resistance 206, 207
epidermis 1–2, 7
epipharynx 31
epitope 5
*Eristalis tenax* 134
*Eutriatoma maculata* 95, 98
*Eutrombicula* 274–275, 275–276
extracellular membrane layers 136

factor III 96
factor V 96, 97
factor Va 96, 97
factor VII 94, 96, 97
factor VIIa 96, 97
factor VIII 96
factor VIIIa 97
factor IX 94, 96, 97
factor IXa 97
factor X 94, 96, 97
factor Xa 96, 97
factor XI 94, 96, 107
factor XIa 97, 107
factor XII 107
Fc receptors 5, 8
feeding mechanisms, arthropod 31–54
feeding strategies, arthropod 63, 66–67, 151–152
feline miliary dermatitis 156
fertility, influence of host on 141
fibrin 97
fibrin-clot microassay 99
fibrinogen 96, 97
fleas *see* Siphonaptera
fleece rot 187
flies *see* Diptera
*Francisella tularensis* 124
fur mite of mouse 276–277

gadding response 178
galea 31
gender and feeding strategy 67
glossae 31

*Glossina*
    hypersensitivity reactions to 179
    saliva 87
    *G. austeni*
        coagulation inhibition 95, 98
        feeding 50
        host preference 141
    *G. fuscipes* 183
    *G. morsitans*
        bloodmeal size *137*
        coagulation inhibition 95, 98
        digestive proteases 139
        feeding 47–48, 50
        peritrophic matrix 134, 135
        vaccine against 296
    *G. pallidipes* 50
    *G. palpalis* 47–48
Glossinidae
    feeding 39, 47–48, 50
    haemolysis 139
    midgut *132*
    peritrophic matrix 134, 135–136
    water excretion 138
guinea pig sensitivity to flea feeding 152, 153

*Haematobia*
    feeding 47
    immune responses in resistant hosts 180
    *H. exigua* 134
    *H. irritans*
        feeding 49–50
        hypersensitivity reactions to 179
*Haematobosca* 47
*Haematomyzus* 33
haematophagy 63, 66
    development 62
    gender differences 67
    life stage differences 66
*Haematopinus eurysternus*
    host response to 165
    population dynamics 166
*Haematopota pluvialis* 45
haemolysis in digestion 139
haemostasis 85
haustellum 33–34

heat-stable antigen 8
HECA-452 12
Hemiptera 150
    digestive enzymes 140
    feeding mechanisms 34, 35, 36–7
    feeding strategies 158
    host responses to 159–163
    midgut *132*
    saliva *89*, 158
heparin 15
*Hesperocimex sonorensis* 159
Heteroptera *see* Hemiptera
Hippoboscidae
    feeding 39, 47, 48, 50
    midgut *132*
    peritrophic matrix 134
histamine 15–16
    in acquired resistance 207
    affinity for nitrophorins 94
    receptors 15–16
HLA-A antigens, host serum levels in *S. scabiei* infection 246
horn fly 49, 50, 179
horse flies *see* Tabanidae
*Hyalomma*
    *H. anatolicum anatolicum* 301–302
    *H. marginatum rufipes* 213
    *H. truncatum* 213
hydrogen peroxide 4
*Hydrotaea irritans* 39
hydroxyl radicals 4
hypersensitivity reactions to haematophagous dipterans 177–180
*Hypoderma*
    egg-laying 178
    host immune responses to 184–186
    vaccines against 177, 189–190, 297
    *H. bovis* 184, 297
    *H. lineatum* 177, 184, 297
hypodermins 184–186, 189–190, 297
hypopharynx 31

ICAM-1 *see* intercellular adhesion molecule-1

immediate hypersensitivity 5, 14, 234
    to *S. scabiei* 235–236, 236–240, 252–254
immune response 3–6, 109–111
    arthropod modulation 113–124
    interaction with coagulation 107–109
    and parasite survival 112–113
immunogens 4, 5
    identification 316–317
immunoglobulins
    in acquired resistance 209
    IgA 5
        host serum levels in *S. scabiei* infection 245–246, 247
        host skin lesion deposits in *S. scabiei* infection 247–248
    IgD 5
    IgE 5, 14
        host serum levels in *S. scabiei* infection 245–247
        host skin lesion deposits in *S. scabiei* infection 247
        in immediate hypersensitivity to *S. scabiei* 236–240
    IgG 5
        in haemolymph 142, 183, 210
        host serum levels in *S. scabiei* infection 245–246, 247
        host skin lesion deposits in *S. scabiei* infection 248
        in midgut 182–183
    IgM 5, 6
        host serum levels in *S. scabiei* infection 245–246, 247
        host skin lesion deposits in *S. scabiei* infection 247–248
immunological control strategies 291
inositol triphosphate 70
integrins 86, 90
intercellular adhesion molecule-1 (ICAM-1) 7
    in acquired resistance 208
    expression
        by eosinophils 18
        by keratinocytes 11–12
        by Langerhans cells 8
interferons 4

IFN-gamma 7, 10, 122
  suppression by *D. andersoni*
    SGE 122, 219
interleukins
  IL-1 7, 121–122
    in acquired resistance 208
    and Langerhans cell viability 9
    suppression by *D. andersoni*
      SGE 122, 123, 219
  IL-1-alpha 9
  IL-1-beta 9
  IL-2 122
    suppression by *D. andersoni*
      SGE 122, 123, 219
  IL-4 122
  IL-5 122
  IL-7 11
  IL-10 9, 10, 122
  IL-12 10
ivermectin 190, 265, 267–268
*Ixodes*
  *I. dammini* 116
    modulation of host immune
      response 116–117, 119,
      217–218
    saliva 89, 90, 116, 217
  *I. holocyclus*
    acquired resistance to 209
    coagulation inhibition 95, 98
    vaccine against 300
  *I. ricinus*
    acquired resistance to 207, 209,
      215
    coagulation inhibition 95, 98
    host immune responses
      to 110–111
    influence on success of infestation
      and pathogen transmission 108
    modulation of host immune
      response 118–119
    vaccine against 308
  *I. scapularis* see *I. dammini*
Ixodidae
  acquired resistance to 205–216
  effects of salivary secretions on
    pathogen transmission 76
  feeding mechanism 51, 52, 53, 54
  feeding strategies 66, 67

  host immune responses
    to 109–111
  modulation of host immune
    response 115–124, 216–220
  pathogen vector strategies 74, 75
  regulation of salivary secretion
    71–74
  saliva 62
  salivary gland morphology and
    pharmacology 65, 67–68,
    216–220
  speed of feeding 108
  vaccines against 299–308,
    309–310

kallikrein 107, 108
keratinocyte growth factor 13
keratinocytes 2, 9–12
  in immunity to *S. scabiei*
    infection 251
  interaction with Langerhans cells 9

L-selectin 20
labium 31
labrum 31
lacinia 31
Laelapidae, feeding 51
LAM 136
Langerhans cells 4, 7–9, 110
  in acquired resistance 207–208
late phase hypersensitivity 234
  to *S. scabiei* 235–236, 240–244,
    252–254
lectins 135–136
*Leishmania* 76, 114
*Leptocimex duplicatus* 159
*Leptotrombidium* 274
leukotriene $B_4$ 15, 16
leukotriene $C_4$ 16, 19
leukotriene $D_4$ 16, 19
leukotriene $E_4$ 16
levamisole 271
LFA-3 8, 12
lice
  cattle
    host responses to 164–166

lice *contd*
    population dynamics 166–167
    chewing 31, 33, 150
    feeding mechanisms 39, 47, 48, 50
    midgut *132*
    peritrophic matrix 134
    sucking *see* Anoplura
*Linognathus vituli* 165
*Lipoptene* 50
locust 69
*Lucilia cuprina* 144, 176
    host immune responses
        to 186–189
    peritrophic membrane 191
    vaccine against 176–177,
        190–192, 297
lumenal apical membrane 136
*Lutzomyia*
    feeding 40
    *L. longipalpus*
        feeding 44
        longevity 45
        modulation of host immune
            response 114, 181
        saliva *88*, 92–93, 108, 181
lymphocyte function associated antigen-3 8, 12
*Lyponyssoides sanguineus* 52
lysozyme 3

Macronyssidae 51, 52–53
macrophages 4, 5, 317
major basic protein 18
Mallophaga 31, 33, 150
mandibles 31
*Manduca sexta* 69
mange mites 259
    demodectic 115, 270–273
    myocoptic 277
    otodectic 269–270
    psoroptic 51, 109, 260–269
mast cells 4, 14–17
    in acquired resistance 206–207
    connective tissue 14
    mucosal 14
maxadilan 92–93
maxillae 31

*Melophagus ovinus* 48, 50, 175
*Menacanthus stramineus* 33
MHC antigens, class II 10
midgut 131–132
    absorption in 140–141
    antihaemostatic factors released
        by 139
    and bloodmeal size 137–138
    digestion in 139–140
    water excretion from 138
    zones 132, *133*
*Morellia horturum* 39
mosquito
    digestive enzymes 140
    effects of pathogen on 76–77
    feeding
        duration 108
        strategies 63
    host desensitization 179
    host immune response to 109
    human hypersensitivity to
        bites 179
    midgut *132*
    peritrophic matrix 134
    regulation of salivary secretion 70
    salivary glands 63, *64*, 66, 67
    vaccines against 182–183,
        292–296
    *see also* Aedes; Anopheles
moubatin 90
mouthparts, arthropod 31–54
*Musca*
    *M. autumnalis* 39
    *M. crassirostris* 39
    *M. domestica* 38
    *M. sorbens* 39
    *M. tempestiva* 39
Muscamorpha, feeding 35, 39, 40, 47–51
Muscidae
    feeding 47
    peritrophic matrix 134
mycetome 141
*Myobia musculi* 276–277
*Myocoptes musculinus* 277

natural killer cells 4, 216, 219

*Nauphoeta cinerea* 69
Nematocera 35, 38–39, 40–45
*Neotrombicula autumnalis* 275
neutrophils 4, 5, 20–21
nitric oxide 4, 90, 91, 93–94
nitrophorins 85, 94, 98
NK cells 4, 216, 219
non-sterilizing immunity 112
northern fowl mite 53, 259, 278–283
*Nosopsyllus fasciatus* 38
Nycteribiidae 39, 40, 47, 48–49, 50–51

*Octobius megnini* 66
*Oestrus ovis* 192–193
opsonization 5, 20
*Orchopea howardi* 89
*Ornithodoros*
    feeding 53–54
    *O. moubata*
      coagulation inhibition 95, 98
      saliva 89, 90, 213
    *O. savignyi* 213
    *O. talaje* 53–54
*Ornithonyssus*
    *O. bacoti* 51, 52–53
    *O. bursa* 53
    *O. sylviarum* 53, 259, 278–283
*Oropsylla bacchi* 89
osmoregulation, role of salivary gland 67–68
*Otodectes cynotis* 269–270
oxygen, singlet 4

P2 189
P-selectin 7, 20
PAF 19, 179
pallidipin 90
palp 31
paraglossae 31
pathogens
    effects of salivary secretions on transmission 76
    effects on vector 76–77
    effects of vector growth stage on 75
    role of salivary gland in transmission 74–77

vector tissue and cell specificity 76
*Pediculus*
    evolution 163–164
    *P. humanus*
      bloodmeal size *137*
      feeding 34
      host responses to 164
pedipalps 31
perforin 4
perimicrovillar membrane 136
periodic ectoparasites 136–137
*Periplaneta americana* 69
peritrophic matrix/membrane 134–136, 191
permanent ectoparasites 137
phagocytosis 4, 20–21
phagolysosome 4
pharynx 31
Phlebotomidae 88
*Phlebotomus*
    feeding 40, 41
    longevity 45
    *P. argentipes* 88
    *P. colabaensis* 88
    *P. duboscqi* 44
    *P. langeroni* 45
    *P. papatasi*
      longevity 45
      saliva 88
    *P. perniciosus* 88
Phthiraptera 31, 33–34, 35, 150
pilocarpine 71, 73
piroplasms
    effect of vector growth stage on transmission 75
    tissue and cell specificity 76
plasma cells 5
*Plasmodium*
    chitinases 143
    *P. falciparum* 142
platelet activating factor 19, 179
platelet aggregometers 86
platelets 85
    aggregation assays 86
    aggregation inhibitors in saliva 86–91
PM 134–136, 191
Polyctenidae 34, 36

*Polyplax*
  *P. serrata*
    acquired resistance to 297–298
    clinical response to 167
    host response to 109, 165
    vaccine against 298
  *P. spinulosa* 167
prekallikrein 107
prestomen 39
proctolin 69
prolixin G 98
prolixin S 94, 98
prostacyclin 217
prostaglandins 68
  assay 91, 92
  in haemostasis 90
  $PGD_2$ 16, 19
  $PGE_2$ 19
    in haemostasis 90
    in saliva 90–91, 93, 120–121, 217, 218
  $PGF_{2\alpha}$ 91, 93
  $PGI_2$
    in haemostasis 90
    in saliva 93
protease IV 139
protease VII 139
protein kinase C 74
prothrombin 96, 97
proventriculus 134
*Psoroptes*
  feeding 51
  *P. cervinus* 260
  *P. cuniculi* 109, 260–269
  *P. equi* 260
  *P. natalensis* 260
  *P. ovis* 260–269
Psychodidae
  feeding 38, 40–45
  midgut 132
  peritrophic matrix 134
*Pulex*
  host desensitization 156
  South America 151
  *P. irritans*
    feeding 37, 38
    guinea pig immune response to 153

  *P. simulans* 153
Pulicidae 151

Reduviidae 150
  cellular responses to 161–162
  clinical responses to 163
  evolution 158
  feeding 34, 36–37
  host immune responses to 160
  immune-mediated behavioural responses to 162
  saliva 89, 160–161
resistance
  acquired 4–6
  model 220–223, *224*
  passive transfer 209
  to tick infestation 205–216
  innate 3–4, 219
  to haematophagous dipterans 180–181
  to *S. scabiei* infection 248–250
Rhagionidae 39, 45–47
*Rhipicentor nuttali* 213
*Rhipicephalus*
  *R. appendiculatus*
    acquired resistance to 209, 212–213
    coagulation inhibition 95, 98
    modulation of host immune response 117–118, 217
    saliva 212–213
    vaccine against 300, *303*, 304, 308
  *R. evertsi evertsi* 213
  *R. sanguineus* 209
  *R. simus simus* 213
*Rhodnius prolixus*
  cell and tissue specificity of pathogens transmitted by 76
  coagulation inhibition 95, 98, 99
  duration of feeding 108
  feeding 36–37
  haemolysis 139
  host immune responses to 160
  saliva 85, *89*, 90, 91, 92, 93–94, 160–161
Rhyncophthirina 33

*Ricinus* 33
rickettsiosis, chigger-borne 274

saliva 62, 85
   and acquired resistance 211–213
   anticoagulation factors 97–99, 181
   platelet aggregation inhibitors 86–91
   processes in production 63
   regulation of secretion 68–74
   secretions and pathogen transmission 76
   vasodilators 91–94, 181
salivary glands 62, 77–78
   alternative functions 67–68
   effects of relationship with host 68
   morphology 63, 64–65
   role in pathogen transmission 74–77
SALT 6
sandflies *see* Psychodidae
*Sarcoptes scabiei* 232
   antigens 233
     immune reactions to 234–244
   cross-reactions with other mites 233–234
   deposits in host skin lesions 247–248
   epidermal and dermal cellular infiltrates in host 240–244
   host immune response to 109
   host serum antibody isotype concentrations 244–247
   model for immune response to 252–254
   modulation of host immune response 115
   resistance to 248–250
   role of epidermal keratinocytes in immunity to 251
   transfer of antibodies to in colostrum 250–251
scabies *see Sarcoptes scabiei*
*Schistocerca gregaria* 69
scrub typhus 274
second messenger activators in saliva 90–91

selectins 7, 18, 20
serotonin 15, 69, 85, 86
sialokinins 92
Simuliidae
   feeding mechanisms 38, 40–45
   feeding strategies 67
   peritrophic matrix 134
   saliva 88
*Simulium*
   feeding 43
   hypersensitivity reactions to 179
   *S. vittatum*
     coagulation inhibition 95, 98
     host immune response to 109
     modulation of host immune response 113–114, 181
     saliva 88, 93, 181
Siphonaptera 150
   feeding mechanisms 35, 37–38
   feeding strategies 151–152
   host clinical response to 155–156
   host desensitization 156–158
   host immune response to 153–155
   host specificity 152
   midgut 132
   saliva 89
   vaccines against 298–289
SIS 6–7
skin 1–2
   immune system 6–21
skin-associated lymphoid tissues 6
slow reacting substance of anaphylaxis 16
snipe flies 39, 45–47
solenophagy 63
*Spilopsyllus cuniculi* 152
spiperone 73
SRS-A 16
stable fly *see Stomoxys calcitrans*
sterilizing immunity 112
*Stilbometopa impressa* 50
stipes 31
Stomoxyinae 39, 47, 49
*Stomoxys*
   midgut 132
   *S. calcitrans*
     absorption of digested bloodmeal 141

*Stomoxys contd*
   bloodmeal size *137*
   digestive proteases 139–140
   feeding 47, 49
   haemolysis 139
   peritrophic matrix 134, 136
   vaccine against 183, 296
Streblidae 39, 40, 47, 48–49, 50–51
stylet bundle 34
superoxide anion 4
symbionts 141

T lymphocytes 12–14
   in acquired resistance 208, 214–216
   antigen-specific 5
   cytotoxic 9, 216
   dendritic epidermal 12–14
   effector 8
   gamma/delta receptor 13–14
   helper 5
      $T_H1$ 9, 10–11, 122, 206, 208, 216
      $T_H2$ 9, 10–11, 122, 216
   in immunogen identification 317
   memory 6, 8, 12
   modulation of responsiveness 218–219
   responses induced by keratinocytes 10
Tabanidae
   feeding mechanisms 39, 45–47
   feeding strategies 67
   hypersensitivity reactions to 179
   midgut *132*
   peritrophic matrix 134
   saliva 90
Tabanomorpha 39, 45–47
*Tabanus*
   *T. abactor* 47
   *T. atratus* 46
   *T. bovinus* 95, 98
   *T. lineola* 46–47
   *T. nigrovittatus* 45, 47
tachykinins in saliva 91, 92
telmophagy 66
temporary ectoparasites 136, 137

Thogotovirus 76
thrombin 86, 96, 97
   activity assay 98–99
   functional sites 98
thrombin-fibrinogen reaction 98
thromboxane $A_2$ 85, 86
thromboxane $B_2$ 19
ticks *see* Argasidae; Ixodidae
tissue factor 96
TNF *see* tumour necrosis factor
*Triatoma*
   *T. barberi* 37
   *T. dimidiata* 37
   *T. infestans* 36–37
      bloodmeal size 137
      clinical responses to 163
      coagulation inhibition 95, 98
      immune-mediated behavioural responses to 162
      saliva 163
   *T. pallidipennis* 90, 161
   *T. protracta* 36–37
      allergens 163
      cellular responses to 161–162
Triatominae
   feeding 34, 36–37
   host immune response to 109
   water excretion 138
*Trinoton anserinum* 33
*Trochiloecetes* 33
*Trypanosoma*
   *T. cruzi* 162
   *T. rangeli* 76
trypsins 139, 140, 143, 144, 188
tsetse flies *see* Glossinidae
tsutsugamushi fever 274
tumour necrosis factor (TNF) 7, 9, 122
   in acquired resistance 208
   suppression by *D. andersoni* SGE 122, 123, 219
*Tunga*
   feeding strategy 151–152
   *T. monositus*
      feeding mechanism 38
      host immune response to 155
   *T. penetrans*
      feeding mechanism 38
      host immune response to 155

Tungidae 38

vaccines 144, 290–292, 308–309, 316–317
   against fleas 298–299
   against haematophagous dipterans 176, 182–184, 193, 296
   against mosquitoes 182–183, 292–296
   against myiasis flies 176–177, 189–193, 297
   against sucking lice 297–298
   against ticks 299–308, *309–310*
   anti-immunosuppressant 302
   stand-alone anti-vector 295–296
   transmission blocking 142–144
vascular cell adhesion molecule-1 7
vasodilators
   peptide 92
   in saliva 91–94, 181
VCAM-1 7
veiled cells 8

warble fly *see Hypoderma*
water excretion 138
*Wohlfartia magnifica* 176

*Xenopsylla cheopis*
   bloodmeal size 137–138
   feeding 37
   host immune response to 154–155
   host preference 141
   saliva *89*

YGGFMRFamide 69